Ukraine's Nuclear Disarmament
A History

Ukrainian Research Institute
Harvard University

Harvard Series in Ukrainian Studies 78

HURI Editorial Board

Michael S. Flier
George G. Grabowicz
Serhii Plokhy, *Chairman*

Oleh Kotsyuba, *Manager of Publications*

Cambridge, Massachusetts

YURI KOSTENKO

UKRAINE'S NUCLEAR DISARMAMENT

A HISTORY

Second, corrected and expanded edition

Edited by
Svitlana Krasynska

Introduction by
Paul J. D'Anieri

Translated by
Lidia Wolanskyj, Svitlana Krasynska,
and Olena Jennings

Distributed by Harvard University Press
for the Ukrainian Research Institute
Harvard University

The Harvard Ukrainian Research Institute was established in 1973 as an integral part of Harvard University. It supports research associates and visiting scholars who are engaged in projects concerned with all aspects of Ukrainian studies. The Institute also works in close cooperation with the Committee on Ukrainian Studies, which supervises and coordinates the teaching of Ukrainian history, language, and literature at Harvard University.

Publication of this book has been made possible by the generous support of Ukrainian studies at Harvard University by the Ostap and Ursula Balaban Ukrainian Fund at the Ukrainian Research Institute, and by generous contributions from Oleksandr Hromyko, Mykhailo Koval, and Bohdan Kozak.

 You can support our work of publishing academic books and translations of Ukrainian literature and documents by making a tax-deductible donation in any amount, or by including HURI in your estate planning. To find out more, please visit https://huri.harvard.edu/give.

Second, corrected and expanded edition

ISBN 9780674249301 (hardcover), 9780674295346 (paperback), 9780674250864 (epub), 9780674250888 (PDF)

Library of Congress Control Number: 2020938440
LC record available at https://lccn.loc.gov/2020938440

Front cover illustration: A Nuclear Missile Is Removed from a Silo, Reuters Pictures. Reproduced with permission.
Cover and book design by Mykola Leonovych, https://smalta.pro, and Mykhailo Fedyshak, bluecollider@gmail.com

Contents

vii   Acknowledgments
viii A Note on Transliteration
  ix   Preface to the Second Edition

   1 Introduction (Paul J. D'Anieri)
  15 Author's Note
  23 *Chapter 1.* An Infant in a Grownups' Game
  75 *Chapter 2.* Ukraine Formulates Its Position on Nuclear Arms
127 *Chapter 3.* Breakthrough
179 *Chapter 4.* The Hawks' Victory
221 *Chapter 5.* Going into Reverse
245 *Chapter 6.* Capitulation
267 Afterword (Paul J. D'Anieri)

269 Appendices
299 Notes
333 Illustration Credits
337 Index

Acknowledgments

The idea of translating *Ukraine's Nuclear Disarmament* into English was born as early as the publication of the Ukrainian edition. Nuclear weapons have been an important global issue, and the Ukrainian language alone does not reach all interested audiences. Thus, I accepted with great enthusiasm the offer of the Ukrainian Research Institute at Harvard University (HURI) to translate and publish this book in English.

From the very beginning, I received tremendous support from George G. Grabowicz, the Dmytro Chyzhevs'kyj Professor of Ukrainian Literature at Harvard University. Serhii Plokhy, the Mykhailo S. Hrushevs'kyi Professor of Ukrainian History and director of HURI, engaged more than once with me in discussions of Ukrainian matters and was most encouraging of the publication of my book. For this support I am very grateful.

The Ukrainian edition conformed to the norms and standards of publication in Ukraine, but required great organizational work and re-editing to overcome discrepancies in approaches with Western practices and to give context to the English-language reader. Dr. Oleh Kotsyuba, HURI's manager of publications, solved these challenges brilliantly, for which I am particularly grateful. My deepest gratitude is also owed to the book's scholarly editor, Dr. Svitlana Krasynska, who spent many an hour with me tracking down original citations, and substantially revising and condensing the manuscript in preparation for translation for a scholarly Western audience. I am also very grateful to Dr. Michelle R. Viise, HURI's monographs editor, for her extensive work on the book, as well as to copyeditor Michael Diem and to Orysia Hrudka, our assistant in Kyiv, who helped obtain permissions for illustrations and image files. They all made a tremendous contribution to making the English edition outstanding.

My special thanks go also to Paul J. D'Anieri, professor of Political Science and Public Policy at the University of California Riverside, for agreeing to read and comment on the manuscript of the English translation, as well as to write the introduction to the first edition and afterword to the second edition of this book. I would also like to thank Thomas S. Blanton, executive director of the National Security Archive at George Washington University and Dr. Svetlana Savranskaya, the Archive's director of Russia programs, who both gave their support to this production of an English-language edition of the book.

Finally, this book would not have come into being in either Ukrainian or English were it not for the dedication and hard work of the journalist Iryna Lukoms'ka, who edited the original publication. She walked the long and thorny road of manuscript preparation with me, from the categorization of the colossal archive of Ukraine's nuclear saga to writing the book itself, for which I offer my heartfelt thanks.

A Note on Transliteration

Personal names have been transliterated in accordance with the Library of Congress transliteration systems for Ukrainian and Russian. Well-known personal names such as Yeltsin, Yushchenko, Nazarbayev, and Yanukovych appear in spellings widely adopted in English-language texts.

Place-names appear in the Library of Congress transliteration (without primes) or in the form sanctioned by the countries in whose political borders they are found today. English-language forms of long standing are used for the most well-known cities, including Moscow (rather than Moskva), Warsaw (rather than Warszawa), and Vienna (rather than Wien). In a small number of cases, place-names identify historical events and treaties. In these cases, we retain the transliterated form for the geographic location (e.g., Chornobyl, Ukraine; Almaty, Kazakhstan) but use the English-language form by which the event is known (Chernobyl, the nuclear accident; the Alma-Ata Protocol).

Preface to the Second Edition

Very early in the morning on 24 February 2022, we awakened to a new twenty-first century reality. The world's second largest nuclear power began a large-scale war against democracy. Although Vladimir Putin, Russia's *de facto* dictator, called this a "special military operation" and said it was limited to the territory of Ukraine, it was immediately clear that this is the latest round in the global redistribution of influence between democracy and autocracy.

Putin's "forced response to the threat of NATO," as he cynically termed his aggression against Ukraine, finally opened the eyes of the democratic world to the fundamental essence of the Kremlin regime. The huge military forces that Russia for years had amassed at Ukraine's borders now invaded sovereign territory, sowing death and committing war crimes and genocide against the civilian population on a scale not seen in this century. The horrors of *russkii mir* (the Russian world) were felt immediately, as the world took in millions of refugees from Ukraine.

However, the threat of Putin's revanchism also rallied democracies. Dozens of countries began to help Ukraine with humanitarian aid, funding, and weapons. Within two months of that first attack, Ramstein Air Base in Germany effectively became the gathering place of an "anti-Putin coalition" involving more than 40 countries. On 9 May 2022, the anniversary of the defeat of Nazi Germany—a day that Russia considers sacred—US president Joe Biden signed a new lend-lease act, establishing a powerful mechanism for supplying modern equipment to Ukraine's forces in order to defeat the aggressor.

Meanwhile, faced with his first military defeats in Ukraine and the determined opposition of the West, Putin began to raise the nuclear stakes—and not just through statements from Moscow's diplomats. At first, the Russian invaders engaged in acts of nuclear terrorism, capturing two Ukrainian nuclear power stations and simulating the bombing of three others with cruise missiles. Then, top officials of the Russian Federation began to threaten that tactical nuclear weapons would be used in Ukraine. Beginning in May 2022, Russia's leadership talked more and more about a nuclear apocalypse.

Under such circumstances, we need to reconsider the documents and decisions presented in this book in an effort to develop more effective mechanisms to ensure global security. *Ukraine's Nuclear Disarmament: A History* offers some key proposals in this regard.

Firstly, this study most extensively captures the deeper political processes that provided the basis for Putin's neoimperialism and gave rise to new threats to democracy. It also outlines clear recommendations for countering Russia's neoimperial ambitions.

Secondly, the case of Ukraine has demonstrated beyond a doubt that the way the West focused entirely on nuclear aspects after the fall of the USSR—and not on the essence of the Soviet totalitarianism that never went away in Yeltsin's or Putin's Russia—has led to the global crisis facing us today. For one thing, Ukraine's legislature agreed to destroy completely the country's nuclear weapons in exchange for swift integration into a democratic Europe. In the meantime, enormous pressure from Russia, its threats against Ukraine, and the West's support for Yeltsin the "Democrat" actually hampered Ukraine's democratic progress and forced the Ukrainian government to turn over its nuclear arsenal—the third largest in the world—to its potential attacker. The consequences of this mistake on the part of the West were the renewal and reinforcement of Russia's imperial ambitions and the emergence of new and more dangerous challenges for the world. Meanwhile, Ukraine, which really did choose democracy, has been denied Euro-Atlantic integration for over 30 years.

Thirdly, the West's mistakes included the decision to "assure" Ukraine's security in the form of the Budapest Memorandum of 1994, which was signed by the heads of three nuclear powers: the US, the UK, and Russia. When the West treated this politically important document as nothing more than a political declaration and not a legally binding act, it allowed Russia to avoid severe sanctions for its annexation of Crimea and the war it started in the Donbas in 2014, although such measures were called for by the Charter of the United Nations and other international treaties. Instead, at Moscow's insistence, we saw the emergence of the Minsk agreements, which went against international practice and were not signed by any head of state. For some reason, they became a kind of "political icon" and, as numerous top officials kept insisting, the only path to peace in Ukraine. This was the very pathway that led to the global threat the world is facing today.

Finally, the history of Ukraine's nuclear disarmament shows convincingly that faulty decisions based on the failure to adequately assess the threat of Russian authoritarianism have led to a worldwide security crisis. The global system of peaceful coexistence among different countries, a system that humanity had built over a long time and with great effort after World War II, has burst like a soap bubble. Paper agreements with authoritarian rulers and democracy's tendency to turn a blind eye to global challenges for the sake of economic dividends do not work in today's world.

At this point, we need a fundamentally different model for world order, based on an ideological unity around democracy and governed by the rule of law and international agreements. At the same time, authoritarian systems and dictatorships need to be restricted in their access to everything that is the fruit of democracy: international organizations and security alliances based on the principles of respect for the sovereignty and territorial integrity of other states; economic, cultural, and social benefits; intellectual and technological achievements; and financial resources. Simultaneously, any and all acts of aggression and threats from authoritarian states against democratic ones must be opposed decisively and effectively, based on the principle of "coercing the aggressor into peace."

Having shown itself to be so fragile and interdependent, the modern world no longer has the right to react in the old ways to the new threats of totalitarianism. We have no alternative but to begin changing the global security system. This book describes not just Ukrainian but world experience of what can happen when such challenges are ignored. Unfortunately, Ukraine has already paid the highest price for volunteering to destroy its nuclear arsenal, while the democratic world has never properly appreciated this action.

Right now, Ukraine is defending democracy against the nuclear madness of the Putin regime and is paying with what is most precious: the lives of its citizens. The democratic world has no moral right to merely watch as this horror unfolds in the twenty-first century. The time has come for us to fight together.

February 2023
Kyiv

Introduction

Paul J. D'Anieri

In January 1994, Ukraine signed the Trilateral Statement with the US and Russia, agreeing to surrender its nuclear weapons in exchange for security guarantees. Having annexed Crimea and occupied portions of eastern Ukraine in 2014, Russia unleashed a massive invasion of Ukraine in 2022. Yuri Kostenko's book is focused on the first event, but the second looms over the analysis. Some believe that Russia could not have behaved as it did in 2014 and 2022 had Ukraine retained its nuclear weapons. Others believe that the security commitments Western states issued at that time should have led to more robust support for Ukraine after Russia's invasion. These beliefs have brought what seemed like a footnote in the history of arms control back to the center of our attention. Kostenko believes things could have been different, and his book recounts the events of that time in order to make that case. This is a first-hand account, but it is also an indictment of Ukraine's policy, along with that of the US and Russia and of many of the people who shaped it.

Ukraine's Nuclear Disarmament is significant in several respects. First, it provides an inside look into the politics that shaped Ukraine's policy on this crucial issue. It affords us a detailed look into Ukrainian politics in the earliest months of independence. Second, it provides a documentary history. Kostenko includes excerpts of debates and documents that have not previously been available in English. It is both a primary and a secondary source. Read as a mixture of memoir, history, and polemic, as it must be, it provides a great deal of insight, as well as a great deal to argue about.

My introduction begins by surveying the events that Kostenko covers in so much detail in the book. It goes on to place Kostenko's account in the context of the very small English-language scholarly literature on Ukraine's denuclearization. I then consider some of the important themes of the book. One set of crucial questions involves arguments about what might have happened had different choices been made. Finally, I consider the fallout thirty years later, as Ukrainians lament the impotence of the guarantees they were given, and other states try to learn lessons from Ukraine's experience.

I. Ukraine's Nuclear Dilemma

Ukraine's July 1990 sovereignty declaration included a statement that Ukraine would become a nonnuclear state and would become neutral. As Kostenko stresses, these declarations must be understood in context: Ukraine was still part of the Soviet Union, and the declaration assumed sovereignty *within* a reformed Soviet Union, not the complete independence that came the following year. In that context, the intent to become a nonnuclear and neutral state was meant to insist that Soviet nuclear forces be withdrawn from Ukraine. It was not intended as a legally binding commitment on an independent Ukrainian state, which at that point was at most a distant hope. After the failed putsch in Moscow in August 1991, Ukraine declared independence, completely changing the situation. While many Ukrainian leaders sought to maintain the policy of denuclearization, others, Kostenko among them, believed that the problem should be reconsidered in the completely new circumstances.

These events took place alongside the broader US-Soviet and US-Russian rapprochement that accompanied the end of the Cold War. Nuclear weapons had always been at the heart of the Cold War, and progress in US-Soviet relations had always been measured in terms of nuclear arms control, from the Partial Test Ban Treaty of 1963 to the Intermediate Range Nuclear Forces (INF) agreement of 1987. The SALT I agreements (1972) and SALT II Treaty (1979), which limited intercontinental ballistic missiles (ICBMs), submarine-launched ballistic missiles (SLBMs), and heavy bombers, were seen as the most important achievements of the "détente" period. Mikhail Gorbachev's "new thinking" in Soviet foreign policy had paved the way to a successor treaty, START I, which Gorbachev and George H. W. Bush signed in Moscow on 31 July 1991. The treaty two sides agreed to deep cuts in numbers of strategic nuclear weapons and launchers, whereas the SALT agreements had only limited the growth of arsenals (and some questioned whether they even had that effect).

On his way home, President Bush stopped in Kyiv, where he gave a speech, later derided as the "Chicken Kiev speech," warning Ukrainians against "suicidal nationalism." While Bush's support for reforming the Soviet Union did not impress Ukrainian nationalists, it reflected the US focus on avoiding the nuclear proliferation that might accompany the collapse of the Soviet Union. Less than three weeks later, Russian conservatives sought to oust Gorbachev; this putsch (or rather its failure) led to Ukraine's declaration of independence.

This chronology explains why the United States was so adamant that Ukraine denuclearize. The START I treaty was both a major advance in US-Soviet relations and an important victory for President Bush, but it now could not go forward if Ukraine did not become party to it. From the beginning, the US applied pressure, one part of which was casting the 1990 Sovereignty Declaration as a legally binding commitment. No element of foreign policy was as important to the US as non-proliferation, and the prospect of adding three new nuclear states appalled US leaders. Russia shared these goals, as well as the goal of using nuclear weapons as a reason to retain a unified Commonwealth of Independent States military, controlled from Moscow, which would limit the sovereignty of the newly independent states. In sum, while Russia had many reasons to want to lean on Ukraine, the US was focused on two policy goals: arms control with Russia and nuclear non-proliferation. The result was a *de facto* US-Russia alliance to convince Ukraine–or to compel it, if needed–to surrender the weapons. This alliance persisted even as the US and Russia increasingly disagreed on topics such as Yugoslavia, and even as democracy in Russia reached a precipice in the autumn of 1993. One cannot mistake Kostenko's smoldering resentment, shared by many in Ukraine, over this state of affairs.

In this unfavorable context, Ukrainian leaders faced a classic version of the security dilemma that features prominently in international relations theory. Facing a palpable security threat in the form of a powerful neighbor that did not accept Ukraine's independence, Ukraine could seek to retain its nuclear weapons. Doing so, however, would almost certainly entail a response from others. Not only would Western states withhold crucial aid, but Russia might attack Ukraine in order to gain control over the weapons. In other words, a measure intended to guarantee Ukraine's security might have the opposite effect. This was the argument made by those who supported rapid denuclearization, and this view was encouraged by US and Russian leaders. Kostenko does not entirely reject this argument, but he finds that a policy of "effective disarmament," by which he means retaining the weapons longer and getting much more in return for them, would have been a better choice than simply "capitulating," as he puts it, to pressure.

Kostenko details two related battles. One was the internal one based primarily in the Ukrainian parliament between people like Kostenko, who wanted to place at least some conditions on denuclearization (such as sending warheads somewhere other than Russia, getting financial compensation, and receiving security guarantees), and those, largely in the executive branch, who believed that Ukraine was best off ridding itself of the weapons as soon as possible, under terms preferred by the West and Russia. The second was the external battle between Ukrainian leaders on one hand, and the US and Russia on the other.

Ukraine's denuclearization became explicitly linked to the START process through the May 1992 Lisbon Protocol, which made Ukraine, Belarus, and Kazakhstan party to the START I agreement. The stakes were raised when George H. W. Bush and Boris Yeltsin signed the START II agreement in January 1993. This made even deeper cuts in the US and Russian arsenals dependent on Ukraine's disarmament. In Ukraine, the debate on denuclearization intensified in 1993 as the need to ratify the START I Treaty brought the parliament, where Kostenko led a group of disarmament skeptics, into the policy process.

This process exposed the political and institutional conflict between the parliament and the executive in Ukraine. As with Russia, the US threw its support behind a president with whom it could work rather than with a parliament that was more recalcitrant. Such institutional conflicts are normal in a checks-and-balances system, but the US essentially took the position that the Ukrainian parliament was obligated to ratify whatever treaty its president signed. It is worth noting that in Ukraine, as well as in Russia, a new constitution was eventually adopted which granted very extensive powers to the presidency, with deleterious effects on efforts to build a functioning democracy.

At the summit in Masandra in September 1993, Russia shut off gas supplies to coerce Leonid Kravchuk to sign on to concessions on both the Black Sea Fleet and nuclear weapons. He was forced to backtrack after an uproar in Parliament. The incident highlighted Ukraine's dilemma: if it did not surrender the weapons, Russia and the US could devastate it economically. But Russia's willingness to coerce Ukraine demonstrated the need for some deterrent or countermeasure.[1]

By this time, the Ukrainian political situation had become more fraught, shaping the dynamics that Kostenko details. Having rejected shock therapy, Ukraine's economy was in freefall, and inflation was spiraling out of control. Street protests led to an agreement to hold early presidential and parliamentary elections in 1994. The run-up to those elections increasingly shaped the Ukrainian actors' calculus. Leonid Kuchma, who as former director of the Pivdenmash (Yuzhmash) missile factory was both knowledgeable about nuclear affairs and supportive of Kostenko's efforts, resigned and became the main challenger to Kravchuk.

At the same time, worried that Russia and Ukraine could not work out Ukraine's disarmament, the US took over the lead in negotiating with Ukraine, turning a bilateral process into a trilateral one. In November, the Ukrainian parliament ratified START I, but with significant reservations that, from the US perspective, amounted to non-ratification.

With the economy reeling and an election coming, Kravchuk needed an injection of aid, and the US made it clear that such an injection was conditional upon ratification of START I. Moreover, in a move with clear echoes in 2019, the US administration refused to schedule a meeting between Bill Clinton and Kravchuk until the deal was done. Only when Kravchuk agreed to sign the new Trilateral Statement with the US and Russia, which filled in gaps in START I, did Clinton agree to meet Kravchuk. Kostenko's account of this meeting is fascinating. The meeting took place at Kyiv's Boryspil airport, and Clinton never even left the airport before reboarding his plane and flying on to Moscow for the treaty signing. Kravchuk, Kostenko reports, was left standing in the cold on the tarmac for an hour waiting for Clinton to emerge from Air Force I. If the goal was to put Ukraine and Kravchuk in their places, the message seems to have been received.

The saga did not end there, for the Trilateral Statement and START I still needed to be ratified by the Ukrainian parliament. Ukrainian political developments made this much easier. The 1994 parliamentary elections led to dramatic turnover, and many of the disarmament skeptics left office. Kuchma became president, and was desperate for economic aid. A new aid package secured ratification.

In December 1994, the Budapest Memorandum was signed, committing the US, UK, and Russia to "refrain from the threat or use of force against the territorial integrity or political independence of Ukraine." The last warheads were moved to Russia in 1996. Kostenko's regret over the events in question stands in stark contrast to the enduring view in the US and Russia that Ukraine's disarmament was an unqualified success.

II. What Have Others Said?

The English-language scholarly literature on Ukraine's disarmament is limited. The memoirs of many participants touch on this process, as do various histories of US-Russian relations in this period. Detailed treatments are far fewer.

In post-Soviet Ukraine, officials relied upon the most accessible Western scholarship, usually published by institutes and think tanks. In 1993 the American international relations scholar John Mearsheimer published an article in *Foreign Affairs* titled "The Case for a Ukrainian Nuclear Deterrent."[2]

> Ukraine cannot defend itself against a nuclear-armed Russia with conventional weapons, and no state, including the United States, is going to extend to it a meaningful security guarantee. Ukrainian nuclear weapons are the only reliable deterrent to Russian aggression. If the U. S. aim is to enhance stability in Europe, the case against a nuclear-armed Ukraine is unpersuasive.[3]

Mearsheimer went on to argue that US policy was futile, because the US could not compel Ukraine to give nuclear weapons "to Russia, the state it fears most." In this second argument, Mearsheimer turned out to be wrong, but the first argument was seized on by Kostenko and like-minded Ukrainians to support their arguments. Mearsheimer later drew considerable attention by arguing, in 2014 in the same journal, that Russia's invasion of Ukraine was "the West's fault."[4] Curiously, he does not cite coercing Ukraine to disarm to support that argument, but rather blames the expansion of NATO.

Mearsheimer's article was followed, in the same issue of *Foreign Affairs*, by a rejoinder by Steven E. Miller. Miller contended that, "When the costs and complications associated with nuclear acquisition are taken into account, the case for Ukrainian nuclear weapons is not compelling."[5] He argued that Ukraine's potential "instant proliferation" would leave it without effective institutional control of the weapons. He also argued that Ukraine would always be outgunned by Russia, and that, ultimately, nuclear weapons would make Ukraine less secure, not more. This essentially was the official US view. Miller argued that the factors that led nuclear weapons to add stability to the US-Soviet relationship did not apply. The argument that what is good for the "great powers" is not good for lesser ones has long irked both the states seeking to obtain nuclear weapons and those who have refrained. It is also worth noting that neither Miller nor Mearsheimer was a specialist on the post-Soviet region, let alone Ukraine. While Kostenko stresses the unpreparedness of Ukraine for independence, the dearth of US expertise on Ukraine at the time was also consequential. Scholars and the government hurried to catch up to a situation that seemed to change faster than we could understand it.

Writing in the late 1990s, I included a chapter on disarmament in my book *Economic Interdependence in Ukrainian-Russian Relations*.[6] This analysis placed the nuclear weapons question in the context of Ukraine's broader effort to establish its sovereignty and autonomy from a country that was both its biggest trade partner and its greatest security threat. From this perspective, Ukraine's acquiescence was driven by both "push" and "pull" factors. The push was Ukraine's economic collapse. Ukraine's disastrous economic policies left it in a weak position with respect to Russia, on whom it depended, and with respect to the US, who held the key to significant bilateral and multilateral aid. The pull was the prospect that the US and Russia would acknowledge, through the Trilateral Statement and the Budapest Memorandum, Ukraine's sovereignty and territorial integrity. While the guarantees were thin, the legitimacy they conveyed was immensely valuable at a time when many still questioned Ukraine's statehood. My book concluded with an argument that still seems relevant today: the reform of Ukraine's state institutions and economy are the keys to effectively preserving independence from Russia.

Christopher Stevens provided a theoretically and empirically novel explanation of Ukraine's disarmament in 2008. He argued that Ukraine's policy was not driven by the issues of compensation and security on which Kostenko and most others have focused, but on Ukraine's identity, and specifically on Ukrainians' conception of Ukraine's relationship with Russia. Based on analysis of Ukrainian news sources and interviews with Ukrainian policy makers, Stevens finds that the long history between Ukraine and Russia left "most Ukrainian citizens socialized not to identify Russia as a real security threat."[7]

Steven Pifer worked extensively on US policy toward Ukraine and its nuclear weapons as a foreign service officer before serving as the US ambassador to Ukraine from 1998 to 2000. His inside account of US-Ukrainian relations includes a chapter on the disarmament process, placing it in the broader context of US-Ukraine relations.[8] His detailed discussion of the internal US policy-making process is analogous to, if much less opinionated than, Kostenko's discussion of Ukraine's process. He acknowledges that the US may not have been sufficiently sensitive to Ukraine's concerns, but while he argues that the process could have been smoother in this respect, he does not find that the outcome could have or should have been dramatically different. Pifer concludes: "From Ukraine's perspective, the trilateral process produced acceptable, if not ideal, outcomes on each issue."[9] Writing, like Kostenko, after Russia's invasion of Ukraine in 2014, Pifer laments that the US and EU did not do more to support Ukraine, but he maintains (along with most other observers) that the more robust security commitments that Ukraine sought in 1993–1994, were simply out of reach: "neither the Bush nor the Clinton administration was prepared to offer that kind of guarantee."[10]

The most detailed scholarly analysis of Ukraine's policy on this issue is Mariana Budjeryn's book *Inheriting the Bomb*, which studies the Ukraine case along with those of Belarus and Kazakhstan.[11] Budjeryn inquires particularly about the impact of the international norm of non-proliferation relative to coercion applied by Russia and the US, finding that the norm and coercion had complementary effects. For Ukraine to keep the weapons, she shows, would be to violate the non-proliferation norm, taking Ukraine outside the behavior of "civilized" states at the very moment it was trying to assert its status as a normal, civilized state. Rooted in constructivist international relations theory, her analysis shows how "going nuclear" contradicted the predominant notion within Ukraine of Ukraine's identity in the international system. She also points out that the disagreements among Ukrainian elites, which Kostenko chronicles in detail, undermined Ukraine's credibility as a bargaining partner with Moscow and Washington.

This theme of internal debate is taken up by Nadiya Kravets in her 2012 dissertation, "Domestic Sources of Ukraine's Foreign Policy,"[12] one chapter of which focuses on Ukraine's denuclearization. Kravets invokes Robert Putnam's concept of a "two-level" game, in which Ukrainian leaders were bargaining internationally with the US and Russia while also bargaining domestically with various constituencies, of which Kostenko's group of skeptics in the parliament was one.[13] She stresses that competition among Ukrainian elites partially accounts for its policy on nuclear weapons, especially as elections loomed in 1994.

This brief survey highlights three points. First, Kostenko's arguments often mesh with those of scholars who have studied the case in depth. For example, Ukraine's internal divisions are a recurring theme. Second, however, there are some key areas of disagreement, most important of which is whether Ukraine could have gotten a fundamentally different deal if not for those domestic divisions. Third, therefore, the extraordinary detail that Kostenko provides will add considerably to our understanding of the process in Ukraine.

III. Kostenko's Contribution: What Might Have Been?

Kostenko's central argument is that, even if Ukraine would not have benefitted from "going nuclear," there were multiple options for how to achieve its disarmament, and the one that was chosen was far from the best. Exactly what was possible and what was not must remain a matter of debate, and it is a debate Kostenko joins with gusto. He makes a compelling case that Ukraine could have reached a better deal. He argues that the outcome was determined primarily by the US and Russia, and that the provisions therefore served the interests of the US and Russia. It is hard to disagree with this view, even if one disagrees about what alternatives were available.

The key question, then and now, is what Ukraine's range of options was. At the time, US and Russian leaders tried to convince Ukraine that it had no choice but to give up the weapons unconditionally. At the extreme, this would mean transferring the warheads to Russia, getting no compensation in return, and with no commitments on Ukraine's future security. The final deals reached indicate that more than that was possible: Ukraine received some compensation and security guarantees that were symbolically important even if insufficient to prevent Russian aggression.

Kostenko believes that even more could have been achieved. Had Ukraine bargained more effectively, he argues, it could have convinced the US to move the nuclear material somewhere other than Russia, to provide compensation closer to the true value of the material, and to provide more meaningful security guarantees–including possibly NATO membership. This is what he calls effective disarmament in chapter 2. The first two of these seem likely. While Ukraine did not have a strong hand to play, one can easily imagine it being played more skillfully. It seems highly unlikely, however, that NATO membership or a security treaty equivalent to the NATO Article V commitment was going to be offered by the US or its allies. Kostenko reports on a conversation between US undersecretary of state Frank Wisner and Ukraine's ambassador to Washington, Oleh Bilorus, in December 1992, in which, he says, Wisner urged Ukraine to join NATO. This was surely a misreading by Kostenko. If Wisner said something to that effect, it was the Partnership for Peace that he was advocating, in keeping with the US and NATO line. Virtually every US source, then and now, says that NATO membership or a treaty committing the US to defending Ukraine were not on the table. Indeed, at this time, the Clinton administration was still a year away from announcing its support for enlarging NATO to include Czechia, Hungary, and Poland.

Complicating the question of security guarantees is the argument, which Kostenko advances, that NATO membership for Ukraine was unacceptable to Russia. The Russian government, as well as many Western critics of US and EU policy, argue that the prospect that post-Yanukovych Ukraine might join NATO was a major factor in provoking Russian attacks in 2014. These arguments highlight the difficulty of Ukraine's security dilemma, given Russia's policies. The very measures intended to protect Ukraine from Russia were seen as potentially provoking a Russian attack.

Whether the US would have agreed to removing the weapons to somewhere other than Russia seems questionable, simply because the warheads were highly toxic and dangerous to move. Putting them on aircraft or ships would have entailed extensive security risks, and required costly measures, compared to trucking them back to Russia. Moreover, it is not clear what the impact of moving the warheads elsewhere would be. While it certainly seemed strange to move such powerful weapons to one's biggest security threat, in practice the consequences would be indistinguishable from moving them elsewhere. If Ukraine were to surrender its nuclear weapons, it would be left with none, while Russia would be left with enough to destroy Ukraine many times over. That would be true whether the weapons were moved to Russia or elsewhere, because they would not make a qualitative impact on Russia's arsenal.

The big question, therefore, was compensation, and here one can certainly make the case that Ukraine might have received much more. Kostenko points to three reasons why Ukraine did not do as well as it could have, and we might add a fourth.

First, an apparent Russia-US alliance left Ukraine in a very weak bargaining position. It is as hard in 2020 as it would have been in 1985 to imagine the US and Russia as allies, but this is what appeared to be emerging after the collapse of communism. Denuclearization of Ukraine was not the only thing on which they collaborated. In 1991, even before the collapse of the Soviet Union, Mikhail Gorbachev acquiesced to the US invasion of Iraq, withholding the Soviet Union's veto in the UN Security Council. In both Washington and Moscow, the prevailing opinion was that communism and the occupation of Eastern Europe had been the main causes of the Cold War, and that, with Russia's rejection of communism and retreat from Eastern Europe, lasting partnership was at hand. The leaderships of the two countries were united in believing that the relationship was on the right track and must not be derailed. Not only would Ukraine's retention of nuclear weapons potentially erode the NPT, it would undermine the most reliable pillar of US-Soviet/Russian collaboration: nuclear arms control.

Even the obvious weaknesses of democracy in Russia did not undermine the Russia-US alliance to force Ukraine to denuclearize. The emergence of a "red-brown" coalition of nationalists and communists showed that the liberal democracy to which Yeltsin appeared to be committed was under threat. Many Western strategists pointed to the importance of Ukraine as a bulwark against Russian aggression. Sherman Garnett, in the first study of Ukraine's post-Soviet security affairs, called it the "keystone in the arch" of European security.[14] Zbigniew Brzezinski, the former US national security advisor and probably the most respected voice in the US on Eastern European affairs, published an article in *Foreign Affairs* in the spring of 1994 asserting that "without Ukraine, Russia ceases to be an empire, but with Ukraine suborned and then subordinated, Russia automatically becomes an empire."[15] Crucially, neither Garnett nor Brzezinski, nor any other mainstream US strategist, advocated that Ukraine retain its nuclear weapons. No one in Russia did either. This helps explain why Ukrainian efforts to link the nuclear weapons question to the broader problem of Russian revanchism fell on deaf ears. The two problems were seen in the West as being of a different order of magnitude. While how to promote Russian democracy in the face of revanchism spurred debate, the need to secure Ukraine's denuclearization was a certainty, and was left undebated. Mearsheimer's argument that Ukraine should be allowed to keep its weapons did not spur a major controversy; it was mostly ignored.

Second, Ukraine was internally divided and utterly unprepared to deal with such a difficult and tricky negotiation. Kostenko stresses the unpreparedness of the Ukrainian government for the negotiations. While many in Ukraine sought greater autonomy from Moscow in 1990–1991, total independence seemed far in the distance. Instead, it emerged suddenly in August 1991. The result was chaos. Kostenko points to several consequences. First, the new Ukrainian leaders did not even know how many nuclear weapons were on their territory, let alone how to control them. Second, there was little apparatus to analyze the issues or to carry out negotiations. If knowledge is power, Ukraine had yet another dramatic weakness compared to Russia and the US. Third, the centralization of the Soviet system, and the subservience it demanded, meant that the new officeholders in Ukraine were largely accustomed to taking orders from Moscow.

Kostenko chronicles the emergence of Ukrainian policy in engrossing detail, and the themes of unpreparedness and division drive the drama and explain the outcome. Kostenko's accounting of the paucity of expertise–or even basic information–that the new Ukrainian state possessed is an important corrective to the assumptions of coherence and rationality we tend to apply. This lack of expertise played into the division of opinion about what options were open to Ukraine and what strategies might work best. Supporters of rapid denuclearization argued that Ukraine could not do anything with the weapons, because they had unbreakable launch codes that Ukraine did not have. To the extent this was true, it made retaining the weapons pointless. This in turn made the threat of retaining them empty. Related was the fear that Russia might actually launch the weapons from Ukraine, making Ukraine an unwilling party to a nuclear attack. Others contended that Ukraine could take control of the weapons, though estimates of how long it would take and what it would cost varied. The truth should have been discernible, but the lack of expertise meant that Ukrainian leaders could not get definitive answers. With competing assertions on these matters, it was easy for those inclined one way or the other to believe whichever technical arguments supported their case.

These strategic divisions in Ukraine seriously undermined Ukraine's bargaining position by reducing the credibility of the negotiators. On the one hand, Kravchuk and the people around him made concessions and commitments that Kostenko saw as foolhardy. On the other hand, efforts by Kostenko and others to slow down the process were seen by Russia and the US as evidence of bad faith. This then made US and Russian pressure seem more justified. Had Ukraine developed a single position–even if it were close to the one Kravchuk sought–and stood behind it with something approaching consensus, the process would likely have turned out better. Such consensus was made impossible by fundamental disagreements on strategy.

Strategic disagreement was compounded by institutional conflict. Constitutionally, Ukraine was in flux, and the prerogatives of the president and the parliament were not clear. Ukraine was not to adopt a post-Soviet constitution until 1996, and even that did not end the institutional battle between the branches. As well as using their institutional power to fight for preferred policies, the two branches were using the policy battles to fight for their institutional power. Thus, the parliamentary leadership was jealous in reserving the parliament's right to ratify–or not ratify–treaties. These divisions allowed Russian and US negotiators to play the various actors off against one another.

Equally consequentially, the institutional conflict prompted the US to come down on the side that would support its goals in the disarmament question, which had long-term institutional ramifications for the nascent Ukrainian state. Because Kravchuk and Kuchma were more amenable to rapid denuclearization than was the parliament, the US acted as though the president had the sole power to make treaties, and that the parliament was obligated to follow the president's lead. One should not overestimate the weight of the US in Ukraine's institutional battles, but the result, as in Russia, was that a hypertrophied presidency, initially supported by the West, later came to threaten the survival of democracy in the country.

Third, the new Ukrainian government was thoroughly penetrated by people who were likely reporting directly to Moscow. As Kostenko notes in chapter 1, "[B]ecause the entire executive branch of independent Ukraine was inherited from the Ukrainian SSR, the new state essentially was being built by those who just a few months earlier were still being paid to ensure that such a state would never exist." Kostenko asserts that systems set up by the KGB to allow Moscow to monitor all phone conversations remained intact. In an interesting aside, he says that this system, not personnel in the presidential security detail, was responsible for recordings made in Leonid Kuchma's office that kicked off the "Kuchmagate" crisis in 1999.

There was a fourth reason, to which Kostenko gives much less attention, why Ukraine could not bargain more effectively for the concessions Kostenko believes were possible. Ukraine's economic mismanagement left it increasingly vulnerable, and US and Russian negotiators took advantage of this. Kravchuk and then Kuchma badly needed an economic lifeline and the political victory that went with it. US leaders correctly surmised, in late 1993 when dealing with Kravchuk and again in 1994 when dealing with Kuchma, that finishing a deal was simply a matter of money (along with some presidential audiences). It was just a question of how much, and in retrospect, one must agree with Kostenko that the amount received–less than a billion dollars–seems pitiful given what was at stake for the US and Russia. Of the many counterfactuals involved in thinking about this case, one that receives insufficient attention is what might have happened had Ukraine embraced economic reform, or at least managed its "go slow" strategy more effectively. Instead, internal division and the oligarchization of the economy then underway made Ukraine easy to divide and conquer. This theme–internal division in the face of powerful external forces–recurs tragically in Ukraine's history.

Related to this, Kostenko asserts that, at several points in the process, decisions for which leaders were not thoroughly prepared put Ukraine at a disadvantage as negotiations progressed. For example, he contends that the provision in the Lisbon Protocol requiring Ukraine, Belarus, and Kazakhstan to join the NPT as nonnuclear states was added at the last minute, and therefore never analyzed or debated. Further, he says, when Leonid Kravchuk verbally committed to George H.W. Bush that Ukraine would denuclearize within seven years, he was committing to more than the Lisbon Protocol required, and thus foregoing important leverage. In a similar vein, he asserts that, by mid-1993, the US was coming to take Ukraine's position on several key issues, but that Kravchuk's concessions at the Masandra summit undermined its position. It is not clear that the US was moving to back Ukraine as quickly as Kostenko hoped, but the Masandra episode embodied many of the problems that he points to: Russia's heavy-handed coercion, Kravchuk's willingness to cave in quickly, the parliament's furious rejection of the president's concessions, Kravchuk's clumsy walk-back from those concessions, and international accusations that Ukrainian leaders were not negotiating in good faith.

IV. Assessing Counterfactuals

Counterfactual arguments are assertions about things that did not happen. Put differently, by "counterfactual," we refer simply to "propositions that take the generic form 'If it had been the case that C (or not C), it would have been the case that E (or not E).'"[16] Such assertions are explicit or implicit in a wide range of historical and political studies. An oft-cited example focuses on what might have happened, had Archduke Franz Ferdinand not been assassinated in July 1914.

Kostenko's argument is that, if Ukraine had negotiated more effectively, it could have received more in return for denuclearization, including security guarantees more meaningful than those in the Trilateral Statement and the Budapest Memorandum. As I have pointed out, especially on security guarantees, at least some observers disagree, arguing that Ukraine could not have gotten more than it did.

Three other counterfactual arguments are worth highlighting. The first was the argument, made at the time and since, that if Ukraine did *not* acquiesce to denuclearization on US-Russian terms, the results would have been disastrous. The second was that, had Ukraine kept the nuclear weapons, its security would have been guaranteed against the sort of aggression that occurred in 2014 and 2022. A third counterfactual is generally left undiscussed: what if Ukraine had surrendered its nuclear weapons as it did, but also embarked on serious political and economic reform, rather than continuously avoiding reform, while Russia eventually rebuilt its might?

What would have happened to Ukraine had it held out for a better deal, and as a result kept the weapons for a much longer period of time? The West was threatening that it would not enter any kind of trade, security, or aid agreement while the nuclear issue remained unresolved. Similarly, Russia used the intermittent cutoff of gas to pressure Ukraine. The West's threat to withhold support may not have been as draconian as it seemed at the time. Ukraine was not getting much from the West, and so did not have much to lose. But it was desperate for aid and symbolic support. The Russian threat was much more significant and much more immediate. The Ukrainian economy was built, under the Soviets, on the premise of heavily subsidized energy. Simply moving to world market prices would have been wrenching. Losing access to Russian oil and gas altogether would have thrown the economy into an even bigger tailspin than the 40 percent fall in GDP that it actually experienced. People both inside and outside Ukraine predicted that this would have led to the collapse of the Ukrainian state. Thus, Kravchuk felt that, by knuckling under, he was preserving Ukrainian statehood. We of course do not know what would have happened. Perhaps a trade war with Russia would have forced Ukraine to redirect trade elsewhere, or to pursue rapid rather than incremental reform.

If Ukraine had kept the nuclear weapons, would Russia have been deterred from attacking? In the long run, the answer is almost certainly yes. Russia incurred very little cost from its attack in 2014, and it believed that it could easily defeat Ukraine in 2022. If Ukraine had nuclear weapons, Russia would likely have been deterred in both cases.

What this argument leaves out is whether Russia would have sat by, once Ukraine showed its intent to control the weapons, and allowed it to happen. As strategic theory stresses, the time of greatest danger is the moment when a potential nuclear state is on the verge of "going nuclear." In other words, if Ukraine were seeking to gain operational control of the weapons, Russia would have had a powerful incentive to destroy them. One must speculate, but the United States may well have endorsed such an attack, or simply stood aside.

Finally, we should ask what would have become of Ukraine's security if it had reformed rapidly and competently after 1991. This might seem the most unlikely of counterfactual scenarios, but it leads us to an important point. Ukraine in the early 1990s got economic shock without much therapy. The military was shrunk, but not dramatically reformed. Had Ukraine's economy actually been reformed and grown, it might have been able to afford to better strengthen its military. Even without more spending, the leftover hardware from the Soviet era could supply a powerful army, if it were reformed. Instead, the Ukrainian army became synonymous with corruption and, when Russia attacked in 2014, there were no major forces based east of the Dnipro River. One can only speculate, but it seems likely that a robust conventional deterrent could have been constructed that might well have prevented or blunted a Russian attack. Kostenko seems to believe that the financial terms of the nuclear deal were so bad for Ukraine that they actually impeded reform, but there is not much evidence of strong impetus for reform at that time.

Rather than pile up more "what ifs," I close with an important point. All the retrospective arguments about Ukraine's disarmament are based on assumptions about what might have happened differently. Thus, how one evaluates

Ukraine's disarmament policy depends to a large extent on what one believes the alternatives were. If one assumes a smooth transition to nuclear status or to much more reliable security guarantees, then Ukraine's policy looks to have been a failure. If one assumes that Ukraine got as much as it could have gotten, the policy looks like a success, and the catastrophes of 2014 and 2022 must be attributed to unrelated factors.

V. Future Implications

Is Kostenko's book only about the past, or is it about the future as well? While the book relates largely to the past, it provokes two important questions about the future. First, should Ukraine now embark on an effort to obtain nuclear weapons? Second, what are the consequences of Ukraine's experience for nuclear proliferation? While Kostenko only briefly addresses these questions, consideration of them shows why his book is so timely in 2023.

Ukraine is at war with Russia, and at the time of writing (February 2023) the war looks far from over. It may end with Russia's occupying significant Ukrainian territory and being poised, after a period of recovery, to try to complete its planned destruction of the Ukrainian state. The West remains deterred by its fear of Russia from fighting on Ukraine's behalf or transferring the most effective conventional weapons. Acquiring nuclear weapons must now look very tempting to Ukraine.

Examining the question helps us understand the dilemma Ukraine faced in the early 1990s. Clearly Ukraine faces an existential security threat from Russia. However, efforts to address it by obtaining nuclear weapons might cause more harm than good. It would take years before a nuclear weapons program would bear fruit and, in the meantime, Russia would have a heightened incentive to attack. The West would likely pivot from supporting Ukraine to sanctioning it, and Ukraine is not in a position to thrive as a pariah.

What about other states contemplating getting nuclear weapons? For Iran and North Korea, Ukraine's experience can only induce additional resistance to abandoning nuclear weapons programs. That Russia invaded Ukraine twenty years after Ukraine gave up its weapons highlights the nature of the commitment problem in nuclear disarmament. A disarmament deal normally trades disarmament for promises not to attack in the future. The problem is that once disarmament is achieved, there is no way of enforcing the commitment to refrain from attacking. Such a commitment has to last forever in order to be worthwhile. Nor is there any way to enforce any third-party security guarantee. In fact, as Robert Powell has shown in general, once a party has agreed to disarmament, it could become rational for another state to violate its commitments.[17] Thus, Powell argues, the security dilemma is hard to resolve because, by their nature, commitments to refrain from future actions can never be fully credible. This is a very theoretical argument. The case of Ukraine has made it a very concrete one.

It would appear that none of the actors involved in Ukraine's disarmament have changed their minds about what happened. Observers in Russia and the West regard Ukraine's disarmament as a great success, see Ukraine as having done as well as it could, and see Russia's invasion and occupation as largely unrelated to nuclear disarmament. Those in Ukraine who supported disarmament do not appear to have reconsidered, Kostenko's discussion shows. Kostenko, who was skeptical in 1993, continues to believe that a much better deal could have been reached.

What is indisputable is that Ukraine disarmed in the early 1990s and later found itself defenseless. While the West has provided crucial aid to Ukraine in its war against Russia, it has refrained from doing all it can to help Ukraine win the war. Russia remains in control of considerable Ukrainian territory, and seeks even more. Ukraine is in mortal peril and it citizens are being killed every day. In these circumstances, one cannot be surprised by resentment among Ukrainian leaders.

Nor can one be surprised at the hesitation of others to take the same deal. Ukraine's fate points to a problem that the great powers advocating nonproliferation have failed to seriously address, let alone solve: How can states be secure without nuclear weapons? It seems like a simple question, but, without a clear answer to it, pushing states to abstain from gaining nuclear weapons is equivalent to asking them to abstain from security. Would Germany and Japan shun nuclear weapons absent their confidence in the American nuclear umbrella? The problem is especially pressing for those, like Ukraine, with clear and present adversaries. The nonproliferation regime set up a world of nuclear "haves" and "have-nots." To the extent that it also sets up a world of security "haves" and "have-nots," it cannot be expected to endure. One imagines that many years from now, when the story of nuclear proliferation is being told, Ukraine's story, as told by Kostenko, will garner much more attention than it has until now.

Kostenko not only dramatically improves our understanding of crucial events that took place a quarter of a century ago. By elaborating how he thinks events could have transpired differently, he compels us to ask about the choices available to the actors today. Are the options really what we tell ourselves they are? Is it again a case where Russia does what it can and Ukraine suffers what it must? Or are there other steps that could be taken, such that the failure to take them will lead to new recriminations in the future? Most importantly, if the question is what was the impact of Ukraine's disarmament, the answer must be that it is too soon to tell. The consequences will continue to emerge.

Author's Note

Unfinished Business

I began working on this book in February 2013. What Ukraine has become after more than twenty years of independence evoked sentiments of indignation, as well as regret, over a missed opportunity. One of the greatest opportunities in modern history was presented to our country through the inheritance of the third largest nuclear stockpile in the world.

During the first months of Ukraine's independence in the early 1990s, I headed the parliamentary commission tasked with developing the national security policy of Ukraine. Later on, I chaired the parliamentary working group (hereafter, the Working Group) that provided the foundation for ratifying the Strategic Arms Reduction Treaty (START I) and the Lisbon Protocol. Additionally, I was the leader of Ukraine's official delegation in negotiations with Russia over the status of nuclear weapons in Ukraine.

My position as a member of Parliament (MP), a member of the Defense Council of Ukraine, and the minister of environmental protection allowed me an insider's perspective on Ukraine's nuclear disarmament process and provided me with insight into the viewpoints of all of the Ukrainian officials involved in the process. My knowledge of English enabled me to communicate with American and European officials, both formally and informally, as well as to gain firsthand information about Ukraine's cooperation with the West, bypassing Russia's persistent interference.

The years covered in this book have left me with more than just memories, emotions, and an understanding of the political actors' views. As a scholar, I have a habit of collecting documents. While working on the nuclear weapons issue in an official capacity, I collected dozens of documents, from publications in the Western press to personal notes and copies of confidential documents.

The idea of recounting the history of Ukraine's nuclear disarmament came to me quite some time ago; yet, with all the preoccupations of daily life, I kept putting the task off. However, the longer this passage of Ukraine's history with its catastrophic outcomes was left unaddressed, the more contemporary developments were driving the country further back into the status of a colony in a new Russian empire.

Prior to Ukraine's Revolution of Dignity (2013–2014) and the events that followed it, monuments to Lenin remained standing in Ukraine, and countless streets still bore the names of Communist Party leaders and Soviet-era events. The Communists felt at ease and remained a political party in the Parliament of Ukraine (*Verkhovna Rada*). Our history textbooks were being rewritten, often jointly with Russia. The border with Russia remained porous. Our army not only lacked fuel, but also quite literally the wheels for service vehicles. The henchmen of Russia's Federal Security Service (FSB) held leadership positions in the security services and armed forces of Ukraine; these included former Russian citizens such as Oleksandr Iakymenko, who was the head of the Security Service of Ukraine (SBU) in 2013. Television programs were mostly Russian sitcoms, radio stations played Russian music, and concert halls heralded Russian pop stars. Monuments to Russia's empress Catherine II and to Joseph Stalin were being erected in Ukrainian cities, while Ukrainian *hetmans* (political and state leaders from the country's history) remained under the anathema of the Orthodox Church of the Moscow Patriarchate. On 9 May, which the USSR celebrated as Victory Day over Nazi Germany, some Ukrainians, with Russian prompting, decorated their clothes and vehicles with St. George's ribbons, seen as a symbol of the Soviet holiday. In the main, government officials attended services in churches of the Moscow Patriarchate.

In 1993, Russia began encroaching upon the city of Sevastopol in Ukraine, and, in 2003, on Tuzla Island. During its *de facto* annexation in 2008 of Georgian territories and creation of new, self-proclaimed states, Russia blatantly demonstrated its disregard for international law. Yet, the Ukrainian parliament and government did not even call it an act of aggression.[1]

Despite Russia's openly aggressive international policy, public opinion in Ukraine still did not support the idea that it was necessary for Ukraine to join the North Atlantic Treaty Organization (NATO).[2] Ukrainian public opinion resisted the heroization of the Organization of the Ukrainian Nationalists (OUN) and the Ukrainian Insurgent Army (UPA) and still considered Russia a fraternal state, while a portion of Ukrainian society even felt nostalgic for the USSR.[3]

In 2014 and 2022, the grimmest predictions we made in the early 1990s became a reality. Russia annexed Crimea and then launched a military incursion into Ukraine's mainland. Back in 2014, I ran into an old acquaintance, an academic turned diplomat in the independence period, and heard him exclaim: "Who would have ever thought that Russia would wage war against us?!" For many deputies who served during the Parliament's first convocation, this possibility was already very real in the early 1990s. However, this bloc of deputies did not become a parliamentary majority until 2014.

Ukraine's disarmament is an episode in the history books in which, presumably, loose ends were tied up in the relationship between Ukraine and Russia during the first days of their existence as independent states after the collapse of the USSR. However, for the better part of three decades, this page in history has not been properly analyzed—not in Ukraine nor globally. The press and academic literature covered this topic mainly by presenting isolated facts and cliché statements about the reduction of the global nuclear threat.

Furthermore, as the years passed, the likelihood of seeing an archive-based analysis of this history continued to wane. The fact was that, until 1994, Ukraine did not even have the proper laws to ensure the adequate safekeeping of archival documents on the topic. Thus, there was yet another inducement for me to write this book, as it seemed to be the only opportunity to publicize a document-based interpretation of the events that I witnessed. My personal archive, which I accumulated during the time period covered in this book, might be of interest to readers, not only because it provides a complex view of the facts surrounding Ukraine's nuclear disarmament, but also because these materials allow for a detailed reconstruction of the country's unique circumstances during the early 1990s, as I briefly explain below.

Firstly, it is important to note here that some of the documents in my possession have no identifying information, including reports and memoranda provided to me by representatives of ministries and governmental agencies, not to mention the information of a consultative nature, which I received from the scientific community. This was not only a symptom of an undeveloped culture of official document circulation in the early 1990s, but also an expression of their authors' untenable position. On the one hand, these individuals wanted to provide me with credible information; on the other hand, they were well aware that their supervisors' stance on nuclear issues opposed my own. Therefore, the answers to my questions in these various documents were often unofficial, albeit in writing.

Secondly, and this was revealed to me in the process of writing this book, some of the translations of Western press publications officially sent by the Ministry of Foreign Affairs (MFA) to Ukraine's prime minister and the head of Parliament in the early 1990s, either did not entirely match the original articles, or the articles to which they referred could not be found altogether. Moreover, often these inaccurate translations or nowhere-to-be-found articles contained viewpoints that were actively used in the Russian press. This could, albeit indirectly, point to the unofficial communication between the Ukrainian and the Russian MFAs. Notably, these were the official Ukrainian sources that served as the basis for nuclear disarmament policy development. Such were the realities in which the president, the parliament, the government, and our working group operated at the time.

Thirdly, as a participant in the events described here, I must underscore the decisive role of personalities as a factor in these historical developments. Every document or event had real people behind it, all with their own convictions, biographies, education, and interpersonal relationships. These factors were pivotal in explaining the actions of government officials in the first years after the dissolution of the USSR. This is why I found it critical to accompany the text of this book with biographical data on the key individuals of the period.

What should be inferred by these biographies? At the time, the entire top leadership of the Ukrainian Soviet Socialist Republic's (Ukrainian SSR) government had worked their entire lives as career communists: they received university diplomas in technical sciences or humanities, and, within one to five years of graduating, obtained permanent positions within the organs of the Communist Party. All Ukrainian SSR officials were confirmed in Moscow, notably, by the KGB's so-called First Directorate. This process not only concerned the appointment of KGB staff, but of all other governmental agencies, particularly the MFA. Foreign affairs professionals, especially those who were sent abroad, were vetted to a high level of scrutiny. The Academy of Sciences of the Ukrainian SSR had the status of a provincial branch of the Soviet Academy of Sciences; Moscow was in charge of approving Ukrainian scholars' doctorate degrees. Many Ukrainian bureaucrats received their degrees in Moscow or other Russian cities, which was also the case for various military personnel. It is true that some of the officials, such as the USSR's last defense minister Yevgeny Shaposhnikov, were educated and performed military service in Ukraine. However, the Soviet Army always followed a unified command, with a complete integration of the Ukrainian and Russian militaries. The Russians were always thoroughly informed of the events taking place in the Ukrainian army, from its structures and weapons, to the identification of specific individuals who were known to be "amenable."

Therefore, the government officials who found themselves in positions of leadership when Ukraine became independent were essentially shaped within the context of the Soviet power vertical, in which Kyiv played the role of a regional center strictly subordinated to Moscow. The overwhelming majority did not comprehend the true essence of issues pertaining to nuclear weapons. However, they did appreciate their role in the Soviet hierarchy. It is understandable why this dynamic could not change overnight, although there were some notable exceptions, including Ivan Pliushch, Anton Buteiko, and Volodymyr Tolubko. At the same time, there was a new generation of politicians and scholars who were not burdened by Soviet baggage; in early days of Ukraine's independence, this new generation included many opponents of the Soviet regime. However, most of these people did not have political leadership experience or specialized knowledge about nuclear weapons, and were easily manipulated.

The events of 2014–2023 have essentially shifted the global dynamics towards an opposition between the West and Russia, something that seemed virtually impossible at the beginning of the 1990s. This English-language edition of my book gives readers throughout the world the chance to see historical events that remain relatively unexplored: that is, the redistribution of nuclear stockpiles after the dissolution of the USSR, and the diplomatic and information wars that followed.

The production of an English-language version of my account, accomplished in collaboration with the Ukrainian Research Institute at Harvard University (HURI), has allowed for preservation of the unique archive that serves as the basis for this book. I am grateful to my colleagues at HURI for this opportunity.

The State of Ukraine at the Onset of the Nuclear Disarmament Process

After the dissolution of the Soviet Union, Ukraine inherited the third largest nuclear arsenal in the world, after the US and Russia. What were Ukraine's internal conditions at the time?

When the USSR collapsed, the Ukrainian SSR held outstanding economic promise— perhaps the greatest of the post-Soviet republics. In the national referendum of 1 December 1991, over 90 percent of Ukrainians supported the country's independence. For the vast majority, the main reason was economic, and was grounded in the hope of a swift increase in living standards. After Ukraine's declaration of independence and the collapse of the USSR, two simultaneous processes unfolded: firstly, the division of Soviet territory and property; and secondly, the formation of the governmental institutions of the new state.[4] In December 1991, the former Soviet republics agreed to divide property proportionately to their contribution to the USSR's gross domestic

1. *Main chamber of the Parliament of Ukraine (Verkhovna Rada), the first convocation, September 1994*

product (GDP). Ukraine was to possess 16.37 percent of all USSR assets, while Russia inherited 69.4 percent. However, Moscow used all methods at its disposal to appropriate as much property as possible. Ukraine's ultimate share in the division of Soviet property became contingent on its ability to appoint a team capable of counteracting Russia and safeguarding Ukraine's interests in the face of its overbearing neighbor.

The formation and staffing of Ukraine's state institutions after 24 August 1991 had a distinct character as compared to those of other East European countries because the entire executive power vertical was inherited from the Ukrainian SSR. In fact, it encompassed the same individuals who had worked in the Soviet system and who were in the habit of submitting to central authorities in Moscow. The new Ukrainian government's only advocates in Parliament were the 130 deputies belonging to the democratic opposition, the People's Council (Narodna Rada), out of a total of 450 deputies; meanwhile, the government held majorities in the local councils of only four out of twenty-six regions (*oblasts*): Lviv, Ivano-Frankivsk, Ternopil, and Volyn.[5] All other governing bodies remained as they had been during the Ukrainian SSR. Democratic-leaning MPs were even rarer in the executive branch of the government.

The old and new generations of Ukrainian politicians functioned in different spheres. The older generation remained entrenched in the Soviet system, where the USSR was held in highest regard and the Kremlin line continued to prevail, while the democratic West remained an enemy against which Russia and Ukraine were to unite. The new political generation focused on the national interests of the newly independent state. They believed that Ukraine's relations with both the West and Russia should be directed exclusively with an eye on the interests of Ukraine. This new group of politicians felt no nostalgia for the USSR; they did not trust Russia and did not carry around any communist ideological baggage.

By comparison, the Baltic countries' democratic oppositions defeated Communist candidates as early as the 1990 elections. This process allowed them to establish majorities within their parliaments quickly and to form democratic governments. Pro-democratic Lithuanians, Latvians, and Estonians severed their ties with the Kremlin during the first days of their countries' independence, and thus were able swiftly to institute political and economic reforms. In Ukraine, the effective control over the economy, the military, foreign policy, and the power vertical of the KGB remained *de facto* subject to the Kremlin.

In 1992, economic crisis and monetary collapse struck: the new Ukrainian currency, the *karbovanets*, severely depreciated in the currency market in the course of a single day, with the state's annual budget barely reaching US $7 billion.[6] The former USSR's economic network was disintegrating and Ukraine's defense industry was grinding to a halt, with 70 percent of the sector dependent on orders from Russia. Dependence on Russian gas and nuclear power plant (NPP) fuel, coupled with severely outmoded energy consumption systems in the industrial sector, pushed Ukraine's economy to the brink of ruin. Yet the government took no action to remedy the situation. On the contrary, it preserved economic dysfunction. Mass industrial closures in Ukraine's east were prompting social unrest by early 1993. Protesters demanded a snap election of the Parliament and the president. Under these circumstances, politicians became preoccupied with how to win elections. All other issues were secondary.

In the first not months but *years* of independence, Ukraine had no realistic knowledge of what it actually owned or how to prioritize its economic development. Thus, there was no development strategy. In early 1993, the Cabinet of Ministers mandated an inventory of all property and of the production output of Ukraine's industries in order to forecast their capacity to produce independently the majority of goods the country needed, and to develop a new economic strategy. However, the ministers never followed through with the mandate; they continued replicating the actions of Russian officials, not because of any formal subordination to Moscow, but for the simple fact that they were too inept to act independently, outside of the former Soviet system.

Likewise, Ukraine did not really know the specific characteristics of the nuclear stockpile (the world's third largest) it had inherited. Moscow was the only proprietor of that information. Nuclear disarmament could have provided the seed capital for the development of a new state. However, a strong Ukraine was of no interest to anyone but the Ukrainians. Russia's preoccupation with discrediting Ukraine's independence, coupled with the West's wager on Russia as the only successor of USSR's inventory of nuclear weapons, not only created hostile conditions for Ukraine's emergence as a new state, but also pitted it against the coalition of the two world superpowers, the US and Russia. These two states possessed experienced diplomatic institutions and powerful information resources; the only thing of interest to them about Ukraine was the presence of nuclear weapons on its territory.

The Ukrainian government had the legal right to dispose of these nuclear weapons in accordance with Ukraine's national interests. However, at the beginning of the 1990s, the state of the Ukrainian government was merely a remnant of the old Soviet system. Metaphorically, the Ukrainian state was the hands and the feet of the greater Soviet organism, whose head was still Moscow. And to that head Ukraine's government continued to yield by sheer habit.

The emergence of a new political generation sparked a kind of civil war within the hallways of the Ukrainian government. The new generation was a resistance movement of sorts inside the old edifice, while our actions were reminiscent of guerilla warfare. Against this backdrop the events unfolded that are the subject of this book: the process of Ukraine's nuclear disarmament.

From the History of the Nuclear Arms Race

The world's first nuclear bomb detonation took place in the state of New Mexico, in the US, on 16 July 1945. It signified the start of the nuclear arms race. The work leading up to this event, the Manhattan Project, commenced in 1939. The fruit of its labor shocked the world on 6 August 1945, when an atomic bomb equipped with highly enriched uranium was detonated over the Japanese city of Hiroshima. Three days later, on 9 August, another one, this time a plutonium bomb, was set off in the air above Nagasaki.

The Soviet Union carried out its own testing a mere four years later. On 29 August 1949, the USSR detonated its first RDS-1 bomb, containing the power of 22 kilotons of trinitrotoluol (TNT), at the Semipalatinsk Test Site. The USSR's production of nuclear weapons officially began in 1945.

On 3 October 1952, the United Kingdom carried out its first nuclear weapon tests over the waters of the Montebello Islands, just off the coast of Australia. Here, the nuclear explosion was equivalent to 25 kilotons of TNT, enough to decimate a a large city.

Next, a host of nations joined in the nuclear arms race. France conducted successful nuclear bomb tests in 1960. China followed in 1964, India in 1974, Pakistan in 1998, and North Korea in 2006. Israel has possessed nuclear weapons for the past half century. The Republic of South Africa also developed six nuclear bombs, but later terminated the program and destroyed its nuclear arsenal. Other large scale secret programs initiated to create nuclear bombs were launched in Iran, Iraq, Sweden, Lybia, Argentina, Brazil, South Korea, and Taiwan among others. Most of these countries abandoned their programs under U.S. pressure, but Iraq came perilously close to building a nuclear weapons before its program was dismantled after its defeat in the 1991 Gulf War. Iran continues to pursue nuclear weapons to this day.

With the goal of regulating the proliferation of nuclear weapons in the world, the US, the USSR, and the UK signed the Treaty on the Non-Proliferation of Nuclear Weapons, or Non-Proliferation Treaty (NPT), on 1 July 1968. The NPT formulated three key principles of non-proliferation: not to manufacture, not to test, and not to transfer nuclear weapons or their components. This is the only international document that provides a definition of a nuclear-weapon state as "one which has manufactured and exploded a nuclear weapon or other nuclear explosive device prior to 1 January 1967."[7]

In 1992, China and France joined the NPT as nuclear-weapon states. By this time, the Treaty had been formally signed by over a hundred countries; an additional thirty countries did not sign it, including those that had their own nuclear programs. The NPT required the holding of review conferences every five years, during which time the participants would decide whether, and for what period, the treaty was to be extended. Although the NPT gave nuclear status to only five countries—the US, the USSR, the UK, France, and China—nearly twenty other countries had come close to producing their own nuclear weapons by the early 1990s, considering it an effective method of deterrence against external aggression. After the collapse of the USSR, the next review conference was scheduled to take place in New York City from April to May of 1995. Unexpectedly, the conference approved an indefinite extension of the treaty.

In 2018, the world's nuclear arsenal consisted of a total of 14,465 warheads, 3,750 of which were deployed. The remainder were stored, in reserve, or retired. The weapons were distributed among the nuclear-weapon states as follows: the US deployed 1,750 nuclear warheads out of its 6,450 total nuclear arms; Russia's strategic carriers contained 1,600 nuclear warheads, with another 5,250 not deployed; the UK had a total of 215 deployed nuclear warheads, with 120 of them placed on strategic carriers. As of 2018, China had 280 nuclear warheads, all of them stored outside of strategic carriers. France was armed with approximately 300 nuclear warheads, almost all of them (280) deployed. Because the following four countries are not officially members of the NPT, the exact data about their nuclear arsenals are unknown. However, it is estimated that in 2018, India's nuclear forces contained 130–140 warheads, Pakistan had 140–150, Israel had 80, and North Korea is believed to have approximately 10–20 nuclear warheads.[8]

After the collapse of the Soviet Union, Ukraine had the world's third largest nuclear arsenal. With the exception of the US and Russia, its stockpile surpassed the nuclear potential of all other countries in the world—not individually, but combined.

CHAPTER 1. An Infant in a Grownups' Game

The Origins of Ukraine's Nonnuclear Status

The very idea that Ukraine might become a nonnuclear-weapon state surfaced before the country declared independence. Transcripts of sessions of the parliament of the Ukrainian SSR provide an accurate account of when the topic was first raised, who introduced it, and under what circumstances this took place.

It emerged unexpectedly. In early July 1990, the parliament of the Ukrainian SSR was reviewing a draft of the first independence document, the Declaration of State Sovereignty of Ukraine. This forward-looking document defined the boundaries of a state that did not yet exist. Some of the Declaration's provisions exceeded the competencies of a Soviet republic, especially the provisions on Ukraine's armed forces.

It was during parliamentary discussions of these provisions on 12 July that Ivan Drach, then the leader of the oppositional People's Movement (Narodnyi Rukh), requested to speak.[1] Beginning his narrative with the time of Ivan Mazepa, Ukraine's political and state leader in the late seventeenth and early eighteenth centuries, Drach concluded by reading the following words from a piece of paper:

> And perhaps it would be best if Ukraine were to declare the following: the Ukrainian SSR solemnly declares its intention to become a permanently neutral state that will not participate in military alliances and will adhere to the three nonnuclear principles of not manufacturing weapons, not distributing weapons, and banning the presence of weapons on its territory.[2]

This extemporization came as a surprise to me, as the idea had never been raised in any of the meetings of the People's Council, even though virtually the entire text of the Declaration had been prepared within its circles. Nor had the People's Movement discussed this issue, even though I was part of its governing body and was aware of all of strategic developments within it.

After Drach's statement, another People's Movement member, Serhii Holovatyi, made a statement about where Ukrainian

2. Ivan Drach, 1996

Ivan Drach (1936–2018), Ukrainian politician, poet, translator, and screenwriter. He was the first president of the People's Movement of Ukraine (Narodnyi Rukh Ukraïny za perebudovu). He led the People's Movement from 1989 to 1992 and was a Ukrainian MP for the periods 1990–1994 and 1998–2006. From the beginning of perestroika, he was one of the leaders of the national democratic movement, drafting its foundational documents: the platform of the People's Movement, the Declaration of State Sovereignty, and the Declaration of Independence of Ukraine. He was awarded the title "Hero of Ukraine" in 2006.

army conscripts should serve, and, as if in passing, gave support to Drach's proposal. Parliament passed Drach's motion by voice vote.

The next day, 13 July, in the course of Parliament's continued deliberations on the text of the Declaration, the situation took an unexpected turn: it became clear that Drach did not entirely understand the motion that he himself had put forward the day before. This is evident from the transcript of the parliamentary session, chaired by Ivan Pliushch:

> S. P. HOLOVATYI: Honorable People's Deputies! This document contains a minor editorial inaccuracy, which has to do with the formulation of the three nonnuclear principles. The thing is, in the development of this proposition, we used a translation from the English version of the Treaty on the Nonproliferation of Nuclear Weapons, and that translation was made somewhat inaccurately. During the break, we clarified the translation, and the language after the colon should be: "to refuse to manufacture, acquire, and test nuclear weapons." Please take this into consideration.

> CHAIRMAN PLIUSHCH: Ivan Fedorovych [Drach], you proposed the text of the motion. What do you have to say in this regard?

> I. F. DRACH: Comrades, I only want to say that the text of the document was thrown together during lunch and that, additionally, I only drafted part of it. It was given to us when we were already in session. Therefore, you must take this document as one in need of careful reading and revision.

> CHAIRMAN PLIUSHCH: Which text? Which text do you mean?

> I. F. DRACH: This one. I am seeing it for the first time. I do not know what is written here and what mistakes were made by the typist.

> CHAIRMAN PLIUSHCH: Yes. Honorable Comrade Deputies! Let us establish the basic reading. Page eight—I am reading the paragraph on page eight: "The Ukrainian SSR solemnly declares its intention to become a permanently neutral state that will not participate in military alliances and will adhere to the three nonnuclear principles." And the following revision, please […]

> I. F. DRACH: "to refuse to manufacture, acquire, and test nuclear weapons."[3]

In other words, the text cited by Drach was missing one of the three principles of a nonnuclear state. As Holovatyi explained it, the mistake occurred in the translation from English.

It was obvious that the idea of a nonnuclear Ukraine was born in a very narrow circle of People's Movement members and was not subjected to any parliamentary procedure. Without any discussion or expert analysis, it was added to the text of the Declaration in apparent haste. However, whose idea it was remained a mystery to me for a long time. As I have learned only fairly recently, the primary author of the idea of nonnuclear Ukraine was the lawyer and international law expert Professor Volodymyr Vasylenko.

The motivation behind this initiative, as Vasylenko personally explained it to me, was the desire to provide in the Declaration the legal foundation for Ukraine's exit from the USSR. That is to say, a nonnuclear-weapon state of

Ukraine could not be part of the nuclear-weapon state of the Soviet Union. As mentioned above, Serhii Holovatyi, whom Ivan Drach consulted before making his statement, joined in the process of adding this provision to the Declaration. Subsequently, this argumentation was used by other influential People's Movement members, including the head of Parliament's Committee on Foreign Affairs, Dmytro Pavlychko.

The extent to which Vasylenko's motive was plausible can be judged from the fact that there were only four republics in the Soviet Union that had nuclear weapons on their territory. Yet, not a single one of the other eleven republics used their nonnuclear status as a pretext for exiting the Soviet Union; all of them declared their independence only after the USSR had already collapsed on its own.

Thus, in this unexpected way, for an ostensibly noble cause, and well ahead of its time (more than a year before declaring its independence) Ukraine began considering the idea of becoming a nonnuclear-weapon state. This formulation in the Declaration would later be used against Ukraine; immediately after Ukraine's independence, the two nuclear weapons states of the US and Russia began interpreting this language systematically as a commitment to immediately disarm. Their diplomatic and information resources, as well as other means of influence, allowed them to contrive and bring to completion the nuclear disarmament of Ukraine in only four years.

The individuals who proposed this language—Vasylenko, Holovatyi and Drach—probably did not realize that they were playing out a scenario already predicted by foreign analytical institutions at the end of 1980s for the eventuality of the Soviet Union's dissolution. The root cause of Ukraine's proclamation of nonnuclear status becomes even clearer when placing these events in the context of preparations by the Soviet Union and the US to execute START I, which involved the liquidation of a substantial part of USSR's nuclear arsenal.[4] By the time the Declaration had been adopted, the START I negotiations, which had lasted nine years, were coming to a conclusion. The US and the USSR managed to sign it on 31 July 1991. However, START I was never ratified and the USSR dissolved quickly thereafter. Four new nuclear states now emerged on its former territory.

Since Ukraine's nuclear forces exceeded the combined strength of France, the UK, and China, it was not a matter of chance that, at the beginning of the process of the USSR's legal dissolution, a new document came into force on 24 October 1991: the Declaration of the Parliament of Ukraine.[5] In the name of independent Ukraine, rather than the Ukrainian SSR, the country officially confirmed the commitment it had made in the Declaration of State Sovereignty to become a nonnuclear-weapon state in the future. Thus, Ukraine's nuclear disarmament process commenced even before Ukraine became a sovereign state. As an irony of fate, this occurred exactly 45 years after the first nuclear explosion in world history, on 16 July 1990.

3. Volodymyr Vasylenko, 2011

Volodymyr Vasylenko (b. 1937), Ukrainian lawyer and diplomat, promoter of the idea of a nonnuclear Ukraine. Vasylenko was a coauthor of the first People's Movement platform and a delegate to the first congress of the People's Movement in 1989. He participated in drafting the Declaration of State Sovereignty of Ukraine. During 1972–1992, he was a technical advisor on legal issues for the MFA of Ukraine. After 1992, he was Ukraine's ambassador to the Benelux countries and later represented Ukraine at the EU and NATO. During the period 1998–2000, he was the ambassador to Great Britain and Northern Ireland. From 1991 to 2009 he participated in the work of the UN General Assembly as an expert, advisor, and member of the Ukrainian delegation. Afterwards, from 2006 to 2010, he was Ukraine's representative on the UN Human Rights Council. In 2009, he was the Ukrainian ombudsman at the International Court of Justice during the case brought by Romania against Ukraine regarding the demarcation of the continental shelf in the Black Sea.

4. Georgii Arbatov, 1983

Georgii Arbatov (1923–2010), Soviet and Russian scholar of international relations and a member of the Russian Academy of Sciences. From 1967 to 1995 he was the director of the Institute for US and Canadian Studies at the Academy of Sciences of the USSR. He was an advisor to secretaries general of the Communist Party of the Soviet Union (CPSU) Leonid Brezhnev, Yuri Andropov, and Mikhail Gorbachev, and to Russian president Boris Yeltsin. From 1991 to 1996, he was a member of the foreign policy council of the Russian MFA. He was a founder of the Russian school of American studies.

The Arbatov Plan

At the time of the USSR's collapse, 69.4 percent of the country's strategic missiles were located in Russia, 17 percent in Ukraine, 13 percent in Kazakhstan, and 0.6 percent in Belarus.[6] The third largest nuclear arsenal in the world was located on the territory of Ukraine.

Russia's nuclear arms strategy was developed by a Soviet intelligence and analytical center, the Institute for US and Canadian Studies, immediately after the August Putsch (1991) and before the legal dissolution of the USSR.[7] The Institute's director, the academic Georgii Arbatov, provided a concise overview of the strategy in a September 1991 document entitled "The Nature of Key Threats to State Security in the Transition Period." It outlined three steps.

Firstly, it prescribed the complete nuclear disarmament of all former Soviet republics other than Russia. In this regard the Kremlin was counting on Western support: "The unambiguous linkage of the diplomatic recognition of former Soviet republics with the declaration by them of their nonnuclear status appears to be a critically necessary component in international relations."[8] This thesis coincided with the US State Department's recommendations not to recognize the new states until they had acceded to the NPT. Indeed, it would be hard to believe that the US and the USSR did not cooperate in authoring these recommendations.

Secondly, Moscow planned to prevent at all costs the post-Soviet states' appropriation of nuclear weapons during the period in which the USSR would transition into independent countries:

> The division of nuclear forces, either territorially or on any other basis, cannot be considered an acceptable option. [...] [T]he goals of international security and stability unequivocally dictate the need to counteract any attempts to "diffuse" control over nuclear weapons. Maintaining a unified command over nuclear weapons would undoubtedly allay the key concerns of Western leaders in this regard.[9]

Thirdly, it entailed maneuvering around the question of legal carryover, and for Russia not to, by circumstance, take upon itself all of the contractual obligations of the former USSR:

> Assuming that, with the consent of all the newly formed states, Russia becomes the sole successor of the USSR, this option

would also be highly undesirable, as in this case the other former [Soviet] republics would "drop out" of the entire system of international treaties and Russia would become responsible for all agreements, including those concerning the reduction of conventional weapons and USSR's external debt.[10]

As for conventional weapons, because the biggest part of the Soviet army was concentrated on the Western border, which was on Ukrainian territory, Ukraine inherited one of the most powerful concentrations of armed forces in Europe, equipped with nuclear arms and modern conventional weaponry and military equipment. At the time of the USSR's dissolution, the Ukrainian army was the largest and best equipped among all those of the former Soviet republics. It comprised 780,000 personnel, 6,500 tanks, 7,000 other armored combat vehicles, 7,200 artillery systems, 350–500 ships and vessels, 1,110–1,600 fighter planes, and, finally, nuclear weaponry. Deployed on Ukraine's territory were one rocket army, three regular (field) armies, two tank armies, one army corps, four air armies, a separate army of the Air Defense Force (ADF), and the Black Sea Fleet.

Taking this inventory into account, Arbatov stressed:

> Significant problems also arise in the field of conventional weapons. In accordance with the Treaty on Conventional Armed Forces in Europe, the Soviet Union can possess no more than 13,300 battle tanks, 20,000 armored combat vehicles, 13,700 artillery systems, 5,150 combat aircraft, 1,500 assault helicopters[…]. If the newly formed states located in the European part of the USSR decide to have their own national guards, the total limits of the Treaty will be exceeded in one or more of these categories of weapons.[11]

Therefore, if Russia were to take on the USSR's contractual obligations under the agreement, it would be left virtually without an army.[12] Thus, Arbatov wrote, "if all newly created states on the territory of the USSR (except for those considering themselves illegally annexed, such as the Baltics) were to become successors to the USSR's international obligations, then, in principle, they could partake in the division of the USSR's nuclear potential."[13] This explains why the Kremlin did not immediately back US President George Bush's unequivocal statement that all Soviet nuclear weapons should belong to Russia, because at the time, there was still no clarity on the division of obligations within the framework of the Treaty on Conventional Armed Forces in Europe, as well as division of the USSR's external debt.

Russia had a different plan: to eventually become the sole successor to all of the USSR's assets, including its nuclear weapons. First, it proposed to the former republics that they voluntarily give up their right of ownership of the nuclear weapons located on their territories. This way, the external debt could be split fair and square, so to speak, and the weapons could be transported to Russia unhindered. If this did not work, then the republics could be recognized as parties to the treaties entered into by the USSR (including START I), which automatically would transfer to them part of the USSR's external debt and other obligations; later, Russia could return to the question of the ownership of the nuclear warheads and transport them onto the Russian territory, preferably without any compensation. Russia anticipated no major problems from Belarus and Kazakhstan in the execution of this plan, and, in fact, this expectation was realized rather quickly.

Nevertheless, the most fundamental question for the Russian leadership was that of Ukraine's nuclear disarmament, which was always seen through the prism of geopolitical interests. Moscow intended to meet several objectives in this phase of Ukrainian-Russian relations: first, to transport Ukraine's most valuable nuclear components—worth in the tens of billions of dollars—without hindrance onto Russian territory; second, to destroy the elite part of the Ukraine's military-industrial complex (MIC); third, to remove a major competitor in the global arms trade; fourth, to destroy Ukraine's defense capability; and fifth, to make Ukraine energy-dependent on Russia.[14]

This political calculation was based in the idea that a nonnuclear Ukraine would be of no interest to the world and thus would be recognized as a country *de facto* located within Russia's sphere of influence. This meant that Kyiv would be left one-on-one with Moscow in the resolution of all "neighborhood" concerns, from the territorial to the economic. John J. Mearsheimer wrote in *Foreign Affairs*, "[N]uclear weapons confer status. It is true that status does matter in the international system and that nuclear weapons enhance a state's status [...]"[15]

Russia's economic calculation was grounded in the desire to monopolize its influence over Ukraine's strategically important industries during the dismantling of the Soviet economic space, including, first and foremost, the energy industry. After all, Ukraine's dependence on Russian gas, oil, and fuel rod assemblies (containing heat-releasing elements [TVEL]) for nuclear power plants, was easy to convert into political dividends, all of which Russia exploited methodically—from the division of the Black Sea Fleet, to negotiating gas deals and inhibiting Kyiv's integration attempts into the larger European community.

Finally, Moscow had one more, perhaps even key, objective: to discredit the very idea of independence in the eyes of Ukrainians themselves. The direct route toward this end was to place the entire financial responsibility for dismantling Ukraine's nuclear arsenal on Kyiv. The Kremlin was keenly aware of the extent of those expenditures and understood that allocating nuclear disarmament costs (in addition to 10–15 percent of Chernobyl-related funding) from Ukraine's meager annual budget would guarantee a decrease in the population's living standard in comparison to that of the Russians. The conflation of impoverishment with independence would extinguish support for the latter, forcing Ukrainians to doubt their need for statehood. This would be a step towards a calculated reincarnation of the empire, albeit now under a different name. "Russian diplomats in Warsaw, Prague and even Kiev have told their Western and Central European counterparts not to bother befriending Ukraine, as the country's independence would not last long," wrote Anne Applebaum in *The Spectator* in June 1993. "Ukraine, to many Russians, is simply not a separate country. Ukraine is part of Russia."[16]

This very strategy was being implemented through Russian diplomacy. One of its main activities was a geopolitical project named the Commonwealth of Independent States (CIS). This structure was intended to become an instrument for the civilized divorce of the Soviet republics, yet was seen by Moscow as the first step towards the restoration of the USSR.

The White House Plan

At the end of the 1980s, the rapid dissolution of the USSR prompted the urgent development of a plan by Washington regarding the USSR's nuclear weapons. The US dreaded the uncontrolled collapse of the USSR. On 1 August 1991, President Bush, in his address to Parliament on the eve of Ukraine's declaration of independence, called striving for independence "suicidal nationalism."[17]

The US was most wary of a possible division of the nuclear weapons. Their discomfort at dealing with fifteen nation states instead of one leader in the Kremlin did not change after 24 August 1991. On the contrary, it lay at the root of the US's ambiguous position towards Ukraine, which remained essentially intact until the 2014 Revolution of Dignity.

The White House had learned not only how to communicate with but also how to influence the Kremlin, as evidenced by the first large-scale agreement obligating it to reduce strategic offensive arms; START I was signed by George H. W. Bush and Mikhail Gorbachev right before Bush's official visit to Ukraine. Building relationships with the other former republics of the USSR would take time and effort. Therefore, the US chose the easiest path. For a country whose world domination is credited, in large part, to the number of its own nuclear warheads, the issue of blocking nuclear armament by other states is a matter of principle. Hence, America's key interest was in preventing an upsurge in the number of nuclear-weapon states. The nuclear club, at least formally, had to remain quantitatively the same as before the collapse of the USSR.

In December 1991, after the execution of the Belavezha Accords, the document that formally terminated the USSR and established the CIS, the Bush administration categorically demanded that centralized control over nuclear weapons be preserved and maintained by Russia. Other countries such as Ukraine, Belarus, and Kazakhstan, were to become nonnuclear weapon states. The Americans saw this as preventing proliferation of nuclear weapons on the territory of the former USSR. The issue was so critical that the US State Department had even advised Bush not to formally recognize Ukraine's independence until it had acceded to the NPT.[18] Nevertheless, the US still recognized an independent Ukraine in December 1991.[19]

Both plans, the Arbatov plan and the White House plan, match the positions referenced above almost verbatim. In this scenario, Russia was assigned the role of a vacuum cleaner for the Soviet nuclear arsenal. Centralization in one country was convenient for the US; it preserved the nuclear weapon status of the former USSR by merely changing the name of the weapons' proprietor from the USSR to the Russian Federation.

The US and Russia's plans converged in their intention to make Ukraine a nonnuclear-weapon state. However, by accepting Russia in its cohort, the West was essentially facilitating Russia's geopolitical goals, or at least not interfering with them. Mearsheimer writes: "An American-Ukrainian confrontation

George Herbert Walker Bush (1924–2018), president of the United States, 1989–1993. On 1 August 1991, Bush delivered his infamous "Chicken Kiev" speech before the Parliament of Ukraine in Kyiv, during which he expressed clear support for Gorbachev and the preservation of the Soviet Union. After the collapse of the USSR, the US supported the newly independent states and established diplomatic relations with them.

5. Seated, left to right, Boris Yeltsin and George H. W. Bush

over Ukrainian nuclear weapons could encourage the Russians to believe that they could destroy those weapons by force without doing much long-term damage to Russian relations with the West."[20] Ukraine became a hostage in a situation in which it was not advantageous for the US to acknowledge any conflicts of interests between former Soviet republics, including the conflict of Ukraine's interests with Russia's. The *Independent* described the situation in May 1993: "Despite pious talk about a 'community' of free states, Western governments never wanted and are still not prepared to accept that the collapse of the Soviet empire is irrevocable."[21]

Furthermore, the *Independent* went on, "the West persists in a policy that merely fuels the imperial addiction of some leaders in the Kremlin. [...] By allowing Moscow to take the lead on this issue, Western governments are reinforcing precisely what they want to discourage: a suspicion among neighbouring states that Moscow's real aim is to reverse the empire's demise."[22]

US policy on the issue of Ukraine's disarmament did not change with the change of US presidents. During the presidencies of the Republican George H. W. Bush and the Democrat Bill Clinton alike, the direction of US diplomacy remained unaltered. Officially, the US used the following argumentation in support of its bet on Moscow and its reluctance to pay heed to Kyiv's demands for security guarantees: Russia had become democratic and rejected any imperial ambitions, therefore, Russia would rightly become the guarantor of security for a nonnuclear Ukraine. The *Washington Post* expressed something to this effect in January 1993: "The government of Ukraine should be reminded that while its security concerns are indeed important, Russia does not have the same ambitions as the Soviet Union did, or even that imperial Russia of the 19th century did."[23] A year and a half later, the *Washington Post* quoted US Defense Secretary Les Aspin reiterating the idea: "I believe a democratic, non-imperial Russia is the best guarantee of security and stability in the new Europe."[24]

US government officials in charge of the country's defense and national security, along with State Department officials, senators, and congressional representatives, with whom I had a chance to meet in 1991–1992, all tried to convince me of the idea that Russia was no longer the same, and that it posed no danger to Ukrainian independence. On the contrary, they maintained, Russia was the guarantor of democratic transformation in the post-Soviet space. I asserted that Russia included not only the formally democratic Moscow and Leningrad, but beyond that, over one-ninth of the world's landmass, throughout which there had not been so much as a whiff of democracy. That is why, I insisted, it was better to bet on Ukraine, which could transition to democracy and become a European nation much faster. Then, through Ukraine, they could increase democratic influence, even over Russia. My views were shared by some Western experts. Accordingly, Jonathan Eyal, director of studies at the Royal United Services Institute in London, insisted that "the West persists in a policy that merely fuels the imperial addiction of some leaders in the Kremlin."[25]

Thus, when it came to gaining control over the former USSR's nuclear weapons, the United States' overriding goal entailed forcing Kazakhstan, Belarus, and Ukraine to declare themselves nonnuclear weapon states and quickly removing all tactical and strategic nuclear weapons from their territories, while making them develop procedures and mechanisms for liquidation programs on their territories independently. This goal corresponded entirely with the draft agreement that Russia brought to the Minsk meeting of the heads of state of the CIS in February 1992, without any prior discussion with its prospective parties. According to this agreement, all nuclear warheads were to be swiftly transported to Russia, while all costs associated with the liquidation of strategic offensive weapons, including delivery systems, command and control, missile launch facilities, and the resolution of socioeconomic issues pertaining to the rocket armies' military personnel, were left to Russia's "fraternal" republics. However, this agreement, which was more reminiscent of trickery than diplomacy, was not supported by the member states.

Then, the Bush administration offered its own version of an agreement, also designed to transport nuclear weapons from Ukraine's, Kazakhstan's, and Belarus's territories as expeditiously as possible. This was a draft of a five-party supplemental protocol to START I, proposed by the US in April 1992. It was signed by all parties in Lisbon on 23 May of the same year. US diplomacy's lightning-fast reactions to the new circumstances arising in the process of Ukraine's disarmament, in contrast to that of Russia's, was exemplary.

In early April 1992, a Ukrainian parliamentary delegation traveled to the US on an official visit. Since no mechanisms existed for the centralized preparation of directive or advisory documents on nuclear disarmament at the time, and there was neither an official agency nor even a national strategy on the issue, the parliamentary group then working on the concept of national security developed a set of special recommendations explaining Parliament's position on nuclear disarmament, and put together a set of proposals for possible collaboration with the US.

The recommendations comprised four key points: first, to raise the issue of providing national security guarantees for Ukraine after it was denuclearized;

6. *The Parliament of Ukraine*

second, to demand that Ukraine be recognized by the US as party to all negotiations concerning nuclear weapons; third, to determine specific possibilities in the US with regard to financial and technical support for the dismantling of the missile systems; and, finally, fourth, to explore the US position on possible assistance reprocessing the nuclear material released from Ukraine's nuclear weapons.

The US leadership understood that Ukraine had not violated any international laws, in contrast to Russia, which had ignored international law in its aim of appropriating all nuclear material. It was the tradition in Russian diplomacy to advance Russia's goals using any means at its disposal, be it force, intimidation, or diplomatic trickery (distorting the contents of international agreements, for instance), and without regard for other parties' positions.

The events unfolding after the Ukrainian delegation's visit to Washington demonstrated that the majority of the White House's initiatives, including the Lisbon Protocol, immediately took all of these signals into account. Yet, certainly, none of these developments could remove America's national interests from the equation. US foreign policy goals regarding the former USSR's nuclear weapons were represented in Bush's official statement of 13 July 1992. They included the following: the accession of the CIS states to START I and the NPT, ensuring the inventory and safety of nuclear warheads, controlling nuclear and missile technologies, dismantling nuclear warheads and controlling extracted nuclear material, and providing occupational training to former Soviet nuclear and rocket technology experts to transition them to civilian professions.

This perpetuated a double-sided game that continued to focus on Ukraine's nuclear disarmament. However, regardless of how monolithic the Russian-American alliance appeared to be at the beginning of this game, further developments have demonstrated one fact: both the emergence of Ukraine's own position, and Ukraine's bet on the clash of Moscow's and Washington's interests allowed Ukraine to reach some noteworthy results, which became evident only in the summer of 1993.

Ukraine's Plan: Seed Capital or Hazardous Legacy?

Unlike Moscow and Washington, when Kyiv declared its intention to become a nonnuclear-weapon state on 16 July 1990, the country lacked not only a nuclear disarmament strategy, but also any unity with regard to this monumental issue. Instead, there were two positions, and they were defined by mutually antagonistic attitudes towards nuclear weapons located on Ukrainian territory. One side maintained that nuclear weapons presented essential seed capital for the new state, and the other asserted that nuclear weapons were a hazardous legacy left by the USSR. Both emerged in the first days of Ukraine's independence.

Immediately after pronouncing the Declaration of Independence on 24 August 1991, the parliamentary Committee on Foreign Affairs (hereafter, the Committee), then headed by poet and People's Movement leader Dmytro Pavlychko, began developing a parliamentary decree on Ukraine's nonnuclear status. The parliamentary opposition, in the minority, which generated most of Parliament's strategic decisions in the first months of Ukraine's independence, took part in the deliberations of the document's provisions. The discussion was very emotional because of the fundamental differences in the participants' viewpoints on Ukraine's nuclear disarmament strategy.

From the onset, most of the participants in the Committee meeting saw nuclear weapons as a hazardous legacy that needed to be got rid of as soon as possible. *Ukraïnska hazeta* described these dynamics in late 1993: "Serhii Holovatyi and Dmytro Pavlychko actively campaigned in support of handing the weapons to Russia. They backed their position with arguments that Russia could capture Ukraine's regions where nuclear weapons are located, without any concessions, and then deploy its occupying forces there for good. That would be it, in other words; it would be the end of independent Ukraine!"[26]

Out of the eighteen deputies in attendance, only three—Serhii Semenets', Tetiana Iakheieva, and I—spoke out against this approach. Ivan Zaiets' hesitated for some time, but he, too, was eventually persuaded by the proponents of immediate disarmament. Among the representatives of the parliamentary (communist) majority, our position was supported by Vasyl' Durdynets', chairman of the Defense and Security Committee. Together, we conveyed the need to use our nuclear weapons as seed capital, gradually transforming them into peaceful dividends for independent Ukraine. However, by a majority vote, the former position prevailed. Its spokespersons were Pavlychko, Holovatyi, and Committee consultant Vasylenko. The Committee finally recommended that the Parliament approve the draft Statement on the Nonnuclear Status of Ukraine.[27]

Nevertheless, our heated discussion bore some fruit in the end. The group reached a compromise on provision 3 of the draft statement, which contained the original language requiring Ukraine to instantly become a nonnuclear-weapon state, and which was now replaced by a provision that allowed the disarmament process to take into consideration national interests. The statement,

7. *Serhii Holovatyi, at the microphone, 1991*

Serhii Holovatyi (b.1954), Ukrainian politician. Coauthor of the first platform of the People's Movement and a delegate to the People's Movement founding congress in 1989. A member of the CPSU from 1977 to 1989 and of the People's Movement from 1989 to 1994. A corresponding member of the Academy of Legal Sciences of Ukraine beginning in 1995 and a Ukrainian MP from 1990 to 2012. He graduated from the Department of International Relations of Shevchenko State University in Kyiv in 1977. In the period 1980–1990 he was a research fellow at the Institute of Social and Economic Issues in Foreign Countries in the Academy of Sciences of the Ukrainian SSR. In 1990, he was elected to the parliament as a proponent of democratization and the rule of law. From 1995 to 2000, and again in 2005, he was a member of the Venice Commission. During the period 1995–2008, he was member of the Ukrainian parliamentary delegation to the Parliamentary Assembly of the Council of Europe (PACE). In 2005–2008, he served as vice president of PACE. In 1995–1997 and 2005–2006, he was the minister of justice of Ukraine.

adopted by Parliament on 24 October 1991, included the following language: "Ukraine's policy will be aimed at a complete elimination of nuclear weapons and all of their corresponding parts located on the territory of the Ukrainian state. It intends to do so in the time frame that is minimally necessary in view of the legal, technical, financial, organizational, and other capacities, and in consideration of environmental safety."[28] Common to the two opposing approaches ("seed capital" versus "hazardous legacy") was that neither position even mentioned the idea of Ukraine actually becoming a nuclear-weapon state. The divergence was in how to handle the disarmament process.

The former approach (using nuclear weapons as seed capital) offered the following strategic pathway: to dispose of nuclear weapons and nuclear material as efficiently as possible, so that the disarmament process could jumpstart the formation of Ukraine's own state policy, which would be independent of the Kremlin, as well as providing an avenue for integration into European and global political, economic, and security structures. This strategy could be termed "effective disarmament." It was being formulated and actively advanced by two parliamentary working groups: one developing the concept of national security and the other preparing Ukraine for ratification of START I and the Lisbon Protocol in accordance with Parliament's decisions.[29] For both of these groups, Ukraine's national interests and a distinct understanding that Russia was not our ally in this process were central to the development of the nuclear disarmament action plan. Thus, the proponents of effective disarmament actively sought strategic partnerships in the West—first and foremost, in the United States.

It is important to understand that, at the time, Ukraine had limitless opportunities for cooperating with the West in the sphere of nuclear disarmament and conversion of nuclear material into "atoms for peace." For instance, in his December 1992 memo to President Kravchuk, the first vice prime minister, academician Ihor Iukhnovskyi, reported on the results of his official talks with the American company General Atomics on the possibility of developing a joint venture that would liquidate nuclear weapons. In the letter Iukhnovskyi proposed the following: first, to announce in the mass media Ukraine's intent to conduct extensive consultations on finding an optimal path to nuclear disarmament; second, to conduct more in-depth talks with representatives of various companies in the US, France, the UK, and Russia, on issues inherent to the processing and storage of nuclear material; and, third, to evaluate all proposals and choose the most effective way for Ukraine to eliminate nuclear weapons.

The strategy emerging from the other approach (nuclear weapons as a hazardous legacy) entailed transferring all nuclear weapons to Russia as quickly as possible, so that Russia would not use them as an opportunity to establish military control over strategic arms in Ukraine, or so that the nuclear warheads located in Ukraine would not cause another Chernobyl. This strategy could be termed "transfer weapons to Russia."

After Parliament's adoption of the Statement on the Nonnuclear Status of Ukraine, the Ministry of Foreign Affairs (MFA) was in the vanguard of implementing the strategy that entailed transferring Ukraine's entire nuclear arsenal to Russia. The Ministry's first practical steps in this direction included Kravchuk's execution of two treaties in December 1991, in Almaty and in Minsk respectively, both of which concerned strategic nuclear arms. Ukraine's foreign policy began to signal that the nation had already made the decision to immediately transfer all of its nuclear weapons to Russia: tactical weapons before 1 June 1992, and strategic weapons by the end of 1994. Henceforth, despite all of the legislature's subsequent decisions aimed at effective disarmament, the MFA's position remained unchanged.

In addition to Ukraine's national interests, the divergence in these strategies encompassed another important issue: its impact on the overall reduction of the world's nuclear arsenal. Since one of the objectives of Ukraine's relinquishing its nuclear weapons was to reduce the number of nuclear weapons in the world, an important criterion for assessing Ukraine's disarmament options ought to have been: Would the number of nuclear warheads on the planet actually decrease as a result of their disappearance from Ukrainian territory? The effective disarmament strategy should have anticipated revisiting options for completely eradicating nuclear weapons (because pacifist motives were at the heart of the decision to attain nonnuclear status in the future), which would have maximal benefit for Ukraine. So, from this point of view, effective disarmament would have been a plan to destroy the nuclear weapons.

The alternative strategy, however, anticipated not the destruction but the transfer of Ukraine's nuclear weapons to Russia. Significantly, none of the official documents executed at the

8. *Leonid Kravchuk*

Leonid Kravchuk (1934–2022), first president of Ukraine (1991–1994), Speaker of the Parliament of Ukraine (1990–1991), Ukrainian MP (1990–1991 and 1994–2006); a member of the CPSU from 1958 to 1991, and a member of the Social Democratic Party of Ukraine (SDPU) from 1998 to 2009. He graduated from Shevchenko State University in Kyiv in 1958 as an economist. From 1960, he worked for the CPSU, including a period as director of the Agitprop ("agitation and propaganda") Department of the Chernivtsi Oblast Committee of the Communist Party of Ukraine. He was a doctoral student at the Academy of Social Studies under the Central Committee of the CPSU. From 1970–1988, he worked in the Agitprop Department of the Central Committee of the CPSU, and from 1988–1990 as director of the Department of Ideology and second secretary of the Central Committee of the CPSU. In 1991, he joined the Politburo of the Central Committee of the CPSU. In 1989, he defended the passage of a failed resolution by the Politburo to ban the People's Movement. In 1993, he agreed to an early election for the presidency and lost in the second round to Leonid Kuchma. In 1994, he was elected an MP. In 1998, he began to cooperate with the SDPU (United). In 2004, he lost his doctor honoris causa from Kyiv-Mohyla Academy for his "uncivil position during the Orange Revolution." In 2009, he called on President Viktor Yushchenko to resign.

The Belavezha Accords were signed by Speaker of the Parliament Stanislav Shushkevych and Premier Viacheslav Kebich on behalf of Belarus, by President Boris Yeltsin and First Deputy Prime Minister Gennadii Burbulis on behalf of the Russian SFSR, and by President Leonid Kravchuk and Prime Minister Vitold Fokin on behalf of Ukraine.

9. The signing of the Minsk Agreement, 8 December 1991

time required any international control over the liquidation of weapons once they were in Russia. This meant that the weapons could have been disassembled in Russian plants or put back into military use, but now in the Russian, instead of Ukrainian, army. William J. Broad of the *New York Times* attested to the reality of this scenario:

> The biggest fear is bomb recycling. Nothing in the arms reduction accords prevents reuse. The United States pioneered this practice by taking warheads from the Pershing 2 missile, retired from Europe, and manufacturing them into weapons for bombers. Those arms, not covered by Mr. Bush's recent weapons withdrawals, are now back in Europe. Some American experts worry that the Russians might do likewise on an expanded scale.[30]

In 1993, the Russian media communicated that the SS-25 mobile ballistic missile systems previously transported from Belarus replaced the obsolete weaponry in the Russian military's strategic units.[31] Consequently, the "transfer of weapons to Russia" strategy, in the context of global nuclear arms reduction, simply meant the relocation of nuclear weapons to another country, rather than their elimination.

The implementation of the effective disarmament strategy (i.e., destroying the weapons, as opposed to "transfer to Russia") entailed developing contacts with US and European companies and politicians, to investigate the possibility of recycling Ukraine's nuclear material. The implementation of the other strategy, "transfer to Russia," involved only arguing why keeping nuclear warheads was a hazard for Ukraine, and why the weapons should be swiftly shipped out to Russia.

The first strategy was pursued by the new generation of Ukrainian politicians; the second—by the Soviet cadre who continued to operate within the bounds of the joint political and economic space of the no longer existent USSR. It is especially noteworthy that the advocates for the "transfer weapons to Russia" strategy continuously altered their narrative in the course of this struggle, ascribing to their opponents the intention to reconsider the country's course toward nonnuclear status and to instead establish a complex and highly costly nuclear weapons production chain in Ukraine. Therefore, "not to transfer the weapons to Russia," "to demand compensation for nuclear material," and "to decommission on the territory of Ukraine," in their interpretation, equated to "delaying disarmament," "postponing fulfillment of commitments," and "intending to create a nuclear weapons cycle in Ukraine." Within the context of the changing narrative, they also used the term "nuclear hawks" to describe the proponents of the effective disarmament strategy. The simultaneous implementation of these two strategies by two different power structures of a single state or, more precisely, their brutal opposition, became the essence of Ukrainian politics on the issue of nuclear disarmament.

The Workings of the Russian Diplomatic Machine

The way Russian diplomacy worked to attain the Kremlin's goals in the first month of Ukraine's independence, as well as the way Ukrainian diplomacy appeared in the context of Russian diplomacy, can be seen vividly in the example of the nuclear agreements signed within the CIS. In the first months after the parade of independence proclamations, these documents were expeditiously signed by the leaders of the newly established states following Russia's prodding.

Among the countries of the former Soviet Union, Russia had a monopoly on international law expertise, as well as on information regarding the USSR's assets from gold reserves to the economy and nuclear weapons. Accordingly, Russia initiated the preparation of documents and brought them to summits with the post-Soviet state leaders. These documents increased Ukraine's and other former republics' obligations with apparent regularity. They also confirmed Russia's intention to remove nuclear weapons from their territories. Thus, the Arbatov Plan was slowly realized, and virtually without resistance. The Belavezha Accords of 8 December 1991, which officially dissolved the USSR and established the CIS, laid the cornerstone for this approach. However, they only vaguely mentioned the issue of nuclear weapons: "The parties shall respect their mutual aspirations of achieving nonnuclear and neutral state status."[32]

Two weeks later, on 21 December, the Alma-Ata Protocol, signed by the presidents of eight other former Soviet republics, had a more precise language with regard to a joint strategy on nuclear weapons, prescribing their complete removal from the territories of the Republic of Belarus and of Ukraine.[33] Thanks to the power of Nazarbayev's personal influence, any mention of the nuclear-weapon state of Kazakhstan was "left out" of the text; Kazakhstan at that point was attempting to bargain with Russia. All signatories agreed to accede to the NPT as nonnuclear states, as well as not to transfer, except to

10. *Alma-Ata Protocol*
From left to right, Leonid Kravchuk, Nursultan Nazarbayev, Boris Yeltsin, Stanislav Shushkevych, 21 December 1991.

the Russian Federation, any nuclear weapons or other explosive devices and technology, or control over them. Consequently, Russia realized through legal means the goal of appropriating all the nuclear weapons without paying for them. This is where a specific timeline for transferring nuclear weapons first appeared: "By 7 January 1992, Ukraine will ensure transportation of tactical nuclear weapons to central factory compounds of the Russian Federation for their dismantling under joint control."[34]

Almost a week later, on 30 December, this timeline was not only repeated in the Minsk Treaty on Strategic Forces, but also was supplemented by additional language, which now stipulated that the dismantling of nuclear weapons located on the territory of Ukraine should be accomplished "by the end of 1994."[35] This provision also stated that in the time period before its complete elimination, the nuclear arsenal located on the territory of Ukraine "shall remain under the control of the Joint Strategic Forces Command," which, of course, was to be headed by Russia.[36] The treaty did not specifically address the elimination of nuclear weapons located in Kazakhstan, which supports the idea that it may have been possible at that time to consolidate the positions of the three post-Soviet nuclear states of Ukraine, Kazakhstan, and Belarus. However, Ukraine's leadership obediently followed the Russian scenario.

The Russian leadership understood that it did not have a great deal of influence over the parliaments of former republics that had, even in the minority, fairly powerful national-democratic forces. Russia implemented its strategy primarily through the former Soviet executive branches, which remained virtually unchanged after the fall of the USSR. The presidents of the new states, accustomed to obeying Moscow's directives, were besieged during high-level official meetings. Most of the time, they were asked to sign numerous documents, which often they had seen for the first time at official summits. For instance, in October 1992, I personally witnessed Russia push through its

11. Boris Yeltsin

Boris Yeltsin (1931–2007), president of the Russian SFSR in 1991 and first president of the Russian Federation (1991–1999). Active in the 1960s, when he worked for the CPSU in Sverdlovsk and Moscow, in the 1980s he became a critic of Gorbachev's policies. In 1987, he was dismissed from all his positions and banned from the Politburo. In 1990, the first Congress of National Deputies of the Russian Federation elected Yeltsin Speaker of the Supreme Council of the Russian SFSR, and in 1991, he became the first democratically elected president of the Russian SFSR. During the Putsch of August 1991, he led the opposition to the coup, making his famous appeal from atop a tank to all Muscovites to defend democracy. His image in the West was that of a democrat, although his policies were aimed at forming a new union under Russia's aegis. In January 1993 he tried to get new statutes of the CIS signed that would have effectively revived the USSR. He used Russia's energy resources and trade wars to manipulate Ukraine politically.

goals, blatantly ignoring standard international relations practices during official talks with state leaders in Bishkek, Kyrgyzstan.

Russia's President, Boris Yeltsin, presided at the meeting with a stack of draft bills in front of him, presenting each of them in turn for discussion. The presidents of the other countries did not have advance copies of many of those documents; they were handed out by representatives of the Russian delegation during the meeting. Yet, not a single foreign minister protested against this Soviet-style method of adopting important resolutions that had a critical bearing on the national interests of the newly formed states.

Therefore, every one of these high-level meetings concluded with mounds of signed agreements, which Russia manipulated afterward in two ways. First, Russia was not concerned with the legal weight of these documents. Most of them had to be ratified by the former republic's parliaments. Two of the three treaties obligating Ukraine to transfer nuclear arms to Russia, signed by President Kravchuk in Almaty and in Minsk, were not ratified by Parliament; in other words, they were not formally put into effect. Nonetheless, Russian diplomats in their official statements regularly declared them to be Ukraine's stipulatory obligations, and the Ukrainian MFA never publicly disputed these allegations.

Incidentally, Russia frequently maneuvered around issues pertaining to agreements that had been signed but not ratified. It did not ratify the Minsk Treaty of 30 December, which referred to the distribution of the USSR's assets; it appropriated all former USSR's external possessions *de facto*, citing the fact that the treaty on the distribution of assets had not been ratified. This act, too, was never protested by the MFA.

Second, Russia actively eroded Ukraine's position, understanding that now corralling the Ukrainian parliament was not as easy as influencing Kravchuk or the MFA, which the Kremlin had treated as subordinates in the Soviet era. This would be achieved by having Kravchuk sign documents obligating Ukraine to terms advantageous for Russia. Then, it would be the president's word against Parliament, which would underscore the lack of a unified position in the country and provide grounds for arbitrary interpretation of the agreements.

The treaties, protocols, and official letters concerning nuclear disarmament signed by Kravchuk in the first months after the dissolution of the USSR created the greatest number of contradictions in Ukraine's nuclear policy. They substantially complicated our external communication, creating immense obstacles in Ukraine's nuclear weapons negotiations with Russia. At this juncture we should also keep one thing in mind: President Kravchuk never decided on his own whether to sign a particular document. Those signatures were always preceded by clearance from the MFA or personally from Minister Anatolii Zlenko.

The problem was not so much in the actual counsel provided by the foreign ministry, but in the fact that the MFA allowed for a review and execution of international documents "on the fly." Under these circumstances, Kravchuk, seeing the documents for the first time and having no chance to garner professional advice upon which to base his decisions, should have refused to sign them, which would have been consistent with international practices. However, how could he possibly have known about international practices, especially when he always had close by him trusted and supposedly experienced diplomats? As a result, the president had no chance of considering alternative points of view to those of the minister of foreign affairs, who accompanied him to all crucial official meetings.

This arrangement became the norm for the MFA in its interactions with Russia. Meanwhile, Russia continued producing additional obligations on nuclear weapons for Ukraine. After a short time the official documents would be signed, not only by Kravchuk, but also by Minister Zlenko.

The US-Russian Uranium Deal

A striking example of the duplicitous political game being played by Washington and Moscow was a secret agreement by which the US would purchase 500 metric tons of the highly enriched uranium (HEU) that was to be obtained in the process of nuclear disarmament during the next twenty years. Although the purchase involved uranium derived from Ukraine's warheads, the agreement was kept secret from Ukraine for over a year. The Security Service of Ukraine (SBU) and the MFA, both of which actively lobbied for the quickest possible relocation of all nuclear weapons to Russia, did not provide any information about the existence of these arrangements either to Parliament, or to President Kravchuk, or to the Working Group.[37]

This information appeared unexpectedly in *Moskovskie novosti* in October 1992. Thanks to my assistant, who dutifully monitored the press, I discovered it before Kravchuk did. As it turned out, US-Russian talks had commenced back in December 1991. Under this proposed agreement, two US companies were to transform 90-percent, weapons-grade HEU into low enriched uranium (LEU), which is used in energy production. They had already entered into preliminary agreements, but then the government decided to take the matter into its own hands. The agreement was initialed on 28 August 1992 by General William F. Burns on behalf of the US Department of Energy, and by Deputy Prime Minister Nikolai Egorov on behalf of the Ministry of Atomic Energy of Russia. The negotiations involved neither the Russian military, nor experts from the joint-stock company Tekhsnabeksport, which had held a monopoly on representing the USSR in the global uranium market for over twenty years. *Moskovskie novosti* reported this sourcing the commander of the Strategic Rocket Forces, Igor Smirnov, and the general director of Tekhsnabeksport, Al'bert Shishkin. The latter noted that the topic was sensitive and that the negotiations were confidential. When communicating with the press, Russia's minister of atomic energy, Viktor Mikhailov, insisted that the agreement was not marked as classified. *Moskovskie novosti* reported, however, that he declined the newspaper's request to peruse the contract for further information.[38]

Then, the confidential status of the uranium deal was unexpectedly breached by the Bush administration. Minister Mikhailov commented on this disclosure: "We did not agree to the Americans' announcing it. However, the president's team likely considered it a powerful political card in the presidential race."[39] Indeed, the deal was beneficial to the Americans, so it is unsurprising that Bush decided to present it to the public as an achievement in the presidential campaign. Aside from its purely political benefits—after all, the deal would result in removing a portion of the nuclear weapons formerly aimed at the US from Soviet territory—there was also an economic advantage. As the Americans explained, the deal would not affect the national budget, because there would be no need for constructing uranium mines or enrichment plants. Instead, the recycling process involved only diluting highly concentrated uranium with natural uranium.

To Russia, which was in the midst of an intense economic crisis, the deal promised several billion dollars, and the Russians were preparing the way to obtain money for the sale of Ukraine's uranium. Furthermore, there was another reason to sell the surplus uranium, mentioned in passing in the *Moskovskie novosti* article: it eliminated the risk of the despoliation of nuclear material warehoused in Russia. "It would also take care of the headache of where to store the nuclear material and how to prevent its accidental use or theft."[40] However, the deal had created a conflict between interested parties in Russia, which had likely prompted the publication of the article in *Moskovskie novosti* in the first place. The Ural Electrochemical Combine, also known as Sverdlovsk-44, the USSR's most powerful manufacturer of enriched natural uranium, had sent three letters to the ministry of atomic energy with strong objections to the deal. *Moskovskie novosti* wrote: "The Ural [specialists] are alarmed at the prospect of fuel derived from military-grade uranium replacing their product in the uranium market, which is already saturated to the limit. They fear losing the dollars they receive from exporting low enriched uranium. Even after all expenses, it amounts to tens of millions annually."[41]

The newspaper voiced still more of the Ural specialists' concerns. "By selling uranium from nuclear weapons for short-term profit, we [Russia] deprive the industry of any prospects for development." A mixture of additional circumstances adverse for production contributed to this forecast: in the aftermath of Chernobyl, nuclear energy development was at a standstill, and now nuclear disarmament, too, was underway. The newspaper article concluded grimly: "Furthermore, the isolation units cannot be shut down, and will deteriorate. In these circumstances, for the nuclear industry to find itself with no market is akin to a death sentence." Underlining the reality of this threat, the newspaper provided some calculations. The 100,000 tons of LEU heretofore accumulated in Western countries were expected to be exploited by 1995. Over the next twenty years, these reserves would have been replenished by 500 tons of HEU. One ton of HEU produced 30–32 tons of LEU for nuclear power fuel. According to the Russian specialists' calculations, this quantity of HEU would be sufficient to fuel all US nuclear reactors for ten years, or the reactors of the entire world for two and a half years. Thus, Sverdlovsk-44 was guaranteed to lose profits.[42]

As soon as I learned about this deal in the next round of negotiations with the Russian delegation in January 1993, I asked about Ukraine's share as compensation for the warheads already taken over by Russia. While we did not receive a response from the Russians, the question was acknowledged by the Americans. By March of 1993 the Ukrainian MFA received a memorandum from the American negotiators. This document officially informed Ukraine about the uranium deal and invited us to divvy up the uranium sales quota with Russia. The Americans then stressed: "The highly enriched uranium mass has an economic value."[43] Shortly afterward, in April 1993, in another memorandum, the Americans proposed that the sales contract provide a share for Ukraine, Belarus, and Kazakhstan. They named a total contract value of US $11 billion and promised not to pay Russia until Russia had reached an agreement on how to distribute it. The Americans also indicated the main reason for "declassifying" these negotiations: during the talks with the Russian Federation the Ukrainian delegation had raised the issue of sharing the profits from the sale of HEU.[44]

Nevertheless, conflicts between interested parties within Russia itself, along with Bush's reelection campaign, enabled us to better understand how to create a formula for converting Ukrainian warheads into nuclear fuel assemblies from Russia, which would fuel Ukraine's own power plants. A group of experts at a February 1993 conference in Kharkiv calculated that there should be 3,400 fuel rod assemblies, worth US $2.72 billion on the international market, although it was valued at only US $901 million on the internal Russian market.[45]

Ukraine, the State as Intended: KGB Reform

In the late 1980s both Moscow and Washington had think tanks that produced forecasts of international developments and worked on national security issues. In 1991 Ukraine, on the other hand, lacked any similar institutions. In fact, no institutions existed to clearly define Ukrainian national interests or develop defense strategies. In 1990–1991, Ukraine was merely the object of larger global events.

After gaining independence, Ukraine had to create governmental structures and formulate policy instantaneously. One of the most fundamental issues for any country is national security, and the development of Ukraine's nuclear disarmament strategy should have been included in its scope.

Within the first days of Ukraine's sovereignty, the KGB fought for exclusive control over the sphere of national security. The Ukrainian KGB's leadership and personnel remained unchanged from the days before 24 August 1991, when the country's independence was proclaimed. Hardened by communist ideology, the old Soviet cadre had their own vision of how to build this new country. Ukraine was not the only country that experienced such takeover attempts by the security services; the KGB was essentially its own state within the state during the existence of the USSR, and the tendency to assert control was neither incidental nor unusual for any of the former Soviet republics. This authoritarian structure remained intact even after the collapse of the USSR.

After the August Putsch, the Ukrainian parliament immediately created an interim parliamentary commission for the reform of the KGB, because the latter was considered the citadel of Soviet totalitarianism. The commission for KGB reform comprised several representatives of the People's Council, including former Soviet political prisoners such as Stepan Khmara, Levko Luk'ianenko, and the Horyn´ brothers, Mykhailo and Bohdan; active People's Movement member Larysa Skoryk, and Communist Oleksandr Moroz, among others. I, too, was elected a member of the reform commission.

During the time of the Soviet Union, the KGB oversaw all security-related issues, internal (including ideological) and external. In line with this Soviet tradition, both KGB reform and the concept of national security were seen by the majority of Ukraine's leadership as one and the same in September 1991; even former political prisoners considered this to be the case.

12. *Andrei Sakharov, 1989*

Andrei Sakharov (1921–1989), political activist, Nobel prize winner, physicist, member of the Academy of Sciences of the USSR. From 1948 to 1968, he worked on the development of thermonuclear weapons. He was one of the founders of the study of controlled thermonuclear reactions but became renowned as a political activist. Sakharov was one of the founders of the Human Rights Committee and spoke out against the testing of nuclear weapons and the death penalty. In 1973, during an anti-dissident campaign, he, along with Alexander Solzhenitsyn, was criticized in the Soviet press for "anti-Soviet actions and speeches." In 1975 he was awarded the Nobel Peace Prize, and in 1980 was exiled to Gorky for seven years. In 1989, he was elected a national deputy of the USSR.

The Ukrainian KGB tried to preserve this tradition by retaining a monopoly over all security matters.[46]

Nevertheless, I already had access to an abundance of literature on contemporary democratic approaches to building systems of national security and understood that internal and external security were two different matters. I had the opportunity to attend a meeting of the people's deputies of the USSR held in Moscow in the spring of 1991. In attendance were members of the parliamentary opposition minority, the Inter-Regional Deputy Group, founded by Andrei Sakharov, Iurii Afanasiev, and Gavriil Popov. The group discussed a new concept of the USSR's national security. At the meeting, Russia was represented by notable democratic deputies, including Boris Yeltsin, Galina Starovoitova, Sergei Stankevich, Anatolii Sobchak, and many others; representatives from Ukraine included Iurii Shcherbak, Volodymyr Cherniak, and Serhii Riabchenko. Afanasiev led the meeting. Academician Iurii Ryzhov gave a presentation on the democratic world's contemporary views on the notion of national security and its key elements.

Shcherbak brought me in as a representative of Ukraine's national democratic movement. On that day, I discovered for the first time the true foundations of modern national security. To my great astonishment, I realized that national security comprised more than military factors; it also consisted of energy, the economy, ecology, finance, information, and many other parts.

When Ukraine and other Soviet republics proclaimed their independence, the USSR continued its legal existence until December 1991. Hence, the Supreme Soviet of the Soviet Union (the USSR's legislative body) likewise remained in operation, and all documents continued to be distributed to the entire deputy body of the Supreme Soviet by their analysts. Among the members of the commission that analyzed these materials and prepared the draft of the USSR's national security policy was Arnold Nazarenko, a deputy from Dnipropetrovsk. In September 1991, he handed over to me a whole package of documents he had managed to collect. Among them was Arbatov's analysis that reviewed strategies for strengthening the Russian Federation's positions after the collapse of the USSR, entitled "The Nature of Key Threats to State Security in the Transition Period." I succinctly presented its contents to the committee on KGB reform. Nevertheless, I was unable to find any support for my viewpoints, not even among the deputies of the democratic opposition. Therefore, on 20 September 1991, Parliament adopted a resolution "On the Establishment of the National Security Service of Ukraine (SNBU)."[47] The title reflected the victory of the Soviet approach.

Although the resolution's first provision officially dissolved the KGB in Ukraine, ensuing events demonstrated that KGB reform ended there. Ievhen Marchuk, a career KGB general, was appointed the head of the newly created SNBU, while staff from the former Soviet Committee on National Security headed its key departments. In the period between September 1991 and

13. *Iurii Ryzhov*

Iurii Ryzhov (1930–2017), political activist, scientist, member of the Academy of Sciences of the USSR and of Russia. From 1954 to 1960, he worked as an engineer at the Zhukov Central Institute for Aerodynamics and the Keldysh Research Center, where he engaged in experimental and theoretical work on the aerodynamics of air-to-air and surface-to-air missiles, and in research in supersonic aerodynamics. From 1989 to 1991, he was elected a national deputy of the USSR and chairman of the Science and Technology Committee. From 1992 to 1998, he was the Russian ambassador to France.

February 1992, the SNBU established its departmental leadership apparatus, regional administration, and the Institute for SNBU Personnel Training, after which it could carry out its staffing process. Needless to say, these were the same people who had stood guard over the ideological foundations of the USSR. Hence, a KGB career officer remained at the helm, while the system of the independent SNBU was fully integrated into the old system of the Soviet KGB.

14. *Ievhen Marchuk, 1995*

The Concept of National Security: the Battle of Two Worldviews

However, surrendering at that point would have been premature. Having analyzed all the documents that Nazarenko had forwarded to me, I went to the Speaker of Parliament. Ivan Pliushch knew better than anyone else that the new Ukrainian state was being built haphazardly. He instantly supported the idea of creating a parliamentary working group that would develop such essential documents as the national security policy of Ukraine.

Although the group had not been formally formed yet, different experts started coming to us of their own volition. However, I must say that they were predominantly military experts, because the concept of national security was still equated with defense. They were alarmed by the political and military aspects of the USSR's dissolution. At the time 70 percent of the Ukraine's MIC was integrated into the USSR's MIC. Thus, after the collapse of the USSR, a crucial question arose: Who would be paying the Ukrainian plants on military contracts? Military representatives also raised the subject of the nuclear disarmament strategy, including the question that seemed obvious to them: How would we be defending ourselves after giving away our nuclear weapons?

It should be mentioned that, at the time, Parliament had become the epicenter for the development of the new state policy. People would appeal to Parliament with innovative ideas and suggestions, because it was the only place where they stood a chance to be heard. In the early 1990s, it was truly the only governmental body receiving an influx of new blood. However, because the entire executive branch of independent Ukraine was inherited from the Ukrainian SSR, the new state essentially was being built by those who just a few months earlier were still being paid to ensure that such a state would never exist.

On 23 December 1991, a decree by the Speaker of Parliament created a new parliamentary working group to work through the issues of refining Ukraine's system of national security. The group consisted of nineteen deputies, representing fifteen

Ievhen Marchuk (1941–2021), first director of the Security Service of Ukraine (SBU) between 1991 and 1994, and member of the CPSU until the August Putsch. He graduated as a teacher of Ukrainian language and literature from the Kirovohrad Pedagogical Institute. He immediately began to work for the KGB of the Ukrainian SSR, ascending the ladder from junior lieutenant to general. In 1994, President Kravchuk issued a decree promoting him to general in the Ukrainian Army. He was deputy prime minister under both Kravchuk and Kuchma, then first deputy prime minister and prime minister. Over the period 1995–2000, he served as an MP; in the 1998 election, he won a seat in the parliament under the SDPU (United) list. In 1999, he became secretary of the National Security Council of Ukraine, and minister of defense in 2003–2004. In 2008, he was an outside advisor to President Yushchenko. In 2014, he headed the International Secretariat on Security and Civil Cooperation between Ukraine and NATO.

(in other words, virtually all) permanent parliamentary committees. These deputies included Henrikh Altunian, Hryhorii Demydov, Ivan Zaiets', Valerii Izmalkov, Anatolii Korzh, Valentyn Lemish, Taras Nahulko, Myroslav Motiuk, Mykola Omel'chenko, Viktor Prykhod'ko, Serhii Semenets', Iurii Serebriannykov, Larysa Skoryk, Volodymyr Slobodeniuk, Volodymyr Stus, Andrii Sukhorukov, Markian Chuchuk, as well as Arnold Nazarenko, who had been a member of an analogous committee in the Supreme Soviet of the USSR.

Although the range of political preferences in this working group reflected the diverse affiliations of the Parliament of the time, from traditional communists to dissidents, it differed from other governmental bodies in being open to considering fundamentally new legislation, rather than replicating Soviet models in republican garb. I was appointed the head of the working group and was tasked with submitting a list of experts for official approval. The group presented its first set of findings in just two weeks.

On 10 January 1992, I wrote an analytical memo to President Kravchuk about the need to shift the emphasis from an exclusively military interpretation of national security to an understanding that included economic, technical, scientific, information, and other aspects of Ukraine's security. With this in mind, I proposed adopting new legislation, implementing changes in existing governmental structures, and creating new ones. This entailed transforming the National Security Council into the National Security and Defense Council; reforming the Security Service (on 13 March 1992 Parliament changed the name of the National Security Service of Ukraine [SNBU] to the Security Service of Ukraine [SBU], signifying the removal of the organization's monopoly on national security); establishing a governmental body within the Cabinet of Ministers for coordinating matters related to economic, energy, technical, scientific, and environmental security; and, lastly, establishing the National Institute for Strategic Studies.[48]

In the context of economic security, I emphasized the need to develop our own industrial production of fresh nuclear fuel and to create safe storage facilities for spent fuel with the goal of maintaining the safe operation of Ukraine's nuclear power plants. This encompassed the production of fuel rods and their utilization, recycling after utilization, and disposal. Our strategy prioritized the country's energy security, which necessitated the development of every phase of production within the energy industry, replacing the Ukrainian SSR's economic focus on raw material production.

Our recommendations regarding the Black Sea Fleet became all the more relevant later, during Russia's annexation of Crimea in 2014. We predicted that the division of the Black Sea Fleet into two parts could lead, quoting from my analytical memo, "to permanent interethnic tensions in Crimea and an increase in separatist sentiment." Additionally, such a division "would be in

15. *Ivan Pliushch, 1990*

Ivan Pliushch (1941–2014), the Ukrainian statesman who, among all the country's top officials, fostered the formation of Ukraine's own strategy for nuclear disarmament and its implementation in the parliament of Ukraine. He was first deputy speaker in the first convocation of the Parliament of Ukraine and became Speaker on 5 December 1991, after Leonid Kravchuk was elected president. An agronomist by education, he worked as a foreman in the council farm and as head of a collective farm. In 1979, he graduated from the Academy of Social Studies under the Central Committee of the CPSU. He worked up until 1981 for the CPSU. From 1984 to 1990, he was the first deputy chairman of the Executive Committee, then the chairman of the Kyiv Oblast Rada. In 1991 he left the Communist Party. Pliushch was parliamentary Speaker 1991–1994 and 2000–2002, and an MP 1990–2006 and 2007–2012.

violation of our principles of neutrality and would worsen our relationship with the West." The memo demonstrated that the framework of the national security policy was ready: it was based on contemporary principles and set the course toward political and economic detachment from Russia.[49]

The memo triggered a massive opposition to our ideological principles in less than two weeks. Kravchuk was slipped a proposal to pass the formation of the national security policy to the executive branch, or, as this was termed at the time, to involve the experts. This opposition to our policy proposals took the form of a parallel process. It was applied to everyone who was not subservient to Russian influence in the Ukrainian government. The ultimate formation of national security policy was a textbook case of this tactic.

On 15 January, Kravchuk issued Resolution no. 41 (Res. 41), which to this day is classified confidential among parliamentary documents, even though it is no longer in effect.[50] Res. 41 formed a presidential administrative committee aimed at developing draft bills concerning the status, action plans, and structure of the National Security Council. The minutes of the 10 February 1992 meeting in open-access sources show that an official decision was made to "prepare the draft National Security Policy of Ukraine by 10 March of this year," laying out key directions for the policy's development and the composition of the groups initiating it.[51] From that moment on, two parallel processes unfolded, embodying two different worlds: the Soviet approach belonging to the past, and the contemporary one belonging to an independent Ukraine with its own national interests. A qualitatively new vision of national security was being developed within our group; in contrast, the hardened Soviet careerists were creating a lifeless philosophical treatise, whose quality can be assessed by accessing the legislative database on Parliament's website.

The key areas to be addressed in developing the security policy according to the aforementioned minutes were the following: external security, internal security, the military-industrial complex, economic security, energy issues, relations with CIS member states, information security, and cultural and environmental policy. Responsibilities for their development were divided among various officials. Three of the groups (external security, internal security, and information security) involved SBU head Marchuk, another group was headed by Foreign Minister Zlenko, and another by Prime Minister Vitold Fokin.

In addition to his decree of 17 February 1992, Kravchuk approved the formation of a consultative group of experts for the development of the national security policy. The group consisted of six people and was headed by Volodymyr Selivanov, who held a doctoral degree in law and was the chief specialist at the Institute of Government and Law of the Ukrainian Academy of Sciences. Other members of the expert group represented the Ministry of Defense (MoD), the SBU, the Ministry of Internal Affairs (MIA), and the Academy of Science and Production. In particular, the group included the following individuals: Colonel Hryhorii Kostenko, deputy head of the Department of Armed Forces Development at the Center for Operations and Strategic Research, which was part of the headquarters of the Ministry of Defense; Major General Ievhen Shvats'kyi, Internal Service, deputy head of the MIA Academy; Major General Volodymyr Sidak, provost of the Institute for Personnel Training of the SNBU; Mykhailo Mytrakhovych, head of the Department of Applied Matters in the executive committee of the Ukrainian Academy of Sciences; and Viktor Moroz, director of the Special Design Bureau "Agropromsystema".

Thus, the process effectively split into two parallel branches: the parliamentary group and the executive-presidential group, each of which was developing its own national security policy. The fundamental incompatibility of their approaches to the process prevented them from developing a joint working document.

The National Security Policy: the Parliamentary Group's Proposal

As the head of the parliamentary group, I presented its draft National Security Policy in the journal *Rozbudova derzhavy*.[52] What follows is a summary of its key provisions and arguments.

* * *

The totalitarian system has established a militarized understanding of national security in our political thinking. That is why the state institutions safeguarding our national interests have been reduced to mainly two elements: an army, which was supposed to protect the state from external threats, and repressive services, whose activities aimed at protecting the state from internal infringement. Within this limited understanding of security, we arrived at a situation where, having accumulated a colossal stockpile of military force and technology, the USSR became immensely vulnerable to internal problems, which eventually led to its collapse. Contemporary concepts of state power are based on the recognition of the growing impact of economic security on the state's defensive capacity and its ability to safeguard its vital interests.[53]

The system of national security is one of the most important standard features of a state-organized society. It is based on state interests. Speaking of Ukrainian-Russian relations, for example, it should be noted that the strategic interests of the two countries differ significantly. Neglecting these differences or delaying addressing them could lead to the loss of the nascent Ukrainian state.

Ukraine's national security is a system of views that are scientifically substantiated and adopted by Ukraine's supreme leadership. Ukraine's national security concerns the concept and contents of the state's vital interests, as well as the goals, structures, and ways and means of safeguarding them. The officially adopted security policy statement is the scientific foundation for the establishment of legislative support and implementation of policies aimed at protecting the essential national values and interests of the Ukrainian state. National security must take into account all factors directly affecting the lives of the Ukrainian people including, foremost, economic, political, military, environmental, and demographic factors.

Economic Security

The Development of Complete Production Cycles. Today, most finished goods are not produced by Ukrainian enterprises. Due to its unbalanced design, our economy aims mainly at the production of raw material and intermediate goods, most of which are exported to Russia and other countries, only to return later to Ukraine in the form of finished goods. This production scheme is highly inefficient. Therefore, without comprehensive structural reforms, Ukraine will be unable to achieve economic strength.

Energy Independence. Another factor negatively affecting Ukraine's economy is the country's complete dependence on supply by energy carriers from other countries, primarily from Russia. Also detrimental is the degree to which Ukraine's strategic industries are integrated into the economies of other former Soviet republics, again, primarily of Russia. This makes Ukraine's national economic system exceptionally vulnerable to external factors. Most notably, this opens opportunities for external players to direct Ukraine's economic strategy for their own benefit, rather than for the benefit of the Ukrainian people, through political pressure or intimidation.[54]

The Assessment of Industries' Competitive Capability. In order to fully grasp the condition of Ukraine's economic security, the following tasks are of utmost urgency: developing methods to assess our industrial systems' competitiveness on the global market; determining the limitations of Ukraine's industrial dependence on other countries; determining technology vulnerability gaps; assessing the presence and stability of the national currency; and monitoring the forward-looking nature of structural reforms. Such national economic development indicators are used for identifying the nature and sources of real threats to vital national interests, as well as options for reducing such threats. The need for a theoretical and practical solution to the problems inherent to Ukraine's economic security is intensified with Ukraine's intentions of developing global economic ties, especially since Ukraine's economy has multiple industries which could eventually be recognized in the global labor market. One of these industries is rocket-building, which should be converted to produce spacecraft, rather than consumer goods.

Therefore, Ukraine urgently needs to develop proper economic security measures. Without them, it will be impossible to also properly identify and solve issues associated with the conversion of the production of materiel. It will also be impossible to determine promising directions for our economy or to establish the optimal proportions in the fruitless debate about what carries more weight in the system of national security: a reliable defense system or an effective national economy and the wealth of society.[55]

Environmental Security

Environmental conditions endanger the survival of our nation. Since 1990 deaths have surpassed the country's births. The uncontrolled expansion of industry in farming areas has turned Ukraine's most valuable fertile "black soil" in the southeastern and central regions into ecological disaster areas. In Mariuopol the level of chemical contamination of the environment is equivalent to the effect of radiation at 180 rem for 30 years. For comparison, 10 rem would be the average dose for 30 years of exposure to natural radiation. Not coincidentally, it was mass public protests in support of environmental rights that laid the foundations for the democratic transformation of Ukraine.

Information Security

Access to information is becoming a commodity equal in importance to energy production; information can be converted into other goods and services. Over 50 percent of the gross domestic product (GDP) of contemporary Western countries is directly related to the production, circulation, and processing of information. In the USSR, information was gauged exclusively through the prism of secrecy and repression for the sake of the regime and as a means of ideological propaganda. Today, Ukraine needs a fundamentally new information policy, including the creation of a nationwide system of information resource management, and new governance mechanisms for the disclosure and protection of secret information.

An important aspect of information security is related to foreign policy. Today, as a result of Ukraine's lack of effective communication with Western countries, the processes underway in Ukraine receive a predominantly pro-Russian interpretation. Russia, with the help of the powerful information infrastructure that it acquired from the former Soviet Union—including its embassies, consular offices, and various foreign policy agencies—is fabricating an image of Ukraine as a neo-nationalist state reaching for the "nuclear button" and imperiling all of humanity.

This distortion of Ukraine's political image significantly affects the state of our economy: lending resources flow predominantly to Russia and businesses are wary of investing in Ukrainian industries. Developing a system of information security is critical and must be directed not only at Western countries, but also at former Soviet republics, where many ethnic Ukrainians reside. The loss of an exchange of information with the Eastern Diaspora (Ukrainians scattered throughout the Russian Empire and the Soviet Union) negatively affects our national security.

Nonnuclear Status as a National Security Factor

Today, the majority of the world does not see nuclear weapons as a means of aggression; they are, rather, considered deterrence against aggression. Having declared its intentions to become a nonnuclear-weapon state, Ukraine has the right to expect adequate support of its denuclearization efforts from the international community.

Complete nuclear disarmament requires significant financial expenditures. Therefore, Ukraine has a reason to expect international financial assistance and preferential loans. However, after Parliament's adoption of the Statement on the Nonnuclear Status of Ukraine, we see that most international loans flow only to Russia.

The next point is crucial: Ukraine needs national security guarantees. Who and what will defend us after we eliminate our nuclear weapons? There are three core national defense systems in the world: the first is military power; the second is economic integration with other countries (for example, any threat to Switzerland, which accumulated colossal levels of global capital, would automatically pose a threat to most other countries); and the third is an advanced scientific and technical capability (such as, for instance, that of Japan, Germany, or South Korea), which in this day and age serves as a defense mechanism no less potent than any military weapon. Unfortunately, none of these security systems apply to Ukraine. Our economy is highly integrated with Russia's, and Russia's potential collapse would threaten any country. Thus, after liquidating its nuclear weapons and without the protection of powerful Western countries, Ukraine is defenseless in the face of military and economic aggression.

* * *

Overall, unless Ukraine's national security system is made to conform to its new political realities, as well as to support the sociopolitical, scientific, technical, environmental, and spiritual development of our society, the Ukrainian people might be deprived of a future. These were the key principles of Ukraine's National Security Policy, developed by the parliamentary group I chaired in early 1992.

The National Security Policy: Operation "Recycling"

Nevertheless, only the proposal of the presidential group of experts made it onto Parliament's agenda. Even so, Parliament only reviewed their draft document for the first time on 19 October 1993. The document was rejected on the grounds of the ambiguity of its provisions and was assigned to Parliament's Defense and National Security Commission, to refine the draft together with the Cabinet of Ministers.

The next review took place another year and a half later, on 24 May 1995, in a changed Parliament. The draft was approved on its first reading, but was again assigned to the same Defense and National Security Commission and Cabinet for further revisions, which were to take into consideration the comments and suggestions made by the MPs and the standing committees of Parliament.

The final version of the document was approved after another year and a half, on 16 January 1997. At the same time, Parliament recommended to the president, the Cabinet, the Defense and National Security Commission, along with other expert committees, that the development and submission of drafts of subsequent, offshoot bills to Parliament be hastened. These included the law on the national security of Ukraine, the law on the National Security and Defense Council, the law on foreign intelligence, and other national programs and legislative acts concerning national security that had arisen from the provisions of the National Security Policy.

However, in the five years that it took to develop the Policy, the essence of the document had been completely diluted. Essentially, it had become an abstract philosophical essay on global development in the coming decade, rather than a policy framework for developing specifically the state of Ukraine under its very specific circumstances. It neither identified the particular threats to the country nor proposed responses to those threats; it did not prioritize any existing problems that the Ukrainian government would need to solve at every stage of the country's development. Furthermore, it did not even mention energy security, which for most of the contemporary world was one the chief foundations of national security, and for Ukraine in the early 1990s was not just a priority, but a goal. The Law on the Foundations of National Security of Ukraine was adopted another six and a half years later, on 19 June 2003.

Considering the speed (or lack thereof) of the adoption of these laws, one cannot help but ask these questions: What had Ukraine been doing to ensure its national security during the twelve years since the establishment of the modern state, when it lacked this essential foundational document? How could a nation exist with no articulated framework of national interests and key threats? The US, for instance, has a national security framework in place that it revisits annually to formulate the most relevant current security threats, from terrorism to Ebola, and then delegates the government to address those threats. In Ukraine, nothing was done regarding national security for over a decade: everyone waited for the official framework. When they finally received it, it had no substance.

In this way, the document that was meant to determine the new state's threats, opportunities, and principles of conduct, using contemporary approaches and guiding principles under the umbrella of Ukraine's national interests, was devitalized. The military doctrine met the same fate, as will be discussed in more detail below.

Our group's version of the national security policy was never brought up for consideration by Parliament. Having prepared a draft of the document in May 1992, I concentrated my efforts on the most critical area of national security at that time: nuclear disarmament. There were two reasons I saw it as futile to continue expending all my time and energy fighting for a modern system of national security. First, a mechanism for blocking any new approaches to governance had been put in place at all levels of the government, and overcoming it was nearly impossible. The minutest of victories could be achieved only by endless office visits and reports, which threatened to take up 90 percent of one's time. The other reason lay in the emergence of a new and critical circumstance. Upon execution of the Lisbon Protocol on 23 May 1992, Ukraine was to initiate the ratification of two complex nuclear treaties: the START and the NPT. Not only did we not take part in the development of these documents, but at the time, we did not even have their Ukrainian translation. Most crucially, and as noted earlier, there was no one center for information gathering or decision-making on this issue.

However, it was already obvious at that time that nuclear arms could be used as significant seed capital for the independent Ukraine. Nuclear weapons were tangible resources that potentially could be taken away from us with no compensation except alluring declarations of peace. Nuclear disarmament had become a hot topic not only in Ukraine but also in the global geopolitics because here the world's third largest nuclear force was at stake.

A Conversation with Brzezinski

In early February 1992, I flew to Washington, DC. Officially, the purpose was to attend the annual meeting of the program entitled Global Learning and Observations to Benefit the Environment (GLOBE). However, two other meetings were organized for me as the head of the parliamentary committee for the development of the national security policy and were more central to my visit. The first meeting took place at the Pentagon, with secretary of defense Dick Cheney. The second one was with the legendary Zbigniew Brzezinski.

At that time Brzezinski was the former national security advisor to the president. However, he endured as one of the most prominent gurus of the American politics, and his opinions were held in high regard by American presidents. It was critical for me to hear his answers to a number of questions arising in the process of formulating Ukraine's national security strategy. Although I understood that US interests would always be his top priority, I also knew that I could count on him to take an objective stance toward Russia and its imperial ambitions.

Brzezinski was greatly sympathetic toward Ukraine and immediately impressed me with his in-depth knowledge of the Ukrainian situation, including the key issues on which the country had to focus first and foremost. He recommended that Ukraine minimize its dependence on Russia over the next two to three years, most notably in terms of the economy and the military. As for the economy, its first steps would be to create Ukraine's own financial and banking systems, and to establish its own national currency. Brzezinski's advice regarding defense was to create a robust nonnuclear army and adopt a doctrine of military defense. In his opinion Ukraine as a nonnuclear state would be seen positively by the West, although other experts, including his own son (who worked on these issues), expressed a contrary view.

Additionally, according to Brzezinski, Ukraine's foreign policy needed to place special emphasis on external communications in order to foster pro-Ukrainian sentiments in the US. Key to this would be the appointment of a dynamic and influential ambassador to the United States, who would not be limited to interactions with the State Department, as had been customary for Soviet ambassadors.

He predicted that Russia's approach to its relationship with Ukraine would involve leaving as many issues unresolved as possible, including Crimea and the Black Sea Fleet, and using these

16. Zbigniew Brzezinski, on a visit to Kyiv in 1991

Zbigniew Kazimierz Brzezinski (1928–2017), American political scientist and statesman. He was on the Policy Planning Council at the US State Department from 1966 to 1968. Between 1973 and 1976, he participated in founding the Trilateral Commission and later became its director. Under President Carter, he was the national security advisor from 1977 to 1981. He served as an adviser on national security and foreign intelligence affairs under subsequent administrations in the 1980s and 1990s.

issues as destabilizing factors. Another lever Russia would try to exploit, he believed, were actions that would weaken Ukraine's economy, so as to provoke public discontent and social unrest.

Upon returning to Kyiv, I immediately prepared a report for President Kravchuk outlining the results of my visit, including Brzezinski's key predictions and recommendations. Their accuracy became evident over the next twenty years, especially in 2014 and 2022 when Russia embarked upon aggression against Ukraine. The transcript of my conversation with Brzezinski, supplied to me by the Ukrainian embassy in the US, documents his recommendations to President Kravchuk with regard to constructing Ukraine's system of national security in the early days of independence. Here are some noteworthy excerpts from our conversation:

BRZEZINSKI: Which security issues are the most important to Ukraine?

KOSTENKO: Our dependence on the Russian financial system is an enormous problem. A year ago we could have resolved this problem, but it is more difficult now. [...]

BRZEZINSKI: Have you seen [Russian ambassador Vladimir] Lukin's new letter, published in *Komsomolskaia pravda?*

KOSTENKO: Yes, he brings up certain issues, reiterating Soviet notions of security from the times of the USSR.

BRZEZINSKI: The letter conveys Russia's resolve to pressure Ukraine in three ways: first, through the Crimea question; second, through the Black Sea Fleet question; and third, through the possible reduction of orders for industrial goods from Ukraine.

This points to Russia's intent to subordinate Ukraine. And the fact that Lukin was appointed the ambassador to the US shows that Russia is hoping the US will support their strategy.

Ukraine's defense minister, Kostyantyn Morozov, has stated that Ukraine needs five years to build a strong sovereign army. I do not think that Ukraine has that kind of time. You have no more than two to three years to create a robust state. After all, Russia can resort to much more forceful attempts to subordinate Ukraine, and it might succeed. Russia can exert military, political, and economic pressure on Ukraine in order to provoke social unrest, strikes in the eastern regions, and so on.

Even Russia's democratic leaders do not believe that Ukraine is its own nation, or that it will endure, or even that it should endure. These sentiments are widespread [...]

KOSTENKO: Can the US help in solving some of these problems?

BRZEZINSKI: To some extent we can. However, many Americans (I am not one of them) tend to underestimate issues like these right now. Many are not convinced that the Ukrainian state is genuinely, historically durable, and that it can survive. It is very important for the United States and Europe to support independent Ukraine, because it will promote Russia's democratic rather than imperialistic, development, in the postcommunist era.

Your task is to establish a genuine state. You have the proper symbols and you can establish a government. But you are not financially independent. You have no military. These two things are critical. Over the next two years you need to achieve financial and military sovereignty.

KOSTENKO: What is your view on Ukraine's nuclear disarmament policy?

BRZEZINSKI: Nuclear arms will not save you. You absolutely need to get rid of nuclear weapons. It will only isolate you from the West. The West sees nuclear weapons as a hazard; it fears the possibility of a nuclear war.

Russia endeavored to create the image of Ukraine as a radical nationalist state. It was a mistake to speak of your armed forces as numbering 450,000 military personnel. I would recommend that Ukraine adopt defensive principles in establishing its military. [...] Show the West that Ukraine wants to be a strong nation, but a nation committed to a defense-based policy. [...] As far as the Black Sea Fleet is concerned, you should tell the Russians that you wish to form a defensive fleet, that you have no ambition to enter the Atlantic or the Mediterranean seas. Ukraine and Turkey ought to be responsible for the security in the Black Sea. For this purpose, you could say that you want approximately 30 percent of the fleet. The Russians could base part of their fleet in Novorossiisk. This is not a battle fleet anyway; it would not be used in war.

Alternatively, you can give Russia temporary use of the base for five years (using the US agreement with Singapore as a model). Charge Russia two billion dollars. What alternatives do you have? You cannot just get rid of them.[56]

KOSTENKO: The issue of the Black Sea Fleet is as much of a matter of principle to us as the Crimea issue. To leave the fleet under the Russian flag would be extremely destabilizing for Crimea. We believe the US should recognize Ukraine as a potentially stabilizing partner in all of Europe.

BRZEZINSKI: There is no need to convince me. Russia's current strategy is to leave every Ukraine-related issue unresolved, to thwart any solution. This will go on until Russia has buttressed its economy and revamped its political situation. Russia will attempt to fuel discontent in Ukraine, to present Ukrainians as chauvinistic, irresponsible and radical. They hope to convince the West of the same.

The Ukrainian people must impart an image of cooperation, responsibility, and commitment to a defense policy, so that the West can see for itself that Ukraine is a durable sovereign state. [...]

The key objective for the Ukrainian ambassador and for the Ukrainian government is to convey to the West that an imperialistic Russia that dominates Ukraine will not be a democratic Russia.[57]

Defense Council Meeting: Kravchuk's First Conversation with Nuclear Weapons Specialists

By the spring of 1992, the so-called experts from various sides had managed to convince President Kravchuk that Ukraine did not need nuclear weapons, because not only could we not use them, but we did not even know how to handle them. From within the democratic camp, one of the People's Movement leaders, Dmytro Pavlychko, who headed the parliamentary Commission on Foreign Affairs, had become a strong influence on the president. He, too, had been persuaded by someone that the sooner we eliminated our nuclear weapons, the sooner we would eliminate the threat of Russian invasion. In other words, for as long as we retained the nuclear warheads, the threat remained that the Russian army would enter to seize control over them. Another potential threat was the possibility that Russia might launch nuclear missiles from our territory, leading to Ukraine's destruction as a result of retaliation.

Therefore, at the 2 April 1992 Defense Council meeting, President Kravchuk showed up with a stack of documents he had signed over the period since late 1991. He was utterly convinced that the nuclear weapons had to be handed over as quickly as possible. The only question remained: How could it all be done?

On the recommendation of Anton Buteiko, who was then in charge of the president's Office of International Affairs and who rendered advice on official decisions, the invitee list included not only various ministers and department heads, but also a number of nuclear weapons specialists. The latter included Iakiv Aizenberh, director of Khartron; Viktor Zelens'kyi, director of the Kharkiv Institute for Physics and Technology; Stanislav Koniukhov, chief designer of Pivdenmash; Viktor Bar'iakhtar, vice president of the Ukrainian Academy of Sciences; Ivan Vyshnevs'kyi, director of the Institute for Nuclear Research; and me, as the head of the parliamentary group developing the national security policy.[58]

However, the meeting took a fundamentally different turn than Kravchuk had anticipated. It ended up being the first high-level discussion with experts about the challenges Ukraine would have to address as soon as it began nuclear disarmament. Moreover, the participants of the meeting presented a plethora of information that demonstrated the following: not only was Ukraine able to adequately handle its nuclear weapons, given proper cooperation with the Russian industrial complex, but it also had the levers with which to induce Russia's cooperation. This meeting signified the end of the political amateurism surrounding the topic of nuclear disarmament that had been fed to Kravchuk, as now actual specialists on the matter finally spoke to the Ukrainian president.

That being said, the first person to present at the meeting was Foreign Minister Zlenko. His main idea was clear and conformed to Kravchuk's general opinion that Ukraine had to give up nuclear weapons as speedily as possible. I watched with interest as the foreign minister made each of his arguments, focused solely on arguing for the transfer of all nuclear warheads to Russia,

17. Dmytro Pavlychko, 1993

Dmytro Pavlychko (1929–2023), poet and political activist; one of the founders of the People's Movement. In 1953, he graduated from the Philology Department of Lviv University, and published a large collection of poetry, as well as literary criticism. In 1990–1999 and 2005–2006, he served as an MP, and chaired the parliament's Foreign Affairs Committee between 1990 and 1994. A close ally of Foreign Minister Anatolii Zlenko, Pavlychko initially supported the immediate removal of all nuclear weapons from Ukrainian territory, but in 1993 changed his position to support the program of effective disarmament. From 1995 to 2002, he was an ambassador to Slovakia and Poland. From 2006 to 2011, he chaired the Ukrainian World Coordinating Council.

without even considering using the nuclear material for Ukraine's own needs. The premise of his presentation was the following: first, Ukraine had no facilities in which to store plutonium and, therefore, we did not need it; second, enriching uranium was easier than downblending it, thus, it was better to hand the HEU over to Russia; third, Ukraine's military doctrine did not sanction heavy assault missiles; and, fourth, dismantling warheads in Ukraine was not possible, because it was classified "secret" by Russia. Whereas the last of these arguments constituted a political question, the first three contained direct factual errors.

Zlenko supported the former two arguments with conclusions of the Academy of Sciences contained, supposedly, in a report authored by Ivan Vyshnevs'kyi, director of the Institute for Nuclear Research. I can confirm that the report, a copy of which I have in my archives, proposed the opposite of Zlenko's claims. For instance, with regard to plutonium, the report highlighted the possibility of using mixed uranium-plutonium oxide (MOX) fuel for the nuclear energy industry. Vyshnevs'kyi wrote:

> Increased attention is being paid to mixed fuel technology involving uranium plutonium oxide; it opens up opportunities for the use of plutonium in already operating thermal reactors after recovering it from spent fuel (Belgium, France, Germany, Japan). Current assessments point to the efficacy of this method, as utilizing the mixed oxide fuel offers significant economic benefits, while not necessitating substantive changes in the reactor control systems. Considering the fact that warheads already contain ready-made plutonium, the benefits of using it in the energy industry are obvious.[59]

As to uranium-downblending technologies, Vyshnevs'kyi wrote that, unlike uranium enrichment, "the process does not pose any kind of serious problem [...]. In any case, I can confirm that

18. Anton Buteiko, 2005

Anton Buteiko (1947–2019), Ukrainian diplomat and advisor to President Leonid Kravchuk's Office of International Affairs between 1991 and 1994. After graduating from Shevchenko National University of Kyiv in International Law in 1974, he entered the MFA of the Ukrainian SSR. From 1980 to 1986, he was a member of the UN Secretariat in New York. Until 1991, he was an advisor to the Department for International Organizations and supervisor of the Contractual and Legal Department within the MFA of the Ukrainian SSR. He was elected an MP in 1994–1998, he became the ombudsman to the Cabinet of Ministries on European and Euro-Atlantic Integration and deputy chairman of the State Commission for NATO Relations from 1994 to 1996. After his November 1992 trip to the bases of the Strategic Nuclear Forces, Buteiko became one of the most active proponents of effective disarmament. He contributed to the coordination of strategic decisions regarding nuclear disarmament between the working group and the presidential administration, blocking of the immediate ratification of START I and the Lisbon Protocol, and working with the working group to draft the text of the ratification resolution of the parliament that was adopted on 18 November 1993. Between 1995 and 1998 and from 2005 to 2006, he served as first deputy foreign minister. In 1998–1999, he served as the ambassador to the US, and from 2000 to September 2003, as ambassador to Romania.

carrying out such a process is far easier than increasing the concentration of uranium-235 in natural uranium."[60]

Zlenko's reference to the military doctrine also sparked confusion, because no such document existed at the time, and it was not even being considered by Parliament. The poor quality of the arguments for nuclear disarmament advanced so vigorously by the Ukrainian Foreign Ministry at that time suggests that their key ideas originated outside of the ministry or, perhaps, even outside Ukraine itself.

The only person who supported Zlenko's arguments was the head engineer of Pivdenmash. He maintained that all of the warheads together contained only ten tons of uranium and plutonium, and that it was better to hand them over to Russia. The figure was a significant underestimation. It is difficult to say at this time whether the head engineer's lack of knowledge was at fault, or it was a case of deliberate disinformation.

Iakiv Aizenberh, director of the Khartron Design Bureau in Kharkiv, presented information on two issues. The first concerned the technological requirements for removing nuclear warheads from strategic delivery systems and the feasibility of their transfer to Russia by 1994 (the schedule Kravchuk had announced earlier). To remove a nuclear warhead from a strategic delivery system, he said, the warhead had to be replaced with a so-called equivalent (an electronic simulator of a nuclear warhead), and those still needed to be manufactured. This was an expensive and rather lengthy procedure. Furthermore, Ukraine did not have sufficient quantities of receptacles for draining out the highly toxic chemicals in the missile fuel, heptyl and amyl, or the technology to recycle them. Thus, Aizenberh concluded, it was technically impossible to complete the transfer of nuclear warheads to Russia by 1994.

The second issue concerned Ukraine's technical capability of creating a so-called lock button: a device that could potentially prevent Russia from launching strategic missiles from the territory of Ukraine. (Khartron specialized in the development and production of strategic control systems.) This issue for some reason worried Kravchuk the most. Aizenberh emphatically maintained that without Russia's consent this kind of interference with the command and control system was impossible. Naturally, Moscow was aware of this conversation, but later it instigated an international scandal, accusing Ukraine of attempting to establish its own missile launch codes for the autonomous use of nuclear weapons.

The most informed person at the meeting on the subject of nuclear disarmament was Stanislav Koniukhov, chief designer of Pivdenmash, the manufacturer of the most formidable strategic missiles, termed in the US the Satan and Scalpel. Each of these carried ten nuclear warheads, and each, in turn, could annihilate a large city. In contrast to many other government officials, Koniukhov had no inferiority complexes regarding Ukraine's nuclear capabilities. On the contrary, he was well aware both of the value of the product manufactured at his plant, as well as of the level of Ukraine's intellectual potential. As a manufacturer, Koniukhov, in contrast to career communists, understood what it meant to compete on the global arms trade and space technology market, something Pivdenmash was already doing. He, therefore, overtly declared: Russia planned to eliminate Ukraine's military-industrial complex.

Koniukhov, like Aizenberh, expressed his doubts regarding the possibility of removing all nuclear weapons from combat duty on a tight timeline. He identified the main problem that would result from a hasty removal: a severe decline in Ukraine's defensive capability, and the absence of an adequate replacement for nuclear weapons in the country's system of deterrence of a military aggressor against Ukraine.

Therefore, on the issue of Ukraine's nuclear disarmament, he proposed the a step-by-step process. First, Ukraine needed to disarm gradually and in unison with the rest of the world. Koniukhov believed that Ukraine did not need to disarm ahead of everyone else, but rather congruently with and proportionally to other countries. Second, the missile systems of outmoded production, i.e., the 130 SS-19 missiles, should be pulled off line and recycled first. Third, all 46 ultramodern SS-24 Scalpel missile systems manufactured in Ukraine should remain, as they constituted more nuclear weaponry than the strategic forces of the UK and France combined. This meeting was also the first at such a high level to introduce the idea of replacing nuclear weapons with precision-guided weapons in Ukraine's defense system. Koniukhov argued that this way, the missiles could still be used as an effective deterrent to aggression against Ukraine. Shortly thereafter, he presented a two-phase program for transitioning the country's armaments to precision-guided weapons, which Pivdenmash was prepared to implement.

Fourth, Koniukhov suggested that the dismantling of warheads could be done in Ukraine, because the country had both the requisite expertise and technical capability. The only obstacle was that Russia would interpret the process as a declassification of nuclear explosives, which it considered secret. Koniukhov's idea was that Ukraine could dismantle the nuclear warheads domestically instead of transporting them to Russia, and use the international disarmament funds to pay Ukrainian, not Russian, scientists.

Stanislav Koniukhov was also the first person to draw the president's attention to the interdependence of Ukraine and Russia relating to the safe operation of strategic nuclear facilities. He conveyed that Pivdenmash's hypothetical refusal to service missile delivery systems located on Russia's territory could potentially result in an unsanctioned launch of SS-18 missiles, with apocalyptic consequences for the entire planet. On the other hand, because of the special safety mechanism built into nuclear explosive devices, the unsanctioned detonation of a Russian-manufactured nuclear warhead on Ukrainian territory would

19. *Anatolii Zlenko, 1991*

Anatolii Zlenko (1938–2021), the first foreign minister of Ukraine. He headed the Foreign Ministry from 1990 to 1994 and again from 2000 to 2003. He was a member of the CPSU until the August Putsch. He graduated from the Kyiv Mining College and worked as a mining specialist. From 1967, after graduating from Shevchenko National University of Kyiv, he worked in the Foreign Ministry of the Ukrainian SSR, as the republic's permanent representative to UNESCO. From 1987 to 1990, he was deputy foreign minister for the Ukrainian SSR. Between 1994 and 2000, he was Ukraine's permanent representative to the UN, then again to UNESCO, and, finally, to the Executive Board of UNESCO. Between 1997 and 2000, he was the ambassador to France and subsequently to Portugal. In 2003, he was appointed President Kuchma's advisor on international affairs, and subsequently Ukraine's representative to the UN Commission on Human Rights.

20. *Pivdenmash*

cause only an ordinary TNT explosion, resulting in the nuclear pollution of a relatively small territory. Many other experts have confirmed this to me, but Kravchuk chose not to believe it.

The latter was a deterrence factor, a lever that Ukraine could use in the event that Russia threatened to discontinue servicing Ukraine's nuclear warheads. However, the Ukrainian MFA never conveyed this information to the global media, even after Russia announced in 1993 that "a second Chernobyl is looming in Ukraine because the nuclear warheads are about to explode." The Ukrainian MFA's subsequent actions made it appear as though Zlenko had heard nothing at that Defense Council meeting.[61]

In my presentation, I drew attention to the rapid changes in the post-Soviet space in light of new threats to Ukraine's security posed by Russia. (Zlenko, for some reason, never mentioned this in his presentation.) In violation of international treaties entered into by Russia prior to that time, the Russian Federation began transferring command and control of the CIS's strategic nuclear forces to the Russian armed forces. To justify its actions, Russia orchestrated several large-scale diversions in the Ukrainian army, including organizing an unsanctioned flight and transfer of Ukrainian fighter planes to Russia, in addition to its persistent attempts to take over the Black Sea Fleet. Furthermore, the mass removal of Ukraine's tactical nuclear weapons and their transport to Russian factories continued without corresponding decisions by the Ukrainian parliament, which had the exclusive power to dispose of state property.

Russia's assertiveness on the nuclear weapons front demanded swift action by Ukraine in the establishment of its nuclear disarmament policy. It was, in its essence, not a refusal to become a nonnuclear state, but rather a search for the most effective path to disarmament. Only Ukraine's interests and national security were to be the grounds for decisions and actions in finding that path. As a first step, I proposed that Ukraine assume administrative control of all strategic nuclear forces located on its territory. However, it was difficult to imagine at the time that Kravchuk would have the courage to do it.

Ukraine Springs into Action

The Security Council meeting resulted in two important documents. The first was the president's decree on the Ukrainian Armed Forces' assumption of command and control over units of the strategic forces, specifically the 43rd Rocket Army and the Strategic Air Forces. The second was Parliament's resolution "On Additional Measures to Ensure Ukraine's Attainment of Nonnuclear-Weapon State Status," approved on 9 April 1992, a mere week after the meeting. Essentially, these were the first specialized actions by the Ukrainian government since the country's declaration of independence, demonstrating Ukraine's resolve on the nuclear issue.[62]

For the first time, Parliament had dared to protest Russia's noncompliance with its official obligations, noting that "the government and the Strategic Forces Command of the Russian Federation did not establish an appropriate system of effective technical control over the non-use of nuclear weapons deployed on its territory, as stipulated by the agreement on joint nuclear-weapon measures of 21 December 1991."[63] The Cabinet of Ministers was assigned to take appropriate measures independently.

The resolution of 9 April 1992 also contained a provision that no one expected from Ukraine at the time "[t]o consider inexpedient the transport of tactical nuclear weapons from the territory of Ukraine before the development and implementation of a mechanism ensuring international control over their destruction, with the participation of Ukraine."[64] With this stipulation, Parliament expressed its intention to seek international assurances that the nuclear arms transferred from Ukraine would actually be demolished, rather than used by Russia. The resolution also declared that Parliament would not consider handing over nuclear material from Ukraine's nuclear weapons arsenal to Russia.

Another provision of the resolution of 9 April 1992 suggested that the international community should not expect a lightning-quick transition, but instead prepare for a long process in Ukraine's becoming a nonnuclear-weapon state. The resolution instructed appropriate committees of Parliament, in consultation with ministry experts, the Ukrainian Academy of Sciences, and if necessary, independent experts, to "consider the complete list of issues associated with nuclear disarmament, including economic, financial, environmental, and organizational aspects of eliminating nuclear weapons, as well as the use of its components for peaceful purposes."[65]

On the same day (9 April), the first vice prime minister of Ukraine, Ihor Iukhnovs'kyi, held a meeting with representatives of the MIC, the Ministry of Defense, and the Academy of Sciences, at which they reviewed a host of technical issues related to disarmament, including questions of strategic systems' security. They proposed creating a special presidential commission to

21. *Ihor Iukhnovs'kyi, 1991*

Ihor Iukhnovs'kyi (b. 1925), Ukrainian theoretical physicist, scholar at the National Academy of Sciences of Ukraine, and an MP during the period of 1990–2006. He was the ideologist behind the independence referendum of 1 December 1991. At his insistence, the Declaration of Independence was brought to the referendum to prevent attempts by future legislatures to revoke it; this gave the Declaration its permanence. He led the legislative opposition as head of the People's Council from 1990 to 1993. He supported the idea of nuclear disarmament as a means of gaining optimal economic and political benefits for Ukraine and the developing the country's own facilities to manufacture nuclear fuel. As first deputy prime minister, he launched negotiations with General Atomics to determine the most effective option for nuclear disarmament. At the time, he chaired the parliament's Commission on Science and Education, and was a member of the presidium. He held the highest-ever post as an MP from the democratic bloc in the first coalition government; he was also the first deputy premier from October 1992 to March 1993. When he resigned in March 1993, there was no one left among top Cabinet officials prepared to support an advantageous national policy of nuclear disarmament in a systematic fashion. From 1990 to 1998, he ran the Western Science Center of the National Academy of Sciences.

coordinate all disarmament activities. Our working group, which was tasked with preparations for ratifying START, later took over this mission.

The next day, on 10 April, Leonid Kravchuk sent a letter to George H. W. Bush. It may have been the only letter in which the Ukrainian president did not accept further obligations, but on the contrary, asked for help in addressing the issues Ukraine would need to resolve before commencing the disarmament process. He asked for US assistance in solving the problem of servicemen being discharged as a consequence of disarmament. This included technical and financial assistance, developing training programs for officers working with nuclear weapons to transition into peaceful occupations, as well as creating jobs and providing housing for them. Kravchuk wrote: "Discussing this with experts, we expect to develop specific, clearly defined objectives and the scope of the US technical assistance program for resolving the most pressing technical issues related to the [disarmament]."[66]

The Kremlin immediately noticed the change in Ukraine's tone. On 11 April, a state delegation with Foreign Minister Zlenko at the helm flew to Moscow for an official meeting of the foreign ministers of Ukraine, Russia, Belarus, and Kazakhstan. I was included in the delegation as the representative of the Ukrainian parliament. The meeting's purpose was to review the status of the implementation of the nuclear disarmament treaties signed in Minsk and Almaty. During the discussion, Russia began discrediting Ukraine's position, including the provisions of the resolution adopted by the Ukrainian parliament immediately prior to the meeting on 9 April. In my report to Kravchuk I wrote in the key conclusions of the official visit: "Taking into account the division of the Black Sea Fleet, Russia is doing everything it can to portray Ukraine as a militant nationalist state eager to get ahold of the nuclear button." I proposed that there was an urgent need "to conduct a comprehensive information campaign explaining Ukraine's nuclear disarmament policy."[67]

A few days after the Moscow meeting, I submitted an analytical memo to Kravchuk, entitled "Certain Remarks on the Implementation of the START Treaty." This was the first thorough summary of information on Ukraine's nuclear disarmament with an attempt to formulate and conceptualize the goals of the process.[68]

I suggested proceeding based on the following facts. First, Ukraine was the first of the former Soviet republics to announce its intentions of becoming a nonnuclear-weapon state. Second, the status of the nuclear weapons located on Ukraine's territory during the period of the USSR was fundamentally different from the status after its collapse. At that point in time, I wrote that there were now four nuclear-weapon states inheriting the USSR's property and legal obligations. Third, Russia was trying to disregard these realities, instead comparing the situation with the basing by the US of its nuclear weapons in Western Europe. It ignored the fact that Ukraine had participated, intellectually and financially, in the development of these weapons, contrasting it with the US case, where its weapons were simply stationed in Western Europe, which was impeding the disarmament process. Fourth, in order to resolve this problem, the US needed to recognize the four nuclear republics as a collective party to START, and acknowledge the fact that the USSR's nuclear potential had been divided among these new states.

As for the technical aspect of liquidating nuclear weapons, I had already written that adhering to some of the conclusions by Ukrainian government officials was unadvisable. The advice from the Kharkiv Institute for Physics and Technology, which recommended ceding warheads to Russia in exchange for nuclear power plant fuel, had used obsolete, Soviet prices. I predicted that, in the absence of uranium enrichment facilities in Ukraine or its own production of heat-releasing elements and recycling of spent fuel for power plants, Ukraine's nuclear energy industry would simply go bankrupt if Russia transitioned to global market prices.

Instead, based on open-source information suggesting that the European Bank for Reconstruction and Development (EBRD) was prepared to write off all of the USSR's debts in exchange for nuclear warheads, I suggested the following scheme for handling nuclear warheads. First, I recommended that Ukraine resume shipping its tactical nuclear warheads to Russia only after implementing an oversight procedure ensuring their ultimate destruction; the oversight was to be carried out by the United States and the CIS countries. At the same time, Ukraine was to ask Russia to return the uranium and plutonium from the Ukrainian warheads that it had already dismantled, or Ukraine was to demand Russia's commitment to supply nuclear fuel for a specified period of time and in quantities proportional to the energy value of the HEU and plutonium.

Second, I suggested simultaneously engaging investors in building a specialized plant for repurposing nuclear warheads in the strategic missiles on Ukrainian territory. Considering the US's eagerness to see Ukraine's swift nuclear disarmament, I wrote, we could build the plant in one to two years. It was already becoming evident then that Russia did not have the capacity to liquidate the USSR's nuclear arsenal within the predetermined time frame. I also suggested engaging Western experts in monitoring all operations associated with the recycling of nuclear warheads. This allowed, first, to eliminate the question of where the nuclear material would go and, second, to force Russia to partake in the process of dismantling nuclear warheads on Ukraine's territory, thus avoiding the declassification of their design.

I also expressed a viewpoint, provocative at the time, that eliminating the forty-six Pivdenmash-produced SS-24 missiles within the framework of START would be disadvantageous and premature. Although Russia had already refused to service weapons on Ukrainian territory and threatened us with the prospect that they would become inoperational and hazardous in a matter of five years, I was convinced the situation could be settled in a mutually beneficial way: Pivdenmash would service the strategic delivery systems of its own manufacture based in Russia and Kazakhstan, while Russian companies would service the nuclear warheads on Ukraine's 46 SS-24 missiles.

Finally, my memo contained one more thought that until 2014 struck those at the top of the Ukrainian government as pure science fiction: fundamentally, Ukraine needed to defend itself from Russia. I wrote that nuclear weapons needed to be a stabilizing factor in our relationship with Russia, and suggested that this could be achieved by retaining the weapons on our territory, under

Ukrainian control, while establishing a blocking system that would preclude the use of the weapons without the consent of both countries.

I had no illusions even then that this correspondence would remain confidential. Subsequent actions by Moscow and Washington were a direct response to the change in Ukraine's attitude and plans. Both understood that the fool's game was over.

The Transfer of Tactical Weapons

Ukraine's new position prompted the Kremlin to adjust its plans, as their leadership realized it had to act not just quickly, but very quickly. By just a month later, in May 1992, all Ukrainian tactical nuclear weapons had been moved to Russia. This transfer of tactical nuclear weapons remains one of the murkiest pages in the history of Ukraine's nuclear disarmament. The only indisputable fact remains that the Ukrainian parliament never granted permission to take the weapons out of the country.[69]

The formal basis for the implementation of the transfer by the Kremlin would be the Alma-Ata Protocol of 21 December 1991. The document obligated Belarus, Kazakhstan, and Ukraine to transfer their tactical nuclear weapons to Russia by 1 July 1992. The treaty was signed by President Kravchuk. However, it never became law because the Ukrainian parliament did not ratify it.

The basis on which Ukraine allowed Russia to take over this highly valuable asset was the first question raised by our working group in preparing the groundwork for the ratification of START. Receiving the official status of affairs at the end of July, the group began gathering relevant information. The Ukrainian MoD, the only source of such data for us, eventually sent us an explanatory brief. The document provided a vivid picture of the chaos prevailing in Ukraine's government structures at the time, as can be judged from its full text:

> On 13 December 1991, the chief of the General Staff of the USSR, on the basis of the Ukrainian parliament's 24 October 1991 declaration regarding Ukraine's nonnuclear status, resolved to transfer the tactical nuclear warheads from Ukrainian territory, beginning on 6 January and continuing to April 1992. Troop commanders of military districts, the Black Sea Fleet and the Air Force, and the Air Defense Force received directives from the appropriate military branch chiefs of the former USSR regarding the planning and transportation of tactical nuclear warheads within predetermined time frames.
>
> On 3 January 1992, the MoD of Ukraine determined the schedule for the transfer of tactical nuclear weapons. The shipment was carried out under the direct control of the chief of the General Staff of Ukraine's Armed Forces, and the sixth branch departments of the armed forces of the former USSR, in cooperation with representatives of twelve chief directorates of the MoD.
>
> Due to the absence of international control mechanisms over the elimination of tactical nuclear warheads, at the end of February 1992, the MoD

of Ukraine suspended the transfer subject to the appropriate decision of the Ukrainian parliament.

After the adoption of the 9 April 1992 Resolution "On Additional Measures to Ensure Ukraine's Attainment of Nonnuclear-Weapon State Status," presidents Kravchuk and Yeltsin signed the agreement "On the Procedure for the Transfer of Nuclear Weapons from the Territory of Ukraine to the Central Production Facilities of the Russian Federation for the Purpose of their Dismantling and Elimination" on 17 April.

After that, from 21 April to 5 May, the shipment was resumed.[70]

Thus, the decision to transfer Ukraine's weapons to Russia was not even made by Kravchuk, but by the general staff of the USSR, an agency of a nonexistent country. Furthermore, Parliament's Statement on the Nonnuclear Status of Ukraine, dated 24 October 1991, was named as the basis for this decision. It is difficult to find expression for the absurdity of this logic. Parliament's Statement of 24 October 1991 mentioned neither the procedure, nor the quantities or nomenclature, nor the time frame for the shipment of weapons.

At the same time, the 9 April 1992 Declaration by Parliament expressly banned the transfer of nuclear weapons from Ukraine's territory. Surprisingly, the Declaration became the basis for the 17 April execution of the intergovernmental agreement on the procedure to transfer the weapons to the Russian Federation. Even though only the Ukrainian parliament had the legal right to dispose of state property, President Kravchuk supported the agreement and the executive branch fulfilled it. Here is another interesting fact. The shipment of state property worth billions of dollars was carried out by agencies with a mysterious status, particularly by "the sixth branch departments" (*shostymy viddilamy*) of the armed forces of the former USSR.

In respect to nuclear threat reduction, the agreement failed to raise the key issue: the question of international control over the destruction of the warheads transferred from Ukraine to Russia. What the Ukrainian government was allowed to observe during the dismantling process through its representatives, of which I happened to be one, was extremely limited. During our official visits to the Russian plant that was dismantling our nuclear weapons, we were shown specific warheads with serial numbers and were told: These are your warheads; we are sending them for dismantling. What would happen to them after that, whether they were really dismantled or put back in use by the Russian army, we had no way of knowing.

These specific developments demonstrated the haste with which Russia endeavored to take advantage of the not-yet-formed governmental structures in Ukraine, as well as dismal levels of professionalism by the latter. Moreover, Russia did this with Ukrainian leadership being virtually unaware of these processes. As an eloquent testimony of this dynamic, Moscow was the first to announce to the world that all tactical weapons were removed from Ukraine, Washington was second, and only third in this line was the commander-in-chief of the Ukrainian Armed Forces, President Leonid Kravchuk.

Why was Russia in such a hurry to take over Ukraine's tactical weapons? The answer was supplied by Russia implicitly, via the Ukrainian parliament's pro-Russian MP, Colonel Valerii Izmalkov. He wrote on 3 November 1992: "Perhaps we could be discussing the issue of relinquishing our nuclear-weapon status if the tactical nuclear weapons still remained on Ukraine's territory. This class of weapons would correspond to the range and power of Ukraine's strategic objectives. However, where were Mykola Porovs'kyi, Stepan Khmara, Volodymyr Tolubko and their supporters at the time? These weapons were shipped to Russia. Now all that is left are ambitions."[71] However, aside from the geopolitical, there was also a purely materialistic aspect of it, as the official daily parliamentary newspaper *Holos Ukraïny* wrote:

> After the tactical nuclear weapons were transferred to Russia, one could encounter with increasing regularity, both on the legal and black markets in the West, emissaries of different Russian small businesses and joint ventures offering for sale [...] weapons-grade plutonium, the payload from our nuclear missiles. Moreover, these deals were made by elements of the Russian government (for example, the Ministry of Energy). This means that we transported our missiles to Russia at our own expense, and now our neighbor shamelessly profited from them.[72]

Finally, the situation took on a legal dimension. If a state has tangible property, there are people responsible for its safekeeping. And should they fail in acting on behalf of the state, these individuals might be punished for neglecting their duties. The names of the individuals who issued and carried out the orders to transfer tactical nuclear weapons are recorded and remain in the archives.

Lisbon Protocol: The Game Begins

As Ukraine began to project an independent position, and Russia attempted and failed at the March CIS summit in Minsk to coerce the new nuclear states into transferring all their strategic nuclear warheads to Russian territory and force their partners to bear the cost of nuclear disarmament, the US decided to correct its approach. In the first half of April Washington offered its own disarmament plan; that is, immediately after Kravchuk had issued the resolution to assume command and control over nuclear weapons on Ukrainian territory, in addition to Parliament's 9 April 1992 Declaration and the visit by Ukraine's parliamentary delegation to the US on 11 April 1992. This was meant to ensure the imminent demolition of nuclear stockpiles of the former Soviet republics, most of which were aimed at the US, a move that would substantially enhance US security.

To achieve this aim, the new nuclear-weapon states of Russia, Kazakhstan, Ukraine, and Belarus, instead of the USSR, were invited to become parties to START I in place of the USSR, with the Lisbon Protocol. To enhance the impression that every state was considered an equal in the process, the US also suggested that the new parties to the treaty negotiate among themselves the exact proportions of nuclear arms reduction on their territories, so that the process would result in the overall liquidation of 36 percent of all strategic delivery systems, and 42 percent of nuclear warheads, as per the previous agreement between the USSR and the US.[73]

22. *Signing of the Lisbon Protocol*

The development was surprisingly advantageous for everyone. First of all, for President Bush and the US this would resurrect START, which was otherwise buried because the USSR no longer existed to implement it. On the eve of US presidential elections in the fall of 1992, START's demise would have been fatal for Bush's campaign.

The plan was lucrative for Russia in that, by recognizing the former Soviet republics as successors to the Soviet Union as owners of nuclear weapons, it also recognized them as successors to the USSR's external debt obligations and to the USSR's obligations under the Treaty on Conventional Armed Forces in Europe (CFE). Once this was completed, according to the aforementioned Arbatov Plan, the task of the Russian diplomatic machine was to arrange things in such a way that Soviet debts would remain with the successor republics, while all nuclear property would be transferred to Russia alone.

Ukraine likewise benefited from the plan. Firstly, Russia now recognized Ukraine's legal rights to the third largest nuclear arsenal in the world. Secondly, the Lisbon Protocol obligated the parties to destroy only a portion and not all of their nuclear weapons: under START I, 36 percent of all Soviet strategic delivery systems and 42 percent of the warheads. As to the reduction of the remainder of the nuclear weapons, the process was to be regulated by other treaties not yet in place. Thirdly, the Lisbon Protocol required parliamentary ratification, which meant that we could carefully study all the nuances of the agreement and adjust the provisions that were detrimental for Ukraine by adding amendments to the ratified document. This was of great importance, as the Ukrainian MFA had allowed two significant details to be added to the brief text of the Protocol (more on this below), both of which essentially nullified all of Ukraine's advantages stemming from the document, and which over the next two and a half years sparked a war in Ukraine's political establishment. Finally, the most important advantage for Ukraine in this arrangement was that the Lisbon Protocol opened doors for us to pursue an independent nuclear policy and negotiate directly with the US, thus bypassing Russia, the traditional intermediary. Ukraine could take advantage of the divergent interests and competition between Russia and the US, which would allow us to advance Ukraine's own national interests.

Therefore, it appears that the signing of the Lisbon Protocol on 23 May 1992 activated a new phase in nuclear disarmament, in which advantage would accrue to the parties most able to take advantage of the opportunities provided by the document. It must be said that both Russia and the US began taking advantage of these opportunities before the Protocol's execution in Lisbon. Since the goals of both involved the complete disarmament of the three nuclear republics, but START only mandated the aforementioned 36 and 42 percent, it was extremely important for them to find other legal avenues to force Ukraine, Belarus, and Kazakhstan to commit to complete disarmament within the next seven years.

On 7 May 1992, two weeks before the Lisbon summit, Leonid Kravchuk unexpectedly wrote a letter to George H. W. Bush in which he jumped the gun by committing Ukraine to actions not sanctioned by Parliament, including to reduce not a portion, but all of Ukraine's nuclear weapons within a seven-year period. As already mentioned, Parliament had the exclusive right to dispose of state property on behalf of its people. Therefore, Kravchuk's actions could be seen either as overstepping the bounds of his authority or as a relapse to Soviet practices, when the general secretary of the Communist Party would make a decision and everyone would automatically support it post factum.

In the letter, Kravchuk referred to the time schedule delineated in the START (although Ukraine was not yet party to it) and for some reason to Parliament's Statement on the Nonnuclear Status of Ukraine, which did not specify any time limits. On virtually the same day as President Kravchuk in Ukraine, the presidents of Kazakhstan and Belarus wrote analogous letters to the US president. The essential congruence of the commitments offered in these letters, along with their contradiction to the text of the Lisbon Protocol and the fact that they were most advantageous only to Russia, suggests that the letters had common authorship. This also points to the influence Moscow still had on the former republics' foreign ministries and presidents.

As I already mentioned, two important details went unnoticed by the Ukrainian MFA in the five brief articles of the Lisbon Protocol. Article II required eliminating not only the nuclear warheads, but also the strategic delivery systems. For the USSR and the US, manufacturers of both the nuclear weapons and their delivery mechanisms, the demolition of a strategic delivery system, a mechanism of delivering a nuclear warhead to its target, automatically meant the reduction in the number of nuclear warheads with target-striking capability. This was the essence of START I, signed by the USSR and the US in July 1991, which aimed at reducing the potential confrontation levels between the two countries. Its implementation was to be assured by tallying the decommissioned strategic delivery systems only, and not the destroyed nuclear warheads.

The USSR and the US had selected this approach because the experts who were working on the development of START I implementation mechanisms could not reach a compromise on the issue of how to control the demolition of the warheads themselves. Ultimately, after a nuclear warhead was dismantled, the HEU and the plutonium removed from it could be reused in a new warhead.

The USSR's collapse had fundamentally changed the situation around the implementation of START I. Since Ukraine, like Belarus and Kazakhstan, was not producing its own nuclear warheads, the path to nuclear disarmament rested in the elimination of the nuclear warheads themselves, rather than in the demolition of nuclear warheads' delivery mechanisms. This fundamentally new circumstance was supposed to be taken into account by the MFA in the course of the Lisbon Protocol's execution.

The second time bomb inserted into the text of the Lisbon Protocol by the US and Russia, which likewise went unnoticed by Ukraine's MFA, was Article V. It emerged unexpectedly and immediately before the Protocol's signing. According to the article, notwithstanding the fact that Belarus, Kazakhstan, and Ukraine were to become parties to START I and obligated to eliminate nuclear weapons in place of the USSR, the three countries were to accede to the NPT as nonnuclear-weapon states within the shortest possible time.

23. *Signatures of the parties to the Lisbon Protocol*

This combination of terms was completely unlawful and showed signs of diplomatic machinations. The Academy of Sciences Institute of Government and Law unequivocally concluded this in its analysis of the Lisbon Protocol, published in the article entitled "Ukraine's Nuclear Status: Legal and Political Issues." First, it wrote, "insisting on Ukraine's ratification of START I fundamentally constitutes the country's recognition as a successor to the Soviet Union's nuclear weapons, including the right of ownership of the portion of nuclear weapons that are located on its territory." This, emphasized the legal experts, "constitutes an indirect recognition of Ukraine's clear nuclear status; otherwise it simply cannot become party to this agreement."[74]

In contrast, demanding that Ukraine join the NPT as a nonnuclear state completely invalidated this right, as it was only possible in two instances: either the non-recognition of Ukraine's succession with regard to international treaties or the non-recognition of Ukraine's succession with regard to property located on its territory, which "unequivocally violates Ukraine's succession rights with respect to the former USSR." The Institute's experts stressed that Ukraine's status "is not provided in the Treaty [NPT]," because the Treaty defined a nuclear-weapon state only as one that has independently manufactured and tested a nuclear explosive device. The Treaty's language also did not provide for instances in which a nuclear-weapon state disintegrated, creating several new states that now possessed its predecessor's nuclear weapons. Therefore, "Ukraine can only join the NPT on the basis of this special status, as a temporary nuclear-weapon state," and this, they concluded, required amending the Treaty to reflect the status Ukraine had at that time, while a "straightforward accession to the NPT, without the corresponding amendments, will be legally insolvent and fundamentally fruitless."[75]

The simultaneous ratification of both treaties, the NPT and the START, along with the Lisbon Protocol, resulted in a diplomatic Catch-22, automatically depriving Ukraine of its legal right

24. *Ministry of Foreign Affairs*

of ownership of the nuclear warheads located on its territory and of any compensation for the nuclear material. By declaring itself a nonnuclear-weapon state, Ukraine would stipulate that it also had no right to nuclear weapons. Therefore, immediately after ratification, Russia would gain the opportunity to declare nuclear weapons located on Ukraine's territory their own, which then would grant the Russian Armed Forces direct command over the military units servicing the nuclear weapons. In this scenario, all other countries would have to accept Russia's actions, as it would have acted within the letter of international law.

Article V of the Lisbon Protocol additionally begs a basic legal question: Whose weapons would Ukraine have to demolish on its territory, if it did not actually own them? The closing questions on this topic are rhetorical: What were the MFA's lawyers thinking when they approved this legal nonsense for official execution? What was the astute diplomat Zlenko thinking when he signed it even though no one had authorized him to do so? As it turned out, the MFA had not yet reached the daily limit for nonsensical behavior.

Lisbon Protocol: Zlenko's Peculiar Statement

Foreign Minister Zlenko, officially representing Ukraine in Lisbon, astounded everyone with the statement he made on behalf of his country, which contradicted not only the Lisbon Protocol, but also itself.

First, in his official statement Zlenko pledged that "Ukraine [would] ensure the demolition of all nuclear weapons, including strategic arms, within a seven-year period." In his words, this would be in accordance with "the provisions of the START treaty." However, as we know, the treaty did not require elimination of *all* weapons. The next point in this statement immediately contradicted itself: "[N]uclear arms reduction will be achieved by proportional and equal elimination of nuclear warheads and their delivery mechanisms, so as to achieve the reduction levels within the timeline set forth by the treaty," in other words, the reduction by 36 and 42 percent of strategic delivery systems and warheads, respectively, but not all nuclear weapons. Second, Zlenko noted Ukraine's status "as a nonnuclear-weapon state" and, contrary to Ukrainian law, affirmed that "Ukraine had voluntarily renounced its right of ownership to nuclear weapons." He topped this off with the following: "Meanwhile, the right of ownership to the former USSR's nuclear weapons, with the express consent of Ukraine and all other successor-states of the former USSR, remains only with the Russian Federation."[76]

From a legal standpoint, this action was criminal. With this arbitrary interpretation, the head of the MFA was violating Ukraine's Law on Enterprises, Institutions and Organizations of the Former Soviet Union Located on Ukraine's Territory, adopted in October 1991, which clearly stipulates that everything located on the territory of Ukraine is its property. Furthermore, according to the Constitution, state property belongs to the people of Ukraine, and Parliament, as its representative body, makes all decisions regarding state property on behalf of the people. No other agency or institution, including the president of Ukraine, has the right to dispose of state property without the authorization of Parliament.

However, in the next point Zlenko's statement once again contradicted itself: "Ukraine cannot recognize the Russian Federation's special status with regard to treaties, right of ownership or other obligations of the former USSR. [...] On no other occasion did the Russian Federation ask for or receive consent from Ukraine or other successor-states, and thus it has no legal grounds for claiming the status of successor to all of the former USSR's property, treaties, or other obligations."[77]

The statement gave the impression of being written by several authors at once, with none of them consulting the others. It is difficult to imagine something like this being composed by the US State Department or the Russian MFA. In this instance the original text was purportedly written in Kyiv, and then amended on the fly in Lisbon, either by Russian or by US officials, or both. Moscow's hand was clearly visible in the parts regarding the seven-year period and the demolition of all weapons, as these terms were persistently used by the Russian diplomatic machine in official documents.

Russia's official statement, presented on this subject the same day, provided a definitive answer as to the origins of the contradictions with national law in Ukraine's statement. The statement had exposed with enviable clarity the kind of developments the Kremlin was expecting from the new nuclear republics: the complete elimination of their nuclear weapons within a seven-year period. The Russian statement referred to the commitments given by the presidents of the former Soviet republics (including Kravchuk) in their letters to Bush. The statement asserted: "The Russian Federation presumes that, at the end of a seven-year period with the execution of START, no nuclear warheads or their strategic delivery mechanisms shall remain on the territories of the Republic of Belarus, the Republic of Kazakhstan, and Ukraine." Additionally, Russia expected Ukraine to accede to the NPT "as a nonnuclear-weapon state."

And in conclusion, it stated that "Russia, as the successor to the USSR, is party to the NPT, acting as the depositary of the Treaty," meaning all weapons of the former Soviet Union had to be transferred to Russia.[78] A comparison of the key theses contained in the two statements, the Ukrainian and the Russian, suggests a common source for their contents' development, and there is no reason to believe it was located in Kyiv.

After the signing of the Lisbon Protocol, a peculiar pattern began to emerge: the MFA's actions on nuclear issues almost always contradicted the decisions of the Parliament. The *Financial Times* revealed this dynamic in January 1993: "The Ukrainian ministries of defense and foreign affairs, by contrast, continue to support removal of nuclear weapons. [...] Ukraine's fragmented legislature has begun to vote with increasing autonomy and is less amenable to control from President Kravchuk."[79]

Having generated these legal contradictions through well-coordinated cooperation, the US and Russia followed up with increasing political, economic and energy pressure, as well as outright threats to the fledgling Ukrainian state. All of this was widely broadcast to Ukraine's leadership by the MFA. In my personal archive I have collected dozens of Western and Russian publications and statements from various officials and even lay people on the dangers of Ukraine's possession of nuclear weapons, as well as the negative ramifications to statehood, should it attempt to become a nuclear-weapon state. Simultaneously, the MFA did nothing to inform the international community on issues related to Ukraine's nuclear disarmament.

It was becoming increasingly evident that Ukraine's foreign policy office was hardly competitive with US and Russian diplomacy efforts in the attempts to safeguard Ukraine's national interests. Had the developments endured in this fashion, that is, had the nuclear disarmament process been driven exclusively by Ukraine's executive branch, there would have been a quick and thorough defeat. This kind of defeat befell Kazakhstan and Belarus, the two other post-Soviet states with nuclear weapons.

25. *Viacheslav Kebich (1936–2020), prime minister of Belarus 1991–1994*

Kazakhstan's and Belarus' Positions

As has already been mentioned, upon the dissolution of the USSR Russia inherited 69.4 percent of its nuclear stockpile, Ukraine 17 percent, Kazakhstan 13 percent, and Belarus 0.6 percent. Belarus, Kazakhstan, and Ukraine each deployed its own strategic rocket forces; Ukraine also deployed a strategic bomber air force.

Belarus had the fewest nuclear warheads: 81 intercontinental ballistic missiles (ICBMs) and the road-mobile SS-25. These were solid-propellant missiles carrying a single 800-kiloton warhead. Kazakhstan deployed three divisions of the USSR's most powerful ICBM underground launchers, the SS-18 missiles manufactured by Pivdenmash. This particular type of liquid-propellant missile carried ten independently-guided warheads, each with a 300-kiloton yield. Kazakhstan's nuclear stockpile comprised 1,100 warheads.

However, Russia wanted to appropriate everything: the 17 percent belonging to Ukraine, the 13 percent stationed in Kazakhstan, and even the 0.6 percent of the former USSR's nuclear weapons inherited by Belarus. The Kremlin led the three post-Soviet nuclear countries towards this goal via a single scenario. At the onset were the CIS treaties, through which these countries declared their intentions to become nonnuclear-weapon states. Then, virtually on the same day in May 1992, the leaders of the three nuclear states wrote letters to President Bush with the commitment to demolish all their nuclear weapons within seven years. Later came the signing of the Lisbon Protocol with Article V, which obligated them to join the NPT as nonnuclear-weapon states and, if ratified, automatically deprived these countries of the right of ownership of nuclear weapons.

26. Nursultan Nazarbayev (b. 1940), president of Kazakhstan 1990–2019 (pictured in 1991)

Finally, there was the Lisbon Protocol, which required Russia, Ukraine, Belarus, and Kazakhstan to negotiate the division of reduction quotas, so as to jointly reduce 36 percent of the delivery systems and 42 percent of the warheads of the former Soviet arsenal, as determined by the START. However, Russia categorically refused to conduct negotiations in a quadrilateral format, instead pressing for the handover of all weapons located on the territories of Ukraine, Kazakhstan, and Belarus.[80]

Despite the old Soviet custom of conceding to Moscow on all matters, the leaders of Belarus and Kazakhstan in the first months of their countries' independence recognized the price of the issue and slightly changed their positions. They were not in any hurry to disarm. Belarus prime minister Viacheslav Kebich declared on 13 February 1992 that "no one is going to talk to a nonnuclear state" and therefore, according to him, the country had completely changed its stance on denuclearization.[81]

Around the same time on 21 February 1992, Kazakhstan president Nursultan Nazarbayev announced in the Russian press that Kazakhstan would become a nonnuclear-weapon state, but that the process would require a substantial amount of time. He stated: "My country does not want to be a nuclear state. However, eliminating nuclear weapons takes time. Kazakhstan's neighbor, China, is a nuclear state, just as are Pakistan and India. Therefore, Kazakhstan's official position is that we will need at least fifteen years to destroy our missiles."[82] On 5 March 1992 he said that he was ready to eliminate nuclear weapons, but only alongside other nuclear states. He essentially repeated the statement made by Yeltsin on 24 February: "Russia is for nuclear disarmament, but not for a unilateral one."[83]

However, Nazarbayev did not maintain this attitude for long. A remarkably quick change of position was announced during a press conference at Vnukovo-2 Airport in May 1992. Before departing from Moscow to Washington on his first state visit, Nazarbayev declared that he would not be asking for anything from the White House, because he saw the existing security guarantees as sufficient, something Russia was apparently also expecting from Belarus and Ukraine. He stated: "As you know, at the meeting of the CIS presidents in Tashkent, we signed a collective security agreement. This, I believe, provides a nuclear umbrella over Kazakhstan." He then made the assurance that Kazakhstan would ratify both treaties, the START and the NPT.[84]

In addition, Nazarbayev criticized Kravchuk's decree signed a month before the Tashkent summit of CIS heads of state, assigning command and control over strategic nuclear forces deployed on Ukrainian territory to the Ukrainian Ministry of Defense. "Does this mean that Ukraine is becoming a nuclear state? I do not understand this position. Russia also intends to create its own nuclear forces. However, I think it is Russia that should remain a nuclear state. The whole world is worried about the possibility of the deadly weapon's spread across the planet," Russian *Izvestiia* quoted him.[85]

The Lisbon Protocol made the three post-Soviet nuclear countries parties to START I, which, consequently, gave recognition to their right to ownership of nuclear weapons. Neither Kazakhstan, nor Belarus, in contrast to Ukraine, had put this right of ownership into practice. They also never raised the question of compensation for nuclear material.

Both Belarus and Kazakhstan surrendered their nuclear weapons quickly and without a fight after joining the Collective Security Treaty Organization initiated by Russia in May 1992, known as the Tashkent Pact.[86] Kazakhstan ratified START and the Lisbon Protocol in the summer 1992, and Belarus ratified both on 2 February 1993. At the end of 1993 they both acceded to the NPT as nonnuclear-weapon states.[87] Incidentally, on 26 March 1993, the *Washington Post* voiced accusations against Clinton for failing to fulfill the administration's promise of assistance to Belarus after it had become a nonnuclear-weapon state, noting that it undermined confidence in the US.[88]

By this time, Ukraine found itself alone in the fight for its property against Russia. It held out until 5 December 1994. On that day Ukraine, the US, Russia, Belarus, and Kazakhstan exchanged START I ratification charters, and Ukraine submitted official documents for its accession to the NPT as a nonnuclear-weapon state.

CHAPTER 2. Ukraine Formulates Its Position on Nuclear Arms

The Launch of the Working Group: Pliushch's Maneuver

Ukraine still had a parliament, and a sizable faction of it was prepared to fight for the country's national interests. However, mobilizing this force would not prove easy. In spring 1992, information about the composition of Ukraine's nuclear arsenal and what should be done with it made its way through various departments and to government officials. The state still did not have a single center for information-gathering or for developing and making decisions on nuclear disarmament. As a result, Ukraine did not have a distinct strategy for tackling this multibillion-dollar question. On the contrary, it was drifting along with the flow, which was directed jointly by the United States and Russia.

The absence of a national strategy led to peculiar situations: the MFA not only refused to lobby Ukraine's national interests on the issue of nuclear disarmament but in fact avoided addressing key questions related to it, such as the right of ownership of nuclear weapons. In April of 1992, Ukraine's Foreign Minister Zlenko, during an official visit to Moscow, held a joint press conference with Russian Foreign Minister Andrei Kozyrev. Because Ukraine's nuclear weapons were making headlines in the world media at the time, the venue was full. However, instead of using this opportunity as a communication platform, Foreign Minister Zlenko declined to comment on Ukraine's nuclear arms policy altogether, asking me, as a representative of the Ukrainian parliament, to field journalists' questions.

The knowledge of nuclear weapons among the MPs and other officials was even more paltry. In May 1992, at the time of the signing of the Lisbon Protocol, Parliament did not even have Ukrainian translations of the START I or the NPT documents, which came to over 600 pages of text.

After the signing of the Lisbon Protocol, I approached the speaker of Parliament, Ivan Pliushch, with the idea of establishing a special working group to develop the groundwork for ratification of these documents. At this point, we ought to give credit

27. Andrei Kozyrev

Andrei Kozyrev (b. 1951), career diplomat and the Russian minister of foreign affairs under the Yeltsin administration. He graduated from the Moscow Institute of Foreign Relations in 1974 and began to work for the MFA of the USSR. In 1989–1990, he supervised the administration for international organizations under the Soviet MFA. From 1990 to 1996, he was the foreign minister. Between 1993 and 2000, he was also a member of the Russian Duma.

28. Stanislav Koniukhov (center) and Leonid Kuchma (right)
Stanislav Koniukhov was the chief designer of Pivdenmash.

to Pliushch's political acumen: despite being fairly distanced from the topic, he was a good manager and understood the need for carefully considering all aspects of the problem before making final decisions in Parliament. Pliushch supported my idea, but to begin with, he suggested creating the Working Group not by a parliamentary decision (which would have required a vote and could be sabotaged) but by the decree of Parliament's speaker, in order to reduce the risks. I prepared the corresponding draft document and held preliminary discussions with other MPs. On 20 July 1992, Pliushch signed the decree that mandated the creation of a working group "for addressing issues related to the ratification of the START by the parliament of Ukraine and to Ukraine's attainment of nonnuclear status."[1]

Thirteen MPs participated in the Group, which included representatives of eight parliamentary commissions. Among the most knowledgeable on the topic were the director of Pivdenmash, the future prime minister and president of Ukraine Leonid Kuchma, and the head of the Kharkiv Military University, General Volodymyr Tolubko. Other members of the Group included Oleksandr Borzykh, Ivan Zaiets´, Pavlo Mysnyk, Borys Mokin, Iurii Serebriannykov, Volodymyr Strelnykov, Andrii Sukhorukov, Vitalii Chernenko, Oleksii Shekhovtsov, and Bohdan Horyn´. I was appointed chair of the Working Group. The group was to work with representatives of the president's administration, the MFA, the MoD, the SBU, the Academy of Sciences of Ukraine, the MIC, and civic organizations. I was tasked with submitting a list of experts for approval and with developing a working plan.

The plan I proposed included the following components: the translation of documents; the creation of expert working groups to investigate the problem's key aspects, spanning the whole range of military, political, legal, scientific, technical, financial, economic, environmental, and social issues; assessment of START I and NPT texts by these expert groups and development of propositions; visits to the 43rd Rocket Army, holding public and closed hearings; meetings with Russian and US representatives; the preparation of a final list of reservations and submission of a draft ratification law for parliamentary review. Pliushch completely agreed with this plan, and we put it into immediate effect.

While preparing these tasks, I worked not only on gaining a detailed understanding of the treaties that were being prepared for ratification, but also on what I believed was most important: to comprehensively study and analyze the nuclear weapons themselves, and to identify key issues Ukraine would face in the process of their liquidation. Because the information related to the USSR's nuclear weapons was in Moscow's sole possession, and Ukraine did not have access to such data, I had to occasionally be both a researcher and a spy. Comparing other countries' experiences with this issue was also important.

Precedents: How the Republic of South Africa Gave up Its Nuclear Weapons

Apart from Ukraine, the only other country in the world that had agreed to reduce its nuclear weapons at the time was the Republic of South Africa (RSA). Therefore, theirs was the only experience we could study in detail.

By the end of the 1980s, the RSA managed to develop only six nuclear explosive devices, yet its nuclear disarmament spanned a two-year period. The official information on the country's nuclear disarmament process was announced by RSA president F. W. de Klerk during his parliamentary address on nuclear nonproliferation on 24 March 1993. Here, in de Klerk's words, is the original motivation behind developing the country's nuclear arsenal:

> At one stage, South Africa did, indeed, develop a limited nuclear deterrent capability. The decision to develop this limited capability was taken as early as 1974, against the background of a Soviet expansionist threat in southern Africa, as well as prevailing uncertainty concerning the designs of the Warsaw Pact members. [...] The objective was the provision of seven nuclear fission devices, which was considered the minimum for testing purposes and for the maintenance thereafter of credible deterrent capability. When the decision was taken to terminate the programme, only six devices had been completed.[2]

With the unfolding of fundamental political shifts in the global arena in 1989, F. W. de Klerk stated that the RSA "decided towards the end of 1989 that the pilot enrichment plant at Pelindaba should be closed and decommissioned."[3] Therefore, in the early 1990s, the RSA developed a detailed step-by-step program to eliminate its six nuclear explosive devices, which also addressed the technological issues that arose in the process. The key issues revolved around what to do with weapons, the nuclear material, and manufacturing equipment, as well as the timeframe for joining the NPT.

The RSA decided that "all the nuclear devices should be dismantled and destroyed"; all nuclear material "be recast and returned" to the Atomic Energy Commission, "where it should be stored according to internationally accepted measures"; and that all facilities of Armscor (a corporation that carried out the RSA's nuclear program) "should be decontaminated and be used only for nonnuclear commercial purposes."[4] Only after implementing these measures would the RSA accede to the NPT, thus placing all of its nuclear material and corresponding parts under international control.

The RSA acceded to the NPT on 10 July 1991, two years after its decision to liquidate its nuclear arsenal of six explosive devices. Ukraine, in contrast, was being required immediately to eliminate all of its 5,000 warheads, without a full understanding of the situation or development of any programs, and also immediately to accede to the NPT.

The Hawkish Strategy

The first practical result of the Working Group's activity was the proposal of a nuclear disarmament strategy. I unveiled it in a programmatic article published in *Holos Ukraïny* under the title "Ukraine's Nuclear Arms: Good or Evil? Disarmament's Political, Legal and Economic Analysis."[5] The article was the first to identify four key questions, answers to which would determine Ukraine's approach to the problem of disarmament. They included: Whose property were the weapons? Who would pay for disarmament? How would nuclear material be used? What would replace nuclear weapons in Ukraine's system of national security?

Arguments collected during the discussion of each of the above questions allowed us to formulate the four chief strategies of effective disarmament (see chapter 1). These strategies provided an ideological foundation for consolidation of a faction of the parliamentary corpus and government officials who were later called the nuclear hawks. These strategies addressed the following: first, Ukraine's right of ownership of nuclear weapons; second, sources of financing for the disarmament process; third, the use of nuclear material in the energy industry; and finally, national security guarantees, along with Ukraine's right to compensation for the tactical weapons already transported to Russia, which people at this point had begun to forget. The hawkish strategy was antagonistic to the "transfer weapons to Russia" approach, organized and gradually enacted by two governmental structures, the MFA and the SBU. Events that unfolded subsequently largely concerned the opposition between these two approaches.

As an aside, I will briefly note here that our logic was supported by an overwhelming majority of the MPs, regardless of their political camps: from the communists to the democrats. This unity allowed Ukraine not only to prevent the immediate and disastrous unexamined ratification of international nuclear treaties, but also to adopt a ratification law at the end of 1993 that at least partially converted nuclear weapons into seed capital for the political, economic, and military development of the fledgling Ukrainian state. The unfortunate way in which the state's supreme leadership sabotaged these prospects by multiple violations of the law will be discussed a bit further on.

Since some of the most fantastic myths began sprouting up around the effective disarmament strategy as soon as it was announced, I will provide the main points of the strategy and outline their supporting arguments.[6]

Point One: Ukraine Is the Owner of its Nuclear Arsenal

In accordance with international law, which includes two Vienna conventions—On the Succession of States in Respect of Treaties of 1978; and On the Succession of States in Respect of State Property, State Archives and State Debts of 1983—all states formed on the territory of the former USSR became its successors in terms of both tangible assets and international obligations.[7] Armed with these laws, Western creditors demanded that not only Russia but also the other independent states, which included Ukraine, repay the former USSR's debt obligations. Therefore, based on these laws, Ukraine, Belarus, and Kazakhstan, by sharing the same status as Russia, became nuclear-weapon states not only *de facto*, but also *de jure*. On the other hand, only Russia possessed the full technological production capabilities for nuclear warheads and, therefore, according to another international document—the NPT—the other three countries were considered nonnuclear states.

The circumstances surrounding nuclear weapons on the territory of the former USSR were fundamentally different from the American deployment of its nuclear weapons in Western Europe; therefore, the US could not be used as a precedent. The US had brought its weapons to Europe already manufactured, whereas Ukraine was directly involved in the creation of the USSR's nuclear arsenal, having invested its own intellectual and material resources into the weapons' production. The signing of the Lisbon Protocol signaled that the US recognized the distribution of the former Soviet Union's nuclear weapons among the newly independent states. Consequently, the immediate adoption of a nonnuclear-weapon state status by Ukraine would have resulted in a legal conundrum: Who owned the nuclear weapons located on the territory of Ukraine?

If Ukraine was considered the owner of the nuclear weapons, it would have to assume all obligations pertaining to them and then be responsible for the process of their destruction. In this case, it would be a logical choice for Ukraine to join START I as an equal party to the agreement. If, on the other hand, Russia was considered the owner of these nuclear weapons, then Ukraine could not take any responsibility for nuclear disarmament on its territory, because it could not destroy another party's property. In this scenario, Ukraine's role within the START would be unclear. Furthermore, the presence of the nuclear weapons of another state (i.e., Russia) on our territory for the duration of the disarmament process, which could take between seven and ten years, would contradict the principles of neutrality already declared by Ukraine. Above all, Parliament never approved the transfer of all the ownership rights of the former USSR's nuclear weapons exclusively to Russia. Therefore, the key political and legal issues associated with the liquidation of nuclear weapons could only be successfully resolved if Ukraine became the owner of the nuclear arsenal located within its territory, and if Ukraine retained the temporary status of a nuclear-weapon state until the weapons were ultimately destroyed.

Point Two: Financing Nuclear Disarmament

Nuclear disarmament entailed considerable costs. Therefore, Ukraine required a technical and economic analysis of possible options for its missile systems. First and foremost, it needed to develop technologies for their safe dismantling and recycling.

Ukraine had two types of strategic nuclear missiles deployed on its territory: liquid-propellant and solid-propellant missiles. Liquid-propellant carriers were filled with highly toxic fuel components, heptyl and the oxidizer amyl. Amyl could be processed at Ukraine's chemical plants, but required significant expenditures. Ukraine had no technology to process and store heptyl, which, by its toxicity, could be compared to chemical warfare agents. Ukraine also did not possess any technologies for the utilization of solid-propellant launchers. However, given Ukraine's significant scientific and technical resources, it had the potential to develop the necessary technologies in the future. Yet, carrying out a large-scale liquidation of the country's nuclear arsenal in the context of an economic crisis without the financial and technical support of the West was nearly impossible.[8]

29. *Protective gear for working with heptyl*

Recycling the key part of a strategic nuclear missile that contained a nuclear warhead was problematic. Although Ukraine did not have the necessary enterprises or experts for their recycling, the Working Group deemed it inadvisable to agree with some of the departments' recommendations to simply hand the warheads over to Russia. The key component contained uranium and plutonium, valued at up to US $100 million per ton of uranium, and between US $500 million and US $1 billion per ton of plutonium on the international market.

Point Three: Recovery of Nuclear Material for Power Generation

Highly-enriched uranium and plutonium are valuable energy sources. Therefore, it would have been most expedient to re-use them for the energy industry. HEU could be blended with natural uranium to decrease its concentration down to three to five percent and subsequently could be used as heat-releasing elements for nuclear power plants (NPPs). Not having its own enterprises with complete technological processing capabilities for enriching uranium or for producing the heat-releasing elements for the NPPs, Ukraine would have been able to substitute the fuel production process with the HEU from nuclear warheads for some time.

The problem of utilizing plutonium extracted from nuclear warheads was more complex. The UK, France, and Germany owned technology that blended plutonium with uranium for the mixture's further use in special reactors; Ukraine did not have such technology. More promising was technology that could make use of plutonium directly, in a new generation of reactors that were projected to be built by 2010–2015. Japan had already launched a plutonium project, intending to produce 400 metric tons of reserve plutonium in the span of thirty years.

Considering that oil and coal deposits were being depleted, plutonium could become one of the most valuable energy-producing elements, worth the price of gold. Therefore, we advised keeping the plutonium that was released in the process of liquidating nuclear arms as a valuable energy source for the NPPs of the future.

For Ukraine, the financial loss in the process could be kept to a minimum if the main components of the strategic rockets carrying nuclear warheads were reprocessed in a certain way. During the first phase, the dismantling of nuclear warheads could be carried out at Russian plants, provided Ukraine received back a portion of the uranium and plutonium. Mutual compensation within this partnership could be made according to the international market price of the nuclear material. This compensation would also include the value of the warheads that had already been transported from Ukraine to Russia. Simultaneously, Ukraine could build its own specialized plants, with full production capabilities for enriching uranium and producing TVELs for power stations, as well as plants for the reprocessing and storage of radioactive waste, including the spent TVELs. Without such technology, the safe operation of our NPPs would become impossible.

Because Russian plants did not have the capacity to dismantle such great quantities of warheads within the timeframes stipulated in international agreements, Ukraine's participation in the process of dismantling and reprocessing nuclear warheads under international control could significantly reduce the workload. There would also be the possibility of maintaining state technological secrets, which concerned the Russians, in the following manner: specific operations could be identified in advance to be carried out only by Russian specialists; other, unclassified activities would be accomplished by Ukrainian and Western specialists.

Ukraine needed to initiate a review of certain provisions to the START. The treaty had been developed in the context of the confrontation between the USSR and the US, while Ukraine took no part in the treaty's development. The political situation had now changed; the USSR no longer existed and many of the requirements had become outdated. Of principal concern was the demolition of strategic missile launch facilities, as the robust engineering structures could be used for agricultural needs and scientific objectives.

Finally, when determining timeframes for the elimination of nuclear weapons, Ukraine needed to make decisions based not on emotions, but on economic considerations and the state's capability to carry out this work while providing for physical and environmental safety.

Point Four: National Security Guarantees

How could Ukraine replace nuclear arms in its system of national security? Nuclear weapons are generally seen not as a means of aggression but rather as a way of deterring it. There are three key factors in the modern context that act as national security guarantees: military strength, economic power, and significant levels of political and economic integration with other countries. Unfortunately, none of these factors fully applied to Ukraine. Therefore, upon complete nuclear disarmament, Ukraine would be in need of protection from more powerful states.

We proposed a pathway for Ukraine's focused political and economic integration with the countries of Western Europe; the degree of this integration would have to be directly related to the pace of Ukraine's nuclear disarmament.

Ukraine's last strategic missile would be destroyed only after the country joined the European community. In this way, Ukraine would gradually replace one of the key factors in safeguarding its national security with another: namely, military strength in the form of nuclear weapons would be replaced without imperiling its national interests by economic and political integration. No other scenario had the potential to compensate Ukraine for the loss to its national security system posed by denuclearization.

The Coordinating Committee for Multilateral Export Controls (CoCom), an international non-treaty organization established with the goal of limiting the spread of nuclear weapons technology, needed to reconsider some of its limitations concerning Ukraine. In practice, CoCom did not have the ability to stop the proliferation of nuclear technology, and its functions were often misused by competitors. CoCom had prevented the introduction of Ukraine's rocket launchers (which were of superior quality) into the international market, even at a time when there was a seven- to ten-year waiting period for launching commercial satellites into Earth's orbit.

Thus, becoming a nonnuclear-weapon state appeared to be a highly complicated and prolonged process that demanded solutions to a wide range of political, legal, scientific, technical and economic problems. Adopting nonnuclear status also demanded new mechanisms for safeguarding the country's national security. Historically, nuclear weapons were created in response to a collective threat, with two opposing sides dragged into the process. At the same time, these weapons presented a threat to all of humanity; surely there was a global interest in their liquidation. It was shortsighted to assign responsibility for the destruction of the entire nuclear arsenal to individual countries.

Ukraine Begins Communicating Directly with the West

The Russian response to Ukraine's attempts to handle its nuclear inheritance independently was characterized laconically by the Russian ambassador Iurii Dubinin, who later headed the Russian delegation negotiating disarmament with Ukraine: "Ukraine's position began causing alarm."[9] The Kremlin became increasingly worried as Ukraine gradually but surely began to function like an independent state. Instead of a quick and complete nuclear disarmament as per the Russian scenario, Kyiv not only began to develop its own strategy, but also took concrete steps towards gaining control of its weapons.

Under these circumstances, it was tremendously important for us to convey Ukraine's position to the West. The world perceived the nuclear developments in the post-Soviet space only through the prism of the Russian media and the Russian MFA's interpretation of events, which was problematic. Ukraine's foreign policy shop at the time conformed almost completely to the Russian scenario, rather than defending Ukraine's national interests in the question of nuclear disarmament before the international community.

30. *Leonid Kravchuk (center, seated) visiting NATO Headquarters, 1992*

However, there was another notable circumstance. Despite US's fundamental support of the idea that Russia should retain the former USSR's nuclear
weapons, there was no actual legal understanding of how to achieve it. Ultimately, the newly independent countries had become nuclear-weapon states
without violating international law, thus, sanctions could not be applied to
them. Sanctions could only be applied to countries secretly developing nuclear
weapons, which violated the NPT.

The United States and its partners launched a diplomatic initiative for
mutually agreeable solutions. Countless groups of government representatives came on official visits to Kyiv and tried to convince Ukrainian leaders to
eliminate their nuclear weapons quickly; they promised that, should this finally occur, Ukraine would be presented with extraordinary opportunities. The
MPs and other officials were deluged with diplomatic invitations to breakfasts,
lunches, and dinners, where the topic of nuclear disarmament would almost
inevitably be raised. This period of time peaked with an array of different
seminars and official meetings, which ultimately concerned security and the
fate of the USSR's nuclear arsenal as key topics of discussion. Between June
and September 1992 alone, three significant international events took place.

31. Ivan Bizhan (left), 2000

Ivan Bizhan (b. 1941), deputy defense minister of Ukraine from 1991 to 2002. He graduated from the Malinovskii Military Armored Forces Academy in Moscow in 1971 and the Military Academy of the General Staff of the Armed Forces of the USSR in 1982. From 1987 to 1989, he was an advisor to the Army commander of the Armed Forces of Cuba, then the commander of the infantry of the Volga-Ural Military District. From April to December 1991, he was the deputy supervisor of the operational administration of the General Staff of the USSR Armed Forces. From 1991 on, he was seconded to the Cabinet of Ministers of Ukraine. From December 1991 to February 2002, he was the deputy and first deputy minister of defense of Ukraine. On 4–8 October 1993, he was the acting defense minister. In 2002, he retired from the military service and entered the reserves, based on his age. He is currently a reserve colonel general.

On 29–30 June Russia's Foreign Affairs Committee held a seminar in collaboration with the Science and Technology Committee of NATO's Parliamentary Assembly. The seminar was held in Moscow and entitled "Issues of Nuclear Security." In July, a non-governmental organization, the Atlantic Council, hosted a seminar entitled "The Relationship between Society and Military in Central and Eastern Europe" at NATO headquarters in Brussels. For the first time in NATO's history, representatives of Eastern European countries, specifically, Poland, Hungary, Czechoslovakia, Bulgaria, Latvia, Lithuania, Estonia, and Ukraine—but not Russia—were introduced to the functioning of democratic institutions in the military in the West, and discussed potential strategies for cooperation after the dissolution of the Warsaw Pact as well as the fate of the USSR's nuclear weapons.[10] Shortly afterwards, a broad discussion about the future of Ukraine's nuclear arsenal finally took place on 15–16 September during a Worldwatch Institute conference in Washington, DC, which was dedicated to the non-proliferation and export control of nuclear weapons.

This was an opportunity for Ukraine to make a breakthrough in the international exchange of information on nuclear weapons, and to clearly communicate its strategy concerning the elimination of the world's third largest nuclear arsenal in a way that also corresponded to its national interests. However, Ukrainian officials, quite to the contrary, tried to minimize Ukraine's participation in these events. Even though the government sent MFA or MoD representatives to these gatherings, it was I, an MP, and not a government delegation official, who was to speak publicly on Ukraine's nuclear disarmament policy.

More than fifty Russian representatives participated in the Moscow seminar, from deputy foreign and defense ministers to leading nuclear industry experts, and almost thirty delegates from the NATO Parliamentary Assembly. Ukraine was represented by just two individuals: deputy minister of defense Ivan Oliinyk and me. During discussions with the NATO general secretary, officers of the Supreme Allied Command from Germany, the UK, Italy, and the US, as well as other key officials of the alliance in Brussels, again, I was the only person who presented Ukraine on security issues. Finally, in Washington, where Volodymyr Tolubko and I were in attendance with representatives of Parliament and other government officials, I again was charged with presenting Ukraine's nuclear disarmament policy.

Given my understanding of the weight of these occasions, I would always emphasize four key points in Ukraine's approach to nuclear disarmament. First, Ukraine's position on nuclear arms was based solely on the decisions of its parliament, and those decisions had to be complied with at all levels of Ukraine's leadership, which included the president and all governmental agencies. Second, Ukraine legally owned the nuclear weapons located on its territory, as stated by international law and Ukrainian national legislation. Interestingly, I had never heard any objections from officials and experts in Brussels or Washington to the notion that nuclear warheads were rightly owned by Ukraine. This idea triggered an indignant response from Russia alone; consequently, the Ukrainian MFA tried to avoid the subject as much as possible. Third, Ukraine saw denuclearization not as an isolated event but rather as a process, which had to be accompanied by an adequate increase in other national security factors, such as the extensive political and economic integration with the West. Fourth, the financial burden of Ukraine's nuclear disarmament should not be borne by Ukraine alone, especially in the context of a mounting economic crisis, or else it would lead to the policy's rejection by the Ukrainian people; Western nations should join in and share the financial burden of Ukraine's disarmament.

The timeline for Ukraine's ratification of START I was a subject routinely asked about to Ukrainian representatives after the signing of the Lisbon Protocol. I stressed that we needed to consider the new circumstances that emerged after the collapse of the USSR, conveying our four key arguments surrounding the ratification of START.

First, Ukraine would never recognize another country's right of ownership to strategic nuclear forces deployed on its territory, a necessary affirmation in the face of Russia's persistent attempts to pressure the leaders of Belarus and Kazakhstan to subordinate their countries' strategic nuclear forces to the Armed Forces of the Russian Federation. Second, as the legal owner of strategic nuclear forces deployed on its territory, Ukraine was prepared to assume obligations stipulated in START I. Third, since Ukraine had not participated in the START I negotiations, and since new political realities had developed, it was essential to revise some of the treaty's provisions, especially those requiring the destruction of missile silos with materials that could be used by Ukraine for peaceful purposes. Fourth, nuclear disarmament was to take place via a careful consideration of the process' political, economic, financial, and environmental consequences, and in alignment with Ukraine's national interests according to the decisions of the Ukrainian parliament.

It was this general position on nuclear disarmament that Ukraine's MFA avoided communicating to the world. Moreover, foreign policy representatives avoided attending these meetings, or kept a low profile, so as to distance themselves from any political intrigue either real or imagined. Ukraine's diplomatic stance allowed Russia to promote its own interests without any challenges.

Even so, our attempts to inform the Western governments and professional communities about Ukraine's stance on nuclear disarmament were not entirely in vain. The first signs of a change appeared in the middle of December 1992, when the American nuclear energy behemoth, General Atomics, approached the Ukrainian government with proposals for utilizing our missiles' nuclear material for peaceful purposes. At the time, we lacked information on what exactly constituted Ukraine's nuclear weapons, and how much money the country needed for the disarmament. Negotiations were impossible without these figures. Therefore, two special trips were made in an attempt to obtain these data. The first was to Ukraine's 43rd Rocket Army, and the second to Russia.

The Visit to the 43rd Rocket Army: Military Estimates

On 18 November 1992, Anton Buteiko, the head of the president's Office on International Affairs, deputy defense minister Ivan Bizhan, and I traveled to the 43rd Rocket Army to conduct what might be called field research. In just two days, we visited command posts and launch facilities in the Vinnytsia, Khmelnytskyi, Kirovohrad and Mykolaïv regions. There, we held a number of meetings with army commanders, experts responsible for the safe utilization of strategic missile systems and nuclear warheads, and officers on combat duty. We were there to gauge their thoughts on Ukraine's nuclear disarmament.

In holding these conversations, we eventually saw a more realistic picture of the magnitude of political problems that Ukraine faced in committing to become a nonnuclear-weapon state. The 43rd Rocket Army commander, Volodymyr Mykhtiuk, presented his preliminary estimates of the financial outlays that Ukraine could anticipate making in order to comply with the conditions of START I. This was an entirely different reality from what we had heard in Kyiv. According to his calculations, Ukraine would need nearly US $700 million annually (and, if we were to include resolving the ecological, scientific, and technical issues, as well as providing social guarantees for the servicemen, the number would exceed US $1 billion) for the duration of all of seven years stipulated in START. It was an inconceivably vast sum for Ukraine's national budget, which at the time consisted of no more than US $8 billion. The US promised to give Ukraine US $175 million for the entire process.

I kept the report provided to our working group by the MoD, "General Conclusions on the Scientific, Technical, Financial and Economic Aspects of Nuclear Disarmament," in my archives. In the report, the MoD speculated that by 1998 Ukraine would need US $4 billion for the implementation of START I provisions, plus another 1.2 trillion Ukrainian karbovantsi.[11] The above sum was broken down as follows: US $2.6 billion were needed for the reduction of 13 deployed missile systems alone, without calculating the cost of the upkeep of the strategic nuclear forces; US $1 billion, for the dismantling of the warheads and the reprocessing of the nuclear material; US $400 million, for

32. *Volodymyr Mykhtiuk, 2001*

Volodymyr Mykhtiuk (1938–2019), commander of the 43rd Rocket Army from 1990 to 2002. He graduated in 1960 from the Black Sea Naval Academy specializing in submarine-based nuclear weapons. He immediately joined the staff of the commander in chief of the Strategic Rocket Army of the USSR. He served in the 50th Rocket Army in posts ranging from the supervisor of technical batteries to commander of a rocket division. In 1979, he graduated from the Dzerzhinsky Military Academy of the Strategic Rocket Forces (SRF) in Moscow (today, the Peter the Great Military Academy of the SRF). In 1980, the commander of the Missile Army, General Volodymyr Tolubko, appointed him commander of the Rocket Division in Barnaul. In 1983, he was appointed chief of staff of the 43rd Rocket Army in Vinnytsia. From 1988 to 1990 he was the commander of the 50th Rocket Army. After the Rocket Army administration was disbanded in December 1990, he was placed in charge of the 43rd Rocket Army in Vinnytsia. In 1991, he graduated from the Military Academy of the General Staff of the USSR Armed Forces. In 1994, he was promoted to colonel general. From May 1996 to August 2002, he was deputy defense minister of Ukraine and commander of the 43rd Rocket Army.

the development of information support systems for the new weapons; and 1.2 trillion karbovantsi were budgeted for the rearmament necessary to maintain defense capacity at the same level.[12] The MoD experts' conclusion read: "Ukraine is not prepared to fulfill the demands of the START I Treaty in the event of its swift ratification, in view of scientific, technical, financial, and economic factors."[13] All three of us, Buteiko, Bizhan, and I, despite our differences in age, professional backgrounds, and political and personal experiences, came to the same unequivocal conclusion. While Moscow and Washington pushed us towards an accelerated removal of nuclear warheads from Ukrainian territory, they left unsolved all of the problems inherent in the process—problems that would cost our nascent state billions of dollars. In our common opinion, this version of nuclear disarmament meant financial ruin for Ukraine.

When we returned to Kyiv, three agencies, the MFA, the MoD, and the State Nuclear and Radiation Safety Committee (Derzhatomnahliad), had already proposed bills for Parliament's immediate ratification of START and the NPT. The three of us persuaded the president, the head of Parliament and the prime minister that we needed to study all provisions and their potential consequences carefully before ratifying START I, along with the Lisbon Protocol and the NPT.

On 20 November 1992, in his brief to Kravchuk, Buteiko appealed for a comprehensive approach to this decision. He stressed that several other ministries and agencies would be involved in the implementation of these treaties, many more than its mere three signatories. He believed the MFA needed to have accurate information about Ukraine's capacity to deliver on these treaties' obligations, and present it before Parliament. Thus he recommended giving explicit instructions to the Cabinet and other ministries and departments to formulate their own proposals for the START ratification bill.

Buteiko stressed that the texts of these treaties, the START and the NPT, were developed without Ukraine's participation, and it was of vital importance to examine them thoroughly.

Parliament's decision regarding these documents, Buteiko wrote, would basically shape Ukraine's subsequent policy on nuclear weapons. So, he recommended that Kravchuk form a separate working group within the executive branch, comprised of representatives of the above-mentioned governmental agencies, to prepare financial and economic projections of the treaty's consequences, as well as to propose a comprehensive plan for their implementation. The group would have to consider the expediency of nonnuclear weapon status for Ukraine within the context of national security.

Even though Kravchuk was already prepared to support the bill proposed by the MFA, the MoD, and Derzhatomnahliad, and to present the nuclear treaties for Parliament's ratification, Buteiko had managed to stop him. This allowed extra time to conduct yet one more important round of research.

Visit to Russia

After our official visit to the military bases, the picture of Ukraine's nuclear disarmament dilemma became clearer. However, our working group was still missing several vital pieces of information needed to formulate our final recommendations on ratification of START I. Unfortunately, no one in Ukraine possessed it. This information concerned the specifics on nuclear warheads located on Ukraine's territory, both tactical and strategic—information classified top secret and judiciously guarded by Moscow.

I needed answers to two questions. My first question was (this worried Kravchuk the most): What was the condition of the nuclear warheads and how long could they safely be kept in Ukraine? Kravchuk was constantly being fed misinformation indicating that the country's nuclear arms were in a terrible condition and could explode at any minute, and that the explosion would be worse than Chernobyl. My second question was: What were the weight characteristics of the nuclear warheads; in other words, how much nuclear material (i.e., highly enriched uranium and plutonium) did they contain? In short, what did Ukraine actually own and what was Ukraine being asked to transfer to its neighbor without any guarantees of compensation or assurances of the weapons' ultimate destruction?

With this in mind, I went to Ivan Pliushch again and was able to convince him that I needed to visit the Russian plant that produced nuclear warheads, which was now dismantling Ukraine's tactical weapons. Officially, the reason provided for my visit was that Parliament had commissioned the government to develop a system of international controls to ensure that nuclear weapons transferred from Ukraine were not used again by Russia. Pliushch telephoned defense minister Kostiantyn Morozov forthwith with a request to arrange a visit with Russia's defense minister. Moscow, albeit without any particular enthusiasm, conceded to the visit.

The trip took place in early December of 1992 and nearly cost me my life. First, on the way from Moscow to the airport in Domodedovo, the brakes on our diplomatic vehicle unexpectedly went out on a dangerous and slippery

December road. Then the plane, which took me from Moscow to Sverdlovsk (now Yekaterinburg), was not able to land until its third attempt. Despite the rough start the results of the trip exceeded all expectations.

First, as one of our official chaperones whispered to me, more than half of this top-secret plants' employees were ethnic Ukrainians, and many of them were ready to return home. In general, much of the USSR's nuclear developments had been carried out by Ukrainians: from uranium mining for the first nuclear warheads to leadership in the USSR's Ministry of Medium Machine-Building Industry (Minsredmash), which was in charge of the organization of the development process and the production of nuclear warheads. At that point, we had a concrete opportunity to send home some of the Ukrainian specialists who, in turn, could then help ensure the safety of Ukraine's nuclear disarmament process. However, for that to happen, the state would need to create the conditions for them to work and live in Ukraine. Unfortunately, no one in our government wanted to take the trouble.

At the time I was the only civilian from Ukraine who had ever visited a plant manufacturing nuclear warheads. So, it was in Sverdlovsk that I finally received the information that our working group needed to formulate our nation's position on nuclear arms. Specifically, I confirmed that a nuclear warhead contained a powerful safety system that would prevent a nuclear explosion unless sidestepped through very specific protocols. I was told: "Yuri Ivanovych, don't worry; nothing will explode over there." This corresponded entirely to what I had heard from Russia's deputy defense minister during a summer seminar on nuclear security. According to his records, there was no precedent for an unsanctioned nuclear explosion in his forty-seven years of working with nuclear warheads, and an accident could not cause a nuclear warhead to explode on its own.

Second, I received data on some of the weight characteristics of a nuclear warhead, which was of utmost interest to me. This information, along with the data that I had received so far, informally and in bits and pieces via international conferences, was finally lifting the veil on what Ukraine was about to lose by handing its nuclear weapons over to Russia without any set obligations in return.

At last, it became clear to me that it would be most beneficial for Ukraine to use the HEU and plutonium obtained from dismantled nuclear warheads for the production of its own nuclear fuel. As almost half of all electricity in Ukraine was produced by NPPs, and all nuclear fuel for them was purchased from Russia, Ukraine's ability to produce its own nuclear fuel would significantly reduce Ukraine's energy dependence.

The picture was becoming clear. At that point, Ukraine was armed with all the necessary expert knowledge not only for nuclear disarmament negotiations, but also for the development of its own peaceful nuclear strategy. Now, it needed to use this knowledge.

What Constituted the World's Third Largest Nuclear Arsenal?

Up until December 1992, Ukraine did not have any systematized information about its nuclear inheritance. Thus, having commenced the nuclear disarmament process, Ukraine, even at the level of state officials, did not have realistic knowledge about the subject of this landmark event. Not a single Ukrainian agency tried to obtain it. President Kravchuk was still more preoccupied with the idea of getting rid of the hazardous nuclear weapons. Ukraine's security services, for their part, instead of trying to collect data on the country's nuclear arsenal, would often try to scare Kravchuk with the horror story that Russia's hypothetical claim to strategic nuclear weapons located in Ukraine would lead to military occupation.

Our working group, which was deputized by Parliament to analyze information and draft documents pertaining to nuclear disarmament, never received any classified data on the characteristics or number of Ukraine's nuclear arms during the duration of its existence, either from the SBU or the MFA. We were essentially excluded from the preparation of expert assessments by Ukraine's Academy of Sciences, even though we engaged their representatives in our working group. Obviously, the situation was further complicated by the fact that Ukraine did not have its own experts on nuclear warhead development or accurate data on the number of warheads themselves.

Thus, our working group turned out to be Ukraine's only official body collecting information on the country's share of the USSR's nuclear legacy. We had to create the informational base almost from scratch, from a variety of fragments, combining and comparing official and unofficial information, from talking with the military, from international documents, press releases, and excerpts from conferences. The information on the weight characteristics of nuclear warheads as we know them today, for instance, are what our working group, against the odds, was able to collect at the time. The following section describes the world's third largest nuclear arsenal—what it looked like in quantitative terms and in terms of the cost of manufacturing the nuclear material in it.

Number of Units

It was known that there were two types of *strategic* nuclear weapons deployed in Ukraine—intercontinental ballistic missiles (ICBMs) and heavy bombers with cruise missiles—as well as a variety of *tactical* nuclear weapons. There were two types of ICBMs; 130 liquid-propellant SS-19s, and 46 solid-propellant SS-24s. These were positioned in 176 missile launch facilities, or nuclear silos, in four divisions of the 43rd Strategic Army. The most powerful missile at the time, the solid-propelled SS-24s of the latest generation, were manufactured by Pivdenmash of Ukraine and deployed in the Mykolaïv and Kirovohrad regions (oblasts). A single missile carried up to ten warheads and was so devastating that even the Americans at that time had no missile with an equivalent payload

33. SS-24 Scalpel

or power. The liquid-propellant missiles, each of which carried six warheads, were stationed with the Rocket Army in Ukraine's Vinnytsia and Khmelnytskyi regions. We did not have exact numbers for any of the other nuclear weapons deployed in Ukraine.[14]

According to the MoD data provided to our working group in early 1993, there were 36 bombers with 274 nuclear warheads located in Ukraine. At about the same time, *Moskovskie novosti* mentioned "670 nuclear warheads on air-launched cruise missiles."[15] Serhii Tolstov mentioned 416 nuclear warheads in his June 1993 article in *The World Today*.[16] *Dzerkalo tyzhnia* provided different numbers. According to the publication, there were 43 long-range strategic bombers with 576 warheads deployed in the 46th Strategic Air Force. These included 14 Tupolev Tu-95MS strategic bombers, each carrying eight AS-15 cruise missiles, and 29 Tupolev Tu-160 ultramodern strategic bombers, able to carry 16 cruise missiles with a range of nearly 3,000 kilometers.[17] William C. Potter, in an article in November 1992 in the *New York Times*, described Ukraine's nuclear arsenal as follows: "At the time, Ukraine possessed 4,000 nuclear weapons. Though the number has been reduced with the transfer to the Russian Federation of all nonstrategic arms, the state still has more than 1,200 nuclear warheads on 176 intercontinental missiles, along with 34 nuclear-armed strategic bombers."[18]

According to the aforementioned data provided by Ukraine's MoD, the strategic missiles and bombers carried 1,514 nuclear warheads all together. However, a report by the Ukrainian MFA, using sources unknown to me, stated that there were 1,656 warheads. In another estimate, a member of the parliamentary Commission on Foreign Affairs, Mykola Balandiuk, maintained that in addition to the 1,656 ICBM warheads there were more than 500 nuclear-armed cruise missiles, as well as numerous nuclear gravity bombs for the arming of 30 bombers[19]—in other words, 2,156 nuclear warheads in total. A different source, publishing in *Dzerkalo tyzhnia,* placed the total number of warheads at 1,837.[20] On 24 June 1993, at the US Senate hearing before the Subcommittee on European Affairs of the Committee on Foreign Relations, Ukraine was said to have 2,000 nuclear weapons.[21]

The quantity of Ukraine's tactical nuclear weapons remained completely unknown to our working group, because they were located in different military units under a different command: the Black Sea Fleet was armed with nuclear torpedoes, nuclear land mines, and nuclear artillery; the ground forces had a significant number of tactical and tactical-strategic missiles equipped with nuclear warheads; and the Air Force and the Anti-Aircraft Defense Force had nuclear warheads, as well.

The MoD never provided us with the total number of nuclear weapons. Data appearing in the *New York Times* on 10 November 1992, stated that overall, Ukraine possessed more than 4,000 warheads, 2,800 of which were tactical.[22] Russian ambassador Iurii Dubinin, in an April 2004 article alluded to "almost 3,000" warheads, and Anne Applebaum wrote in the *Spectator* on 26 June 1993 that Ukraine "had already sent at least 4,200 tactical weapons to Russia."[23]

The exact numbers were in the exclusive possession of the General Staff of the Armed Forces of the USSR, and they were not accessible to the Ukrainian parliament. Table 1 provides data on Ukraine's nuclear arsenal, summarizing information from various sources.

Table 1. Ukraine's Nuclear Arsenal at the Time of the Dissolution of the USSR[24]

Type of nuclear weapons	Quantity
Intercontinental ballistic missiles (ICBM)	176
SS-19 (liquid-propellant)	130
SS-24 (solid-propellant)	46
Heavy bombers	30–43
Nuclear warheads on strategic arms	1,514–2,156
Tactical nuclear warheads	2,800–4,200

Nuclear Material Stockpiles

The UN Security Council and the International Atomic Energy Agency (IAEA) had developed methods for making an approximate assessment of the amount of nuclear material contained in the tactical warheads transferred from Ukraine to Russia.[25] Mark Shteinberg wrote in *Dzerkalo tyzhnia* that according to UN document No. 6858, the minimum amount of fissile material required for the production of a nuclear device was 8 kg of plutonium (95%), 25 kg of uranium-235 (90–95%), and 8 kg of uranium-233. These parameters were established by the IAEA.[26]

In multiplying the above numbers by 3,500 tactical warheads, we reach the following: 87.5 metric tons of uranium-235, 28 metric tons of uranium-233, and 28 metric tons of plutonium. However, our working group received information unofficially about which kinds of nuclear warheads were in our tactical weapons. There were nine: their uranium-235 weight ranged between 7 and 11 kilograms, whereas the plutonium weight ranged between 7 and 16 kilograms. We made our calculations on the basis of these data, to determine that our tactical nuclear weapons alone contained between 30 to 40 tons of HEU, and between 30 and 55 tons of plutonium over 95 percent.

Strategic weapons contain two to three times more nuclear material than the tactical weapons. The content of HEU (90–95%) in a mid-yield strategic weapon warhead varies from 15 to 35 kilograms, and the same proportion applies to plutonium. Since Ukraine's strategic carriers were loaded with high-yield nuclear warheads, calculations suggest that they contained 50–68 tons of uranium-235, and 27–40 tons of plutonium. According to the data provided by the Ukrainian MoD, the 130 liquid-propellant strategic missiles contained 40.25 tons of HEU, and the entire strategic nuclear arsenal of Ukraine contained 64 tons of HEU.

A report provided to our working group by the Kharkiv Institute for Physics and Technology confirmed the accuracy of these calculations. The Institute's experts conducted consultations with their Russian counterparts regarding compensation to Ukraine for the nuclear material from its strategic weapons, and used Russian sources in their calculations.[27] When calculating the cost of converting HEU into fuel rod assemblies (TVELs) for Ukrainian NPPs, the Russians recorded an inventory of 64 tons of HEU. Consequently, Ukraine's tactical and strategic weapons could have contained between 80 to 108 tons of HEU and between 57 and 95 tons of weapon-grade plutonium.[28]

34. *SS-19 engine case*

Nuclear Material's Value

The biggest difficulty for our working group was finding data about the value of nuclear material. Because the selling of material used in manufacturing weapons of mass destruction was banned by the NPT, its market value simply did not exist. Thus, we sought data on the cost of its production, in order to understand what kind of compensation for HEU and plutonium Ukraine could propose.

In an attempt to raise the stakes during the US-Russian Uranium Deal negotiations (which were still being kept a secret from Ukraine in early 1992), the Russian security services would broadcast figures demonstrating the high cost of nuclear arms. In February 1992, *Komsomol'skaia pravda* published an article entitled "We Are Not Rich Enough To Be Selling Warheads," and cited the following numbers: "One ton of weapons-grade uranium costs $100 million, one ton of weapons-grade plutonium costs between $500 million and $1 billion." The article continued, "Even with conservative estimates, the value of the 5,000 to 6,000 nuclear warheads from the decommissioned tactical weapons equals to our [the USSR's] entire external debt!"[29] The Soviet Union's external debt at the time equaled US $60 billion.[30]

In late 1992, *Holos Ukraïny* circulated information that the US would obtain US $50 billion-worth of nuclear fuel for its NPPs by dismantling nearly 6,000 US strategic warheads. According to the Americans, Ukraine could get US $8–10 billion-worth of nuclear fuel for the HEU removed from the nearly 5,000 nuclear warheads alone.[31]

In October 1992, another relevant piece of important information surfaced. In a *Moskovskie novosti* article mentioned above, "Whom Will Enriched Uranium Enrich?", Vladimir Kiselev revealed a secret agreement between Russia and the US regarding the sale of 50 tons of HEU obtained in the process of nuclear disarmament, "enough to fuel all US nuclear reactors for at least 10 years, or the entire world's reactors for two and a half years."[32] As a side note, the US had 180 nuclear reactors at the time, while Ukraine had only 15. It was clear that Ukraine's uranium was included in that 50 tons; the secret agreement gave justification to our working group and the Ukrainian government delegation to raise the question of compensation to Ukraine for the transfer of its nuclear warheads to Russia. Ukraine's share in this deal constituted between 20 and 40 percent, according to a *Radio Free Europe/RL Research Report* published in February 1992.[33] Subsequent events confirmed the accuracy of the RFE estimate of Ukraine's share of the deal for the HEU. In June 2001, the Group of Seven (the G7, consisting of Canada, France, Germany, Italy, Japan, the UK and the US), agreed to provide US $20 billion to Russia for the reduction of its strategic nuclear arms under START II.[34]

For some reason, in estimations of the value of nuclear material and in considerations of possible compensation, plutonium was almost always excluded, even though it was the most valuable of the nuclear materials. In the transfer of 80–108 tons of HEU and 57–95 tons of plutonium to Russia, Ukraine handed over nuclear material valued at US $8–10.8 billion worth of HEU, and US $28.5–47.5 billion of plutonium. The value of the plutonium was calculated according to the most conservative pricing of US $500 million per ton of plutonium, whereas the total cost of production constituted US $57–95 billion. In other words, the nuclear material in Ukraine's weapons constituted a minimum total monetary value of US $36.5–57.5 billion, and up to US $65–105 billion. Table 2 presents a summary of this information.

Table 2. Estimated Quantities and Cost of Production of Nuclear Material Contained in Ukraine's Nuclear Weapons

Nuclear material	Type	Number of weapons	Total tonnage	Production cost per ton (billion USD)		Total cost of production (billion USD)	
				min	*max*	*min*	*max*
HEU	Tactical	30–40	80–108	0.1		8–10.8	
	Strategic	50–68					
Plutonium	Tactical	30–55	57–95	0.5	1	28.5–47.5	57–95
	Strategic	27–40					
Total						**36.5–57.5**	**65–105.8**

Sources: Based on various sources obtained in the press and official documents.[35]

Consequently, the value of material in Ukraine's nuclear weapons was estimated at over US $100 billion. As already mentioned, Ukraine's entire national budget in 1992–1993 constituted no more than US $8 billion. The calculations have demonstrated that Ukraine had good reasons to fight for its right of ownership of the nuclear weapons. The amounts in question also remove any doubts about the Russian diplomats' reasons for launching an aggressive campaign during the negotiations with Ukraine in late January 1993 for the transfer of the weapons to Russia.

However, when I communicated to the president that Ukraine should demand of Russia no less than US $10 billion in compensation for the HEU from our warheads, Kravchuk looked at all these data with little confidence, and said it was utter madness to ask for that kind of compensation with a national budget of US $8 billion. Yet even the figure that so startled Kravchuk accounted only for our HEU and was nowhere near the total amount for the weapons we possessed; our warheads also contained plutonium worth over US $90 billion.

35. *Types of missiles used by the USSR (1)*

Repurposing Nuclear Material: The General Atomics and Westinghouse Proposals

Having unraveled what exactly constituted the third largest nuclear arsenal in the world, we faced another question that was no less complicated: What do we do with nuclear warheads and how can we use nuclear material for the needs of Ukraine's economy? The nuclear warheads and nuclear material dilemma posed a complex problem for Ukraine, as numerous scientific, technological, and economic factors were involved, in addition to unknowns that were classified secret by other countries. Not having its own analogs, Ukraine was forced to use information from other countries—information that reduced the reliability of the calculations Ukraine was using in its decision-making. Therefore, Ukraine needed to widen the network of countries with which it could collaborate in eliminating its nuclear weapons.

Protecting Ukraine's national interests entailed using nuclear material for the country's economic needs. First and foremost, it would reduce Ukraine's energy dependence on Russia. It would also make the country's nuclear energy industry safer and more cost-effective; after all, more than half of Ukraine's electricity production depended on it. I brought questions before the Working Group, keeping in mind Ukraine's use of nuclear material obtained as a result of disarmament.[36]

36. *Types of missiles used by the USSR (2)*

The realization of this task demanded, first, the development of a Ukrainian strategy for initiating its own nuclear fuel production cycle, and second, the assessment from the point of view of that strategy of all proposals put forth by other countries regarding the utilization of nuclear material. In contrast to Ukraine's own institutions, the major American companies had a lightning-fast reaction to Ukraine's new position on nuclear arms, which was divergent from Russia's position. They immediately approached us with official offers, addressing the first vice prime minister, academician Ihor Iukhnovs′kyi, and me, as the head of the Working Group.

I have openly stressed on more than one occasion that Ukraine needed to seek alternatives to the Russian proposal for

37. Soviet air defense rockets

dismantling its warheads and reprocessing weapons-grade uranium and plutonium. It was becoming obvious to me that everything transferred to Russia was disappearing without a trace. Furthermore, our direct interactions with the West would signal that Ukraine was distancing itself from Moscow, both economically and politically; that, I was convinced, was the number one objective for the young Ukrainian state. Iukhnovs'kyi publicly backed my position, which is why we were chosen as the addressees for American companies' official offers of cooperation.

The first to approach us was General Atomics (GA), who wrote to us in December 1992. Because Ukraine did not have an official strategy on what it was planning to do with nuclear warheads at the time, GA's first proposal was based on information from international seminars, where we previously expressed Ukraine's aspirations to use nuclear material for peaceful purposes. First and foremost, this related to the opportunity to produce nuclear fuel for NPPs inside Ukraine's borders. GA vice president James J. Graham then sent Iukhnovs'kyi a letter in which he officially introduced the possibility of converting the HEU obtained from nuclear warheads to LEU for use in Ukraine's energy industry.[37]

On 17 December 1992, we held our first meeting with GA representatives. Kuchma, who then was prime minister, supported the idea, and then delegated Iukhnovs'kyi and me to lead the meetings. GA presented some concrete propositions, the chief of which was to develop an enterprise converting HEU to LEU in Ukraine within one year. This would necessitate a small production site measuring approximately 13 x 13 x 16 meters, and would cost US $20–$30 million. After completing the HEU to LEU conversion program, the site could be repurposed for other needs in Ukraine's nuclear fuel production cycle. One of the short-term benefits of the program, Iukhnovs'kyi and I stressed in our report to Kravchuk and Kuchma summarizing the outcomes of the meeting, was that it would eliminate the need and cost of transporting nuclear warheads to Russia for dismantling (for which Russia charged millions of dollars, as will be discussed further). The program's long-term benefits included cooperation with global leaders in the field of nuclear energy.

Overall, the program would take 10 years. GA also presented a possible mechanism to finance the program. It proposed ways in which Ukraine could convert 60 metric tons of nuclear material into fuel without investing a single penny by using the HEU as collateral.

Meanwhile, GA, in seeing a promising new market development opportunity proceeded without delay. A mere two weeks later, in early January 1993, we held another meeting in Kyiv. After the meeting on 13 January, in a letter to Iukhnovs'kyi, GA wrote that in collaboration with the International Commodities Exchange Corporation (ICEC), an affiliate that worked with HEU, Ukraine could establish its own nuclear fuel production in a matter of four years. This would mean breaking away from Russia's monopoly by creating critical elements of Ukraine's own nuclear fuel cycle, which would include initiating the independent production of TVELs and ensuring the safe storage of spent nuclear fuel and other highly radioactive nuclear waste, thus liberating Ukraine from dependence on Russia in this hazardous operation.

I had a young scholar working on my team, Kostiantyn Rudia, a doctoral student and a former operator at the Chernobyl NPP. Thanks to him, we had all the latest information about global development trends in the nuclear energy industry. Because we did not have the internet then, his research demanded that he sacrifice himself sifting through paper archives and library periodicals, all in the name of science and our country. I can attest to the fact that no one within Ukraine's governmental agencies conducted any such research. The Ukrainian State Committee for Nuclear Power Utilization (Derzhkomatom), whose cadre remained virtually unchanged from the time of the USSR's collapse, oriented itself towards Russia as a power center, and awaited instructions from Moscow on what to do.[38] The fact was that Ukraine's strategy on the development of the nuclear power industry simply did not exist, and to make matters worse, there was no one to even formulate it.

As I realized the situation was dire, I continued communicating with GA on my own initiative. On 15 January GA sent official answers to my questions. I was most concerned with the legal aspects of our prospective collaboration and asked whether GA had a legal right to work with the NPT. Their answer was affirmative, as Graham wrote: "GA has extensive experience working with countries that have signed agreements under the NPT[...] [and] GA has experience in working with US and international agencies, such as the IAEA, which are involved with administering the NPT."[39]

The second important question I had, was whether GA had any experience in dismantling warheads. This was a fundamental question, given Russia's insistence that it was indispensable in the disassembling process, using it to coerce Ukraine into following its scenario. I heard this argument from the Russian delegation during negotiations in Irpin; they even propagated the idea that only the one who assembled the warhead could disassemble it.[40] Ukraine's high officials repeated this thesis in official communications. As it turned out, General Atomics also had experience in working with enterprises in this line of work. Thus, we could manage without the Russians in the eventuality that they refused to dismantle Ukraine's warheads.

38. *R-12 Dvina, a Soviet medium-range single-stage liquid-fueled surface-to-surface ballistic missile*

Yet another problem we faced was the use of plutonium, the estimated value of which in strategic weapons alone was up to US $40 billion. At this point, because we had information about a development program in Japan, I asked whether GA cooperated with Japan on the issue, and also for their view on the benefits of switching to a mixed uranium-plutonium fuel. GA also confirmed their expertise in this subject area.

This was not even all of it, because General Atomics clearly saw a mechanism for reclaiming nuclear material from Ukraine's tactical nuclear weapons that were transferred to Russia under dubious circumstances in 1992. By that time, we already knew about the Uranium Deal, through which Russia intended to sell 500 tons of HEU released from nuclear weapons, which included Ukraine's tactical weapons, to the US. Since the US had officially announced that it would start paying Russia only after Ukraine and Russia agreed on how the money would be divided, we could in theory get paid, not only for the HEU, but also for the plutonium from Ukraine's tactical weapons, the total cost of which was estimated at US $60 billion.[41]

Following that letter, we received another three on 15 March, 26 April, and 3 May, and each contained concrete proposals on the topics about which I had inquired in my previous correspondence to them. The first proposal entailed the construction of a site in Ukraine for converting HEU to LEU, which upon completion of its main function, as already noted, could be used as a component in the nuclear fuel cycle, as well as for the subsequent disposition of low-level radioactive waste.

GA's other proposal included creating a joint venture with Ukraine for the organization of the dismantling process of Ukraine's nuclear weapons, and the subsequent conversion of HEU to LEU for use as fuel for NPP reactors. Graham wrote: "The proposal as described has the advantage of reducing the time and cost of handling HEU derived from the nuclear warhead dismantlement by eliminating the need to build a dilution facility."[42]

39. *Kh-22NA cruise missile*

GA suggested that Ukraine should store its plutonium in the US or in one of the European countries for further reprocessing into fuel. Plutonium was the most problematic element. As already mentioned, Russia insisted in its reports that the use of plutonium in nuclear energy was impossible, and that Ukraine's best option was to hand it over immediately and incur financial expenses "only" from disassembling and transporting it to Russia, which would then assume the "burden" of its subsequent ownership. After all, Russia insisted, it would not be able to return the weapons-grade plutonium to Ukraine, as that would violate the NPT.

On 3 May Graham sent an addendum to GA's earlier proposal for creating a joint venture "designed to assist Ukraine in realizing the full commercial potential of nuclear material stored in Ukraine while providing long-term fuel requirements for Ukrainian reactors and to render required services associated with the decontamination of nuclear warhead facilities."[43] He suggested, instead of building conversion plants for depleting HEU, to only build oxidizing sites. After reprocessing the HEU at these sites, the material could then be transported to the US by air, where GA would produce uranium hexafluoride pellets and would ship them back to Ukraine in special containers for Ukraine's further production of fuel rods (TVELs) for its NPPs.[44]

In August 1993, I was invited to visit the key nuclear power centers in the US through a democracy development support program with the goal of helping Ukraine develop its peaceful nuclear strategy. During my 30-day stay in the US, I visited most of the top-secret laboratories, including Pacific Northwest National Laboratory, the Hanford Site, Los Alamos National Laboratory, Sandia National Laboratory, and Lawrence Livermore National Laboratory, among others. These were the key institutions where the US was developing its nuclear might, and in which, after the dissolution of the USSR, the US began focusing increasingly on "atoms for peace" and on solving issues of radioactive contamination.

This immensely informative visit resulted in a series of propositions I submitted to the government and different governmental agencies, including to

40. *Model of a warhead*

the aforementioned Derzhkomatom and the State Committee for Safeguarding the Population from the Effects of the Accident at Chernobyl NPP (Derzhkom-chornobyl). The two latter agencies worked directly on nuclear issues, one in the nuclear energy sector, and the other in the domain of the safe utilization of nuclear material on Ukrainian territory, considering the experience of the 1986 accident at the Chernobyl NPP. My proposal focused on the following issues: the optimal strategy for the peaceful utilization of nuclear material removed from the warheads in the process of nuclear disarmament; optimal avenues for the development of the equipment for the nuclear fuel cycle, including, above all, the production of our own nuclear fuel; strategies for radioactive contamination cleanup in the Chernobyl zone and for solving the Sarcophagus problem; and alternative energy solutions, among numerous other issues.[45]

However, by then, the situation in the executive branch had fundamentally changed. Iukhnovs′kyi had already left his post as vice prime minister; Vasyl′ Ievtukhov, who took over this part of Iukhnovs′kyi's responsibilities, took no interest in the issue; and Prime Minister Kuchma, who had expert knowledge of nuclear weapons, was getting ready for his official resignation. I remained the only official who maintained communication on disarmament strategies with the United States, instead of with Russia, and thus forwarded all their proposals to specialized governmental structures—that is, to the Derzhkomatom for the development of specific measures aimed at Ukraine's realization of its own nuclear program, and to the MFA for use in nuclear disarmament negotiations. However, it became clear soon enough that all these ideas had been halted.

Interestingly, the Americans continued approaching us with proposals for creating a nuclear fuel cycle in Ukraine, even after the country completely disposed of its nuclear weapons. In the spring of 1997, we were getting ready for the meeting of the Kuchma-Gore Commission. On 8 May Westinghouse, a US company, in its letter to then-president Kuchma proposed a broad cooperation program, which considered nuclear fuel.[46] Given Ukraine's complete dependence on the Russian Federation for NPP fuel, Westinghouse offered the following key directions for cooperation, which included: diversification of nuclear fuel supply into Ukraine; introducing Western safety standards with regard to VVER-1000 (Ukrainian: *vodo-vodianyi enerhetychnyi reaktor*) nuclear power reactors; increasing the efficiency of Ukraine's NPPs, and replacing the energy produced by the Chernobyl NPP, which was being shut down. They even were ready to assist Ukraine in getting financial support from the US for these purposes.

Despite the apparent lack of interest on the Ukrainian side, Westinghouse sent Kuchma a letter after the meeting in Washington with the hope that "Ukraine would move towards energy independence." [47] In this scenario the earlier propositions from General Atomics provided answers to the following questions: how to dismantle nuclear warheads without taking them out of Ukraine; how to develop Ukraine's own nuclear fuel cycle using materials removed from nuclear warheads, including mechanisms of financing the process; what to do with plutonium; and how to receive compensation from Russia for the already transferred tactical nuclear weapons. Thus, Ukraine could have achieved the following by cooperating with the US on nuclear material utilization:

1. compensation by Russia of the cost of the nuclear material in tactical weapons transferred from Ukraine to Russia, valued at nearly US $60 billion.[48]
2. the establishment of Ukraine's own nuclear fuel cycle by using HEU from its strategic weapons. This would have meant departing from Russia's control in operating Ukraine's NPPs from nuclear fuel supply to nuclear waste disposal, as well as an independent market entry into the global system of peaceful production and use of nuclear energy, thus foregoing Russia as an intermediary.
3. the utilization of plutonium as a promising energy source in Ukraine's economy, including using a financial guarantee mechanism under the IAEA's control. Plutonium in Ukraine's tactical and strategic weapons transferred to Russia was valued at US $95 billion.
4. creating a geopolitical option. By cooperating with the US, Ukraine had the chance to begin departing from its political dependence on Russia, thus commencing the process of European integration. Instead, our nation made the choice in Russia's favor in the early 1990s, and Ukraine found itself unprepared for Russian military aggression in 2014 and 2022, both economically and politically.

These potential outcomes allow us to evaluate what Ukraine lost by rejecting proposals from the US and choosing the Russian scenario. Instead of Ukraine's independent use of nuclear material under international control, the Russian scenario envisioned a unilateral transfer of all weapons to Russia, in exchange for so-called peaceful dividends, that is, compensation in the form of TVELs for NPPs and in amounts determined by Russia itself. The US and Russian proposals presented the possibility not only of the optimal use of an immense economic resource, but also of a new geopolitical posture.

What to Do with the Nuclear Material: Ukraine's Readiness for Independent Decisions

The executive branch was responsible for formulating a strategy for developing Ukraine's nuclear fuel cycle, as well as for making decisions on how to utilize nuclear material to bolster Ukraine's economy. These strategies were to be driven by the decisions made in Parliament. However, at the onset of the disarmament process, the executive branch began arbitrarily interpreting parliamentary decisions, which in essence, was ignoring them. As such, Parliament's Statement on the Nonnuclear Status of Ukraine of 24 October 1991 had declared Ukraine's intention to destroy its nuclear warheads, because it was through the elimination of nuclear warheads that the nation was to become a nonnuclear-weapon state. This was the way in which the RSA attained its nonnuclear status, as discussed earlier in this chapter.

Nevertheless, in December 1991, a mere two months after the adoption of the preceding statement, President Kravchuk and Ukraine's MFA, signed the Minsk and Alma-Ata treaties on behalf of Ukraine as party to the CIS. The documents included language on transferring the tactical and strategic weapons to Russia. At the same time, neither of the treaties obligated Russia to destroy the transferred warheads, nor did they make any mention of the procedure ensuring international control over the reprocessing of nuclear material into peaceful fuel. As some of Ukraine's experts warned as early as spring 1992, such an approach exposed the possibility of turning Ukraine's nuclear disarmament into Russia's nuclear rearmament. One of the expert reports provided to our working group stated:

> There are no guarantees that Ukraine's HEU would necessarily be converted to LEU and nuclear fuel. Previous consultations with Russian colleagues attest to the fact that HEU from Ukraine's warheads could be put in storage, and after certain reprocessing, could be used again for military purposes; thus, Ukraine, having received a certain amount of fuel for its NPPs in return, would become a country that enabled a violation of NPT.[49]

The two CIS treaties, Minsk and Alma-Ata, were never ratified by the Ukrainian parliament. However, it is through these treaties that Moscow for the first time had officially declared its plan with regard to Ukraine's nuclear weapons, a plan it was pursuing for the next four years until its victorious end: to transfer all warheads to Russia for which Ukraine was to receive the so-called peaceful dividend of 3,400 TVELs, and which supposedly covered the value of the HEU from the strategic weapons. As it turned out, Russia had no intention of compensating Ukraine for the nuclear material from the tactical weapons, or for the plutonium from its strategic weapons.

Since the nature of our nuclear problem required multidisciplinary expertise and information, Parliament's Resolution on Additional Measures to

The silo was a highly protected and fortified structure made of a monolithic, reinforced concrete, hermetically-sealed cylinder with a hardware compartment housing the land-based equipment for launches and the equipment supplying power for long-term support of the missile's launch readiness, with a 120-ton protective sheath to ensure that the shaft was sealed and the missile protected. In launch phase, the missile rose in six to eight seconds.

41. Missile silo for the SS-24 Scalpel

Ensure Ukraine's Attainment of Nonnuclear-Weapon State Status of April 9 1992 required that "the entire complex of nuclear disarmament issues" be reviewed with the participation of "experts from ministries, state agencies, and Ukraine's Academy of Sciences."[50] What actually happened was something quite different.

The two main Ukrainian agencies that were supposed to conduct analysis, craft a strategy for the development of the nuclear fuel cycle, and offer an action plan for the utilization of HEU and plutonium in the Ukrainian economy, were the Academy of Sciences and the Derzhkomatom. The information enabling these activities was to be supplied by the MoD and the SBU. To prepare recommendations to Parliament on this topic, our working group also created a special expert group comprised of representatives of different governmental agencies.

As the unfolding of events further demonstrated, the Academy of Sciences and the Derzhkomatom had effectively removed themselves from the strategy development process, limiting their contribution to a few short reports that offered information we already knew. The SBU provided no information officially, and the Defense Ministry just insisted on transferring everything to Russia.

The only thing Derzhkomatom managed to deliver to our working group was the agency's opinion on the expediency of exchanging nuclear warheads for NPP fuel, informing us of the nuclear fuel needs for the ten VVER-1000 reactors that produced the highest volumes of electrical energy for Ukraine. At the same time, all of the proposals that the Derzhkomatom had received from American and European companies on opportunities for using nuclear material from the warheads for the development of Ukraine's nuclear energy industry were passed over without analysis for the country's highest leadership as well as the Working Group.[51]

The Academy of Sciences had not presented even a hypothetical plan for creating a nuclear fuel cycle for Ukraine by using nuclear material from the warheads. The Academy's Institute for Nuclear Research provided only the generally known information that converting HEU to LEU posed no problem and that plutonium was a promising energy carrier. There was no analysis of how to reduce Ukraine's energy dependence on Russia, or of the political and economic consequences of such dependence, or of any methods of using nuclear material to change the situation in collaboration with leading Western companies. Neither the Derzhkomatom, nor the National Academy of Sciences provided any such analysis to the Working Group.

As for the security and defense agencies, it was obvious they were limiting themselves to the former USSR's power vertical, and had no interest in pursuing alternative expert information from abroad. Most of the data provided by Ukraine's MoD was distant from the defense of Ukraine's national interests. In one of its reports, the MoD cautioned that since the Russian factories were developing the warheads' main components without their consent, we could not engage any international organizations, especially the IAEA, in the development of technologies for the warheads' utilization in Ukraine. The MoD was thus forcing onto the Ukrainian leadership the idea that Russia had an exclusive right to dismantle warheads. The MoD blankly argued that plutonium should only be considered a substance to be destroyed, because pure plutonium was not used as nuclear fuel except in certain types of submarine reactors.[52]

The SBU provided no reports or analytical data to the Working Group. However, *Molod Ukraïny*, citing the SBU as a source, published the following figures: "The 176 nuclear missiles will provide, according to the SBU, roughly 10 tons of explosives. The bulk of it is plutonium, while no more than four tons is highly enriched uranium. Supplying all of Ukraine's nuclear power reactors requires 1,500 tons of fuel." Without citing any sources, the newspaper continued: "The highly enriched uranium from warheads is unsuitable for peaceful energy use. It needs to be 'diluted.' This requires considerable financial outlays and special equipment. The costs clearly outweigh the gains, with the latter amounting to four tons of uranium fuel in the form of energy."[53] Yet, according to various expert estimates, Ukraine's tactical and strategic nuclear weapons contained not four, but between 80 and 108 tons of HEU. Furthermore, one ton of HEU could generate 30 tons of LEU (according to GA's calculations). Therefore, Ukraine could have potentially produced 2,400–3,024 tons of fuel for its NPPs.[54]

During this time period, the majority of the executive branch, much like Ukraine's scientific community, did not yet operate in the realm of Ukraine's national interests, or see a fundamental difference between Russian and Ukrainian state development strategies. They proved unprepared for the role of a central authority of a new state and continued operating like republican powers of the former USSR. Thus, most of the reports and proposals presented to our working group regarding nuclear material at the time had to do mainly with argumentation in support of handing all nuclear warheads to Russia, rather than seeking an independent path to disarmament that would be optimally beneficial to Ukraine.

Utilization of Nuclear Material: The Russian Strategy

As soon as the Working Group was launched, I held a conference with representatives of various governmental agencies, the Academy of Sciences, and the MIC and its research branches, with the goal of developing Ukraine's options for the optimal use of its nuclear material, as decreed by Parliament. However, the general attitude among scholars and government officials at the time was shaped by the notion that they lacked, and were totally convinced that they would never have, sufficient information to formulate their own recommendations on the issue, since all state and military secrets were possessed exclusively by Moscow. Therefore, they stated diplomatically, these issues required further study, and only preliminary conclusions could be drawn at the time. As a result, instead of strategy recommendations and cost-benefit analyses (as was provided by GA, for instance), our working group received different technical calculations, and even those were only estimations, with their reports containing a massive amount of unnecessary detail. This became clear in late February 1993, and only steered the Ukrainian state's decisions in the direction of Russia's interests.

Until early 1993, most government officials and scientists did not dare to openly recommend that all nuclear weapons should be transferred to Russia. However, after the 15 January 1993 summit in Moscow, where Kravchuk and Yeltsin moved the nuclear negotiations into a bilateral format, and mutually agreed to the Russian disarmament scenario of transporting all warheads to Russia's territory, the entire executive branch and the scientific community instantaneously shifted their positions to reflect that of the president. Everything we received after that openly referred to information from "Russian colleagues," omitting entirely any offers from other countries.

The experts from the Academy of Sciences' Institute for Nuclear Research and from the National Science Center of the Kharkiv Institute of Physics and Technology (KIPT) were the most knowledgeable on nuclear material issues in Ukraine at the time. KIPT members, specifically, took part in designing some of the nuclear weapons' components in the past. I communicated with them the most. They formed the bulk of the Working Group's expert committee tasked with advancing opportunities for utilization of nuclear material extracted from Ukraine's warheads. The committee was headed by the Institute's director, academician Viktor Zelens'kyi.

42. Viktor Zelens'kyi

Viktor Zelens'kyi (1929–2017), academician specializing in nuclear physics and energy industry. He was the director of the National Science Center Kharkiv Institute of Physics and Technology (KIPT) from 1981 to 1996. Zelens'kyi was also a highly respected specialist in solid-state physics, radiation phenomena, and radiation material science. In the early 1990s, he chaired the working group to determine Ukraine's compensation conditions for nuclear material. Based on his proposal, the decision was made to request Russia to supply nuclear fuel for Ukraine's NPPs worth over US $901 million as compensation for nuclear material contained in warheads transferred to Russia from Ukraine. He has written nearly 370 scientific papers and has 70 discoveries to his name. For many years, he was director of the Southeastern Research Center of Ukraine's National Academy of Sciences.

43. *Transporting the SS-24 to its launch position on a modified MAZ-537, designated for transporting containers with missiles and loading them into a silo*

From the onset of our exchange in 1992, the KIPT leadership promoted the idea that Russia was Ukraine's best and sole option for collaborating in the sphere of nuclear material utilization. Of all the analytical reports they provided, none paid any attention to strategic issues emerging from Ukraine's energy security needs or trends in the global nuclear energy industry. Instead, there was a sea of information about the technological side of nuclear plant operations: constructing warehouses and storage facilities, manufacturing containers, transportation, HEU and plutonium storage, among others. In other words, they chiefly raised technological issues concerning the existing plants that manufactured nuclear weapons, rather than providing expert opinions on what Ukraine actually needed for the conversion of nuclear warheads into the "atoms of peace." Even the recommendations by them concerning the conversion of the weapons-grade uranium and plutonium provided no understanding of which option was optimally beneficial for Ukraine, either strategically or economically.

The reports our working group had received also contained countless financial estimates. They were convoluted, relying on different calculation methods: part of the data was provided in Russian rubles, another part in Ukrainian karbovantsi, neither of which had a stable exchange rate, and yet another part, in US dollars.

This "smokescreen" of minutiae did not only fail to provide any answers to the questions of strategy, but also diverted attention from the topic of plutonium utilization, and later from the topic of plutonium in general, the value of which in Ukraine's strategic weapons was estimated in tens of billions of dollars. All financial estimates we received focused exclusively on the use of HEU, and its worth in Ukraine's strategic weapons alone was estimated at up to US $6.8 billion, as shown above in table 2. What kind of information did Ukrainian scientists give to us about the two immensely valuable elements contained in Ukraine's nuclear weapons, in order to inform the country's subsequent nuclear strategy?

How to Handle the Plutonium

While the need for converting HEU into fuel rods was evident to everyone (more on this later), there was no consensus on what to do with plutonium. When analyzing information supplied by different scientific institutions, the Working Group revealed that there was no unified position among Ukrainian scientists on what to do with the artificial material produced in special heavy hydrogen reactors from uranium-238. In fact, expert opinions on this issue were diametrically opposed. KIPT experts from the onset considered the plutonium conversion program inadvisable. This was the logic and data behind their assessment.

First, plutonium is toxic and difficult to store; it is a highly hazardous substance, both from the military perspective, and also in terms of its potential effect on human beings. Dispersing one to two grams of plutonium renders one square meter of land unsuitable for human habitation.

Second, weapons-grade plutonium has no market value, as its sale is banned by the NPT. At the time, reactor-grade plutonium was valued at US $30,000 for one gram (for comparison, reactor-grade uranium was valued at $0.65 per gram). Zelens'kyi's group wrote that pure plutonium could only be stored in Russia. Interestingly, without naming the actual amounts of plutonium, the Russians were prepared to charge Ukraine for its storage, as Zelens'kyi conveyed: "Russian experts cannot reveal the amount of plutonium contained in strategic nuclear weapons located on the territory of Ukraine at this time, as it is classified information. However, it was verbally confirmed that the cost of storing all plutonium removed from these weapons would amount to 400 million rubles per year in fourth quarter 1992 prices."[55]

Third, using plutonium for our own needs in the form of mixed oxide (MOX) fuel, which is produced by blending plutonium and uranium, would require a significant refitting of the existing reactors, which would make the resultant fuel four to six times more expensive than the conventional uranium fuel. Therefore, the Kharkiv experts concluded the best option for Ukraine was to transfer all warhead plutonium to Russia without any conditions, and only in exchange for their disassembly of Ukrainian warheads, or, simply stated, to get rid of it.

The Working Group based its position on a much more comprehensive collection of information that gave us reason to believe the situation was not quite so clear cut. Back in June 1992, at the aforementioned seminar in Moscow, I heard the opinion from Vladimir Sukhorukin, the laboratory head at the Kurchatov Institute (Russia's leading institution in nuclear energy research and development), that accumulating plutonium required significant sacrifice, literally and figuratively, and to simply dispose of it would be a crime.

The most promising method of handling weapons-grade plutonium after extracting it from a nuclear warhead at the time was storing it without conversion, and then gradually turning it into an energy source. Russia took this path. While trying to prove Ukraine's incapability of storing plutonium, Russia demanded money from the US government for the construction of a long-term plutonium storage facility; the US Embassy confirmed that the funds were eventually supplied.[56] This meant that the US could have potentially provided

44. *Mock-up of a Unified Command Post (UCP)*

The purpose of the UCP was to allow for directing and controlling the launch of SS-24 Scalpel missiles, maintaining constant temperature and humidity, and protecting personnel from harm. The UCP weighed 125 tons, was 108 feet (33 meters) long and 10.8 feet (3.3 meters) in diameter. When staffed and equipped, it could operate autonomously for up to 45 days.

similar assistance to Kyiv for the subsequent storage of plutonium on Ukraine's territory under international control.

Additionally, plutonium could have been used as collateral, which was suggested by GA. This would have necessitated neither expenditures for storing plutonium elsewhere, nor the building of its own storage facilities, yet would have provided Ukraine with resources to create its own nuclear fuel cycle.[57] However, Ukrainian scientists never even mentioned this as a possibility.

At this time, in the early 1990s, the global nuclear energy industry was already using plutonium in the form of MOX fuel in France. France also offered CIS countries its services in building nuclear arsenal dismantling facilities in the span of one to two years, as I wrote to President Kravchuk on 22 April 1992.[58]

The fact that Belgium, France, Germany, and Japan were already using mixed-fuel technology was confirmed in a report produced by the Academy of Science's Interdepartmental Commission for Communication with the IAEA, signed by its head and director of the Institute for Nuclear Research, academician Ivan Vyshnevs'kyi, on 6 April 1992. Vyshnevs'kyi wrote that mixed-fuel technology opened up possibilities for utilizing plutonium in the existing nuclear reactors and, therefore, he considered this direction rather expedient; he wrote: "given that the warheads already contain the ready-made plutonium, the advantages of using it in the nuclear energy industry are obvious."[59] However, Ukrainian scientists did not elaborate any more on this statement.

By then, we also had information that Japan had begun accumulating reactor-grade plutonium for use in special reactors that were being constructed in collaboration with the US in the early 1990s. This new type of reactor was to operate using plutonium. According to the press, in 1992, Japan had already accumulated 400 tons of plutonium.[60] Ukraine had the opportunity to join the US and Japan in the process of developing methods to use weapons-grade plutonium for peaceful means. As I already mentioned, in January 1993 the GA approached Ukraine with a proposition for potential collaboration in this field.

According to a 27 January 1993 article in the Japanese newspaper, *Nihon Keizai Shimbun*, for instance, Japan's MFA and the Japan Science and Technology Agency were seeking collaboration opportunities for pursuing further development in the area of storage and utilization of plutonium released in the process of disarmament; in theory, Ukraine, as an owner of plutonium, could also have been involved in the process.[61] However, again, these opportunities were overlooked by Ukrainian scientists and the executive powers. No options other than to transfer everything to Russia were ever mentioned.

How to Handle the HEU

In 1992, the first working documents on nuclear material supplied by KIPT at my request outlined four options for handling Ukraine's HEU.[62] The simultaneous use of dollars and rubles in their calculations makes it impossible to thoroughly evaluate their recommendations. As with plutonium, the Institute provided no comparative evaluation of foreign companies' proposals with regard to the HEU; it only mentioned services offered by Russian companies. The only conclusion that could be drawn from these reports was that everything needed to be transferred to Russia, and no other options were viable. These documents contained the following in a nutshell.

Option one suggested reprocessing the HEU into fuel rod assemblies. The 40,256 kg of HEU contained in Ukraine's 130 strategic liquid-propellant SS-19 missiles, which were the first to be eliminated according to MoD's recommendations, were used as the basis in KIPT's calculations. Out of this quantity, Ukraine could receive 2,119 fuel rod assemblies produced in Russia, which would cost 236 billion rubles. However, since Ukraine would simply act as a purchaser of reprocessing services, it would also have to assume all cost of transportation and insurance for the HEU and LEU. Thus, the total cost associated with this option would be roughly 345 billion rubles. However, producing the same number of fuel rod assemblies from Russia's natural uranium (rather than from Ukraine's HEU as in the above estimate), would cost 322 billion rubles.[63] It was impossible to grasp Ukraine's potential advantage in processing 40,256 tons of its HEU in Russia from this analysis. Similarly, the report provided no understanding regarding the balance between the value of the material, which was estimated at roughly US $4 billion by other experts, and the cost of production and transportation from this report. Finally, in reading this report one could make no comparison of the Russian propositions to any others because they simply were not mentioned.[64]

Option two proposed Ukraine's own conversion of HEU to LEU, and its subsequent sale on the international market in accordance with US-Russian Uranium Deal prices for the afore-mentioned sale of 500 tons of HEU over a twenty-year period. According to Zelens'kyi's report, the price of one kilogram of LEU was US $650; the cost of processing warheads and HEU would be 25.3 billion rubles, or roughly US $39 million; storing HEU would cost additional 10.6 billion rubles. Overall, the revenues from the sale of LEU would constitute roughly US $639 million. However, by taking into account insurance expenses, the sum would be cut in half to about US $320 million.[65] The report did not explain the point of selling LEU on the international market, when there was a need for NPP fuel domestically.

Option three provided cost estimates for building Ukraine's own facilities for HEU-to-LEU conversion. According to KIPT data, a plant with an annual processing capacity of 30 tons of HEU would cost US $150 million to build. This, by the way, contradicted the data we received from GA, which was offering to build the same facility for US $20–30 million. Therefore, the report concluded, developing such a program would be impractical: with Ukraine's mere 64 tons of HEU, it would not be competitive in the already saturated nuclear fuel market.[66] Again, why Ukraine would want to sell the fuel, while needing it for its own NPPs, was not explained in the report. There was one more curious detail. In this report the amount of HEU in Ukraine's warheads was listed as 64 tons. However, in the final documents the amount inexplicably turned into 60 tons.[67]

Incidentally, at this point, we already saw publicly broadcast opinions that selling nuclear material to other countries was inadvisable. These were the inferences drawn in the Russian press in response to EBRD president Jacques Attali's proposal to exchange the reduced warheads for cancellation of the USSR's debt of US $60 billion. Thus, the Russian press concluded, selling nuclear material from warheads was disadvantageous.[68] A year later, this conclusion was confirmed to Zelens′kyi's expert committee by its Russian counterparts: "The Russian Federation [...] holds that the sale of HEU is exceedingly inadvisable. It should instead be used in nuclear energy development programs, or left for the country's descendants to use in the future, when fuel resources become limited."[69] However, as has already been mentioned, Russia had just signed a deal with the US to sell 500 tons of HEU.

Option four anticipated the possibility of storing HEU in Ukraine, but with a caveat: this could happen only if there were a well-defined program for its use within five to ten years, stressing that the idea of HEU and plutonium utilization required additional research.[70] The scientists once again ignored the fact that the storage of nuclear material under the IAEA's control could have been used by Ukraine as collateral for credit to finance the development of its own nuclear fuel cycle.

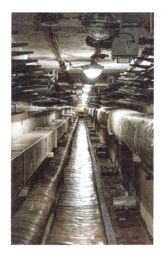

45. Passageway to the UCP, lined with conduits for power, temperature control, communications, and other infrastructure

Joint consultations between Ukrainian and Russian experts became formalized and intensified after the first round of their respective government delegations' talks on nuclear weapons and after the creation of expert groups. Summarizing the results of the consultations held in Moscow on 16–18 February 1993, Zelens′kyi sent a document to our working group, entitled "Report on the Results of Negotiations with the Russian Delegation on Issues of the Dismantling, Transportation, and Elimination of Ukraine's Strategic Nuclear Warheads." In the accompanying note, he stated that during the negotiations, the Ukrainian delegation received "calculations (made by Russian experts) of the cost of the reprocessing of strategic nuclear weapons" and that "the calculations offer two options: first, the reprocessing of HEU to produce TVELs and fuel rod assemblies for VVER-1000 reactors; and second, the storage of HEU."[71]

The report specified that both sides "had come to an agreement to consider the following framework for resolving issues related to the elimination of nuclear warheads and the use of material." Peculiar in the schemas' presentation was the fact that each proposal opened with language reflecting exclusively Russia's position, such as: "the Russian Federation determines," "the Russian Federation provides," "the Russian Federation agrees," "the Russian Federation deems." The essence of the proposal was that Russia would determine the amount of nuclear material in Ukraine's weapons; it would charge Ukraine for the transportation and dismantling of the warheads; plutonium would be transferred to Russia for storage, but without the right of its

46. *Map of 43rd Rocket Army unit locations*

return to Ukraine; and finally, that Ukraine would be "advised" to join Russia in its nuclear energy industry development program.[72]

On 23–25 February 1993, an interdepartmental conference took place in Kharkiv. The expert committee formed a final set of recommendations for handling Ukraine's nuclear material based on the results of this conference. The attendees from various departments and agencies reviewed all information provided by the Russian experts and then discussed three possibilities for using nuclear material from Ukraine's strategic weapons. As a side note, we already had propositions from General Atomics, but as the recommendation document attested, they did not make it onto the committee's tender list. Thus, this time around the following three options were presented as a result.

The first option was to use the HEU to produce nuclear fuel for Ukrainian NPPs. According to Derzhkomatom's report, Ukraine would need to import 576 fuel rod assemblies for its ten VVER-1000 reactors from Russia annually. An additional three VVER-1000 reactors were under construction, and they would require a further 489 fuel rod assemblies to become operational. Russia was offering 3,400 fuel rod assemblies for VVER-1000 reactors in exchange for Ukraine's HEU extracted from strategic weapons. Conference attendees concluded that having assessed the value of fuel rod assemblies offered by Russia against global market prices, the added revenue for Ukraine from selecting Russia's offer would amount to US $1.8 billion.[73]

The second option was to store HEU in Russia. Conference attendees concluded that HEU was a strategic dual-purpose material. As for peaceful uses, this was an excellent material that could be used for developing new types of fuel for various kinds of reactors. Consequently, the most practical option for Ukraine could be to use HEU without downblending it. However, given the state of Ukraine's economy and in the absence of a well-defined nuclear industry development strategy, compounded by the expenses incurred from Russia's

47. *Residential quarters of UCP personnel*

warhead dismantling operations (costing eight billion rubles) and storage of HEU (costing an additional three billion rubles annually), the committee unequivocally recommended following the aforementioned proposal of exchanging HEU for fuel rod assemblies, with the possibility of reserving a portion of the HEU for the time when Ukraine finally determined its strategy in the sphere of nuclear energy.[74]

The third option entailed assessing the sale of HEU on the global market. Considering the fact that only reactor-grade uranium (U-235 of less than 20%) could be sold on the global market, financial calculations were provided for the sale of LEU (4.4% U-235), rather than HEU. As a result of all operations associated with the conversion of HEU to LEU, as well as the sale of nuclear fuel on the global market, Ukraine could only receive approximately US $1 billion for its HEU. However, as I noted earlier, this is where the 64 tons of HEU referenced in the original set of recommendations had turned into "nearly 60 tons." In the following text, the document reiterates the notion of the extra revenue Ukraine would receive by selecting Russia's offer to manufacture fuel rod assemblies, as presented in the document's first option. Accordingly, Russia's price for one rod assembly constituted US $265,000. To purchase the same assembly on the global market would cost Ukraine US $800,000. Therefore, if Ukraine opted for the 3,400 fuel assemblies to be produced in Russia, the order's total price would be only US $901 million; however, the same order on the international market would cost Ukraine US $2.72 billion. As a result, by ordering fuel rod assemblies from Russia, Ukraine would achieve US $1.8 billion in hypothetical

48. *Control panel of the security and defense shift operator in the central missile launch complex*

savings. The committee recommended that the Ukrainian leadership use Russia's services for Ukraine's nuclear power industry needs.[75]

Every conference attendee, without exception, had signed the meeting protocol, agreeing with the committee's recommendations without reservation.[76] Thus, the plan which the scientific community and other experts recommended to Ukraine's executive branch entailed transferring all nuclear warheads to Russia in exchange for 3,400 fuel rod assemblies.

The Final Touches

All of the reports concerning the use of nuclear material that were produced by the KIPT (with Russia's direction) and provided to our working group essentially documented Russia's aspiration not only to appropriate Ukraine's nuclear warheads, but also to reduce the material compensation it owed to Ukraine to a symbolic minimum. Here are two examples:

When discussing its capacity for reprocessing Ukraine's HEU into fuel, Russia confirmed that it could supply the production of up to 400 fuel rod assemblies per year, the first of which would be delivered in the two to three years after the warheads' disassembly. All the while, the Russians also noted, as if in passing, that "the expenditures associated with plutonium operations during the warheads' disassembly shall be added to the cost of the fuel rod assemblies."[77] As is well known, plutonium was not used in the production of fuel assemblies for Ukraine's NPPs. Moreover, in all of its reports, the Russians were persistently persuading Ukraine that plutonium was not a viable energy source; therefore, they had no intention of compensating Ukraine for its value,

49. *Transport and trailer with container for SS-24 missile*

or of returning it in the form of weapons-grade plutonium, as Ukraine was a nonnuclear-weapon state. Hence, along with planning to appropriate the most valuable material from Ukraine's nuclear weapons arsenal, Russia also intended to let Ukraine assume all expenditures related to the removal of plutonium from the strategic warheads.

Another example was the supposed error in at least six tons of HEU. On 16 February 1993, Deputy Minister of Energy Vitalii Konovalov, the head of the working group of the Russian Federation's delegation wrote that the HEU removed from Ukraine's strategic nuclear weapons according to their data amounted to 60 tons, even though in their 1992 reports the number was 64 tons, which could produce 3,400 fuel assemblies for VVER-1000 reactors by enriching the 4.4-percent uranium.[78] However, the 23 April 1993 report provided by Viktor Havryliuk, the deputy director of the Academy of Sciences Institute for Nuclear Research, shows different calculations; namely, that the production of 3,400 fuel rod assemblies required that only 54.1 rather than 60 tons of U-235 uranium be added to the natural uranium. Furthermore, if Ukraine's own natural uranium were used in the production of these 3,400 fuel rod assemblies, instead of Russia's natural uranium, the production cost would be reduced by 105 billion rubles, or an equivalent of roughly US $182 million.[79] Once again, these significant details were overlooked by the Ukrainian specialists.

Thus, despite the obvious contradictions in Russia's proposals, the Kharkiv experts had all agreed that the Russian Federation was the party determining the strategy of utilization of nuclear material in Ukraine's nuclear weapons. The Russian Federation was also to estimate the cost of transportation and storage of nuclear material, including the storage of plutonium that was never to be returned to Ukraine, and the warhead disassembly and liquidation of certain components. All expenses associated herewith were to be incurred by Ukraine. This was the essence of the vision for the processing of nuclear material, which was shared by the majority of Ukrainian scientists and nuclear experts.

What to Do with Nuclear Material: Conclusions

Summarizing the total course of events related to the development of proposals by Ukrainian scientific institutions on the subject of what to do with Ukraine's nuclear material, the following conclusions could be drawn:

1. Ukraine's executive branch did not aim to develop its own strategy for the elimination of Ukraine's nuclear weapons or the use of nuclear material for the needs of the country's economy, contrary to the decisions of Parliament.

2. Ukraine's Academy of Sciences, despite its significant intellectual resource base, did not manage to form a vision of its own regarding the fate of the world's third largest nuclear arsenal, but instead accepted the Russian plan. This was different from what the RSA did, when it independently destroyed its nuclear weapons.

3. Derzhkomatom failed to develop a national strategy for the advancement of Ukraine's nuclear power industry, a strategy that would seek to reduce the dependence of Ukraine's NPPs on Russia. Subsequently, this dependence opened doors to Russia's persistent coercion, forcing Ukraine to develop its nuclear power industry according to Russia's scenarios and political interests.[80]

4. In the process of evaluating options for Ukraine's utilization of nuclear material, all departments and state agencies limited themselves to communication within the old Soviet hierarchy, and therefore only focused on Russia's proposals. Consequently, Ukrainian institutions, whether deliberately or unintentionally, became the conductors of Russian interests, and thus advocated for the transfer of all warheads to Russia.

5. The effectiveness of proposals put forth by Ukraine's executive and scientific institutions can be assessed as follows:

   5.a. The ultimate transfer of all nuclear weapons to Russia had led to significant economic losses for Ukraine. Giving up nearly US $100-billion worth of nuclear material, including HEU and plutonium, Ukraine received compensation in the form of 3,400 fuel rod assemblies worth US $901 million.

   5.b. The supposed advantages of cooperation with Russia in the energy production industry proved to be illusory during the energy crisis of December of 1993. Russia's threat to terminate the supply of fresh nuclear fuel and the acceptance of spent fuel was used as a lever to pressure Ukraine to lift its reservations to START I, to sign the Trilateral Statement, and to join the NPT as a nonnuclear-weapon state.

50. Viktor Chernomyrdin (right), 1994

Viktor Chernomyrdin (1938–2010), Russian prime minister between 1992 and 1998. He was also Russia's ambassador to Ukraine from 2001 to 2009. After graduating from the Kuibyshev Industrial Institute (now Samara Polytechnical Institute), he worked in the Orsk Municipal Committee of the Communist Party of the USSR from 1966 to 1973. Prior to moving to Moscow in 1978, he directed the Orenburg Gas Processing Plant. He then worked in the CPSU Central Committee. From 1985 to 1989, he was the minister of gas industries of the USSR, and in 1989–1992 he was head of the Gazprom. While prime minister, he was once made acting president of the Russian Federation for several hours by special decree while President Yeltsin was undergoing heart surgery. In 1995–1999, he led the political bloc Our Home–Russia (NDR). In 1999, he was elected to the State Duma. From 2001 on, he was the ambassador to Ukraine and President Putin's special representative for developing trade and economic ties with Ukraine. After 2009, he was responsible for economic cooperation with CIS member countries.

In the end, Russia got what it wanted, which is what it had been trying to achieve ever since its first mention of 3,400 fuel rod assemblies as compensation for all of Ukraine's warheads in early 1992. Moscow saw the goal, but unfortunately Ukraine did not create obstacles to its realization, either on the part of Ukrainian scholars or in the executive branch.

Confidentiality in the Hallways of the Ukrainian Government

In a recent interview, a journalist asked me whether Moscow had any knowledge of Kyiv's attempts to develop its own disarmament strategy, so that it could keep a step ahead of the process, so to speak. I replied that in the early 1990s, Ukraine did not have any systems ensuring confidentiality of any information, whether it be decisions, or plans, or negotiations. My answer surprised the journalist. However, in the first years of independence, those were Ukraine's realities.

After the USSR's collapse, the question of confidentiality emerged almost immediately. Ukrainian KGB officers were the first to raise the subject. The parliament had created a commission to reform the KGB into the SBU, and I took part in that process. (See chapter 1.) From the beginning, some of the patriotically-inclined members of the security services came to us, stressing that the first thing we needed to do was to cut off the special telecommunication channels with the Kremlin, because

51. Protests against President Kuchma, 2001

all information was leaking into Moscow through them. They did not elaborate on technical details, but simply recommended cutting it off, for only then could we count on a modicum of confidentiality in our state politics. However, this was not accomplished until approximately the year 2000.

This special telecommunications was a government wire telephone connection with a built-in safeguard to prevent third-party tapping. All high-ranking and even some lower-ranking government workers had official telephones on their desks, from the country's president to the Cabinet of Ministers staff. However, this official communication system also had a reverse function; it was a wiretapping mechanism that recorded all conversations taking place in the offices. During the Soviet era this system was naturally connected to Moscow, and nothing had changed since Ukraine declared its independence.

The fact that Russia essentially controlled all conversation taking place in Ukraine's governmental offices through this special communication system was confirmed for me, ironically, through Leonid Kuchma, immediately after his resignation from the office of prime minister in September of 1993. He reminded me of one of our conversations, when we had discussed creating at least some sort of adequate control system over nuclear warheads maintenance work on Ukrainian territory. Immediately after I left his office, Kuchma said that Russian Prime Minister Viktor Chernomyrdin called him on the phone, and in some colorful and blunt language, demanded that we not even attempt it, because before we could even think of anything, Moscow was already one step ahead of us.

Another confirmation of Russia monitoring the highest Ukrainian officials' offices was the so-called cassette scandal. In September 2000, an audio recording from Kuchma's office was released, most likely not by Kuchma's bodyguard with Major Mykola Melnychenko's recorder, as was popularly accepted, but through the special governmental telecommunication system, which had not yet been disconnected from Russia.[81]

Whereas all verbal information was being transmitted through wiretapping, the monitoring of documents, including those marked confidential or classified, was accomplished through a cadre policy. All offices, especially those of the highest-ranking politicians, had specially trained personnel who remained there after the fall of the USSR. It is a known fact that the prime minister's and the presidential administration's secretariat stayed virtually unchanged at least until 2004 or 2005. The front offices were staffed exclusively by these men, and the staffing had not changed since the early 1980s—in other words, the Soviet era. Thus, whatever Kyiv was developing in the sphere of nuclear disarmament was already known to Moscow in real time. Therefore, it was very easy for Russia to neutralize any plans or prevent any movement by Ukraine that went against the plans of the Kremlin.

Russia Lays Its Bets on Missile Forces

By spring 1992, it had become obvious that the idea of the United Armed Forces of the CIS had ceased to exist. Marshal Yevgeny Shaposhnikov, who had been serving as the commander-in-chief of the United Armed Forces since early 1992, acknowledged the impossibility of maintaining this structure in his address to the Russian parliament. His statement indicated that Russia was accepting the inevitable truth that there would be multiple armies on the territory of the former USSR, instead of one army under a unified command.

The new situation elevated the significance of two critical issues for the Russian Federation. The first had to do with the division of the former USSR's military assets (above all, the Black Sea Fleet's nuclear weapons), since the word *ours* was left behind in the past, along with the United Armed Forces of the CIS. The second concerned the formation of the Russian Army. Shortly thereafter, the Russian parliament was presented with a new military doctrine, as well as a plan to create the new Russian army, consisting of five types of forces, before the year 2000.

Russia placed its main bet on the strategic missile troops. In early November 1992, the Russian newspaper *Izvestiia* announced, "Russia is commencing a reconstruction of its strategic forces," considering them the cheapest and most effective.[82] Naming the first deputy minister of defense, Andrei Kokoshin, as a source, the newspaper termed the missile troops "the most economical type of armed forces," for they were capable of resolving sixty to 100 percent of strategic combat tasks, yet required only five to six percent of the total defense budget.[83]

At the same time, in view of the ongoing reductions in strategic arms, the Russian Federation's MoD expressed its intention to review the country's entire defense system and to come up with a new development plan for the strategic missile forces, making them the foundation of national security, including determining directions for research and design. The strategic forces' overhaul would also be complemented with the SS-25 mobile ballistic missile systems, which were the very units being transferred from Belarus. That is to say, Belarus was handing over its missiles to Russia for destruction, but instead Russia was using them to equip its own army. Subsequently, the Russians publicly announced that they developed "a fundamentally new standardized missile of the next generation."[84]

A scandal broke out in spring 1992 surrounding Russia's supplying rocket engines to India. The international community considered this a violation of nonproliferation principles, which had to do with the technology aimed at the production of weapons of mass destruction. On 14 May, *Izvestiia* essentially published Russia's official position on the matter. This article with an unambiguous title, "We Do Not Have to Ignore Military Purchase Orders," was signed off by a representative of the Russian MFA's legal department. It showcased a rather skeptical attitude towards the idea of nonproliferation, calling to review it within the context of the shifting times, hinting that some ideas simply became obsolete.[85]

Likewise, the Ukrainian military understood the cost-effective nature of the missile forces. This information was actually found in the Ukrainian press a number of times. For instance, in November 1992, Major General Tolubko wrote:

52. *Yevgeny Shaposhnikov*

> At first glance, nuclear arms and their delivery mechanisms may appear rather expensive and exorbitant for the budget of our young state. However, a basic calculation of the missile systems' manufacturing and operating costs attests to the fact that the political and economic benefits of having nuclear weapons are far greater. The cost of maintaining strategic nuclear forces constituted 6–8% of the Soviet Armed Forces' total budget. At the same time, the annual expenditures on operations and combat training for the general armed forces exceeded similar expenses for the maintenance of strategic nuclear forces several times over.[86]

General Tolubko also emphasized the start-up opportunities Ukraine had at the time, contending:

> Ukraine has first-class missile technology and strategic aviation right now. It has a cadre with high professional competencies in the use of these weapons. We have plants producing missile equipment using the most advanced technologies. To get rid of all this for political gain would be, at the very least, an extremely careless and premature decision.[87]

In the Russian Federation's estimation, using Belarus's strategic missiles instead of destroying them or selling rocket engines to India were not actions undermining global peace; however, Ukraine's intention to develop its own missile industry and replace nuclear weapons in its system of national security with precision-guided munitions was cause for increased concern. *Moskovskie novosti* wrote at the time: "According to Ievhenii Sharov of the Ukrainian Institute of World Economy and International Relations, the local defense industry and the state leadership are seeing an overall strengthening of the position of the hawks. This concerns the creation of nonnuclear precision-guided ballistic missiles, as well as the development of space programs.[88]

Yevgeny Shaposhnikov (1942–2020), aviation marshal (1991), the last minister of defense of the USSR (August–December 1991), commander in chief of the Joint Armed Forces of the CIS (February 1992–September 1993). He graduated from the Hrytsevets Higher Military Aviation School in Kharkiv in 1963 and from the Gagarin Air Force Academy in Moscow in 1969. From 1971 to 1975, he was deputy commander of the Air Regiment for Political Issues and subsequently a regiment commander. Starting in 1975, he served in the Prykarpattia and Odesa military districts. In 1990 he became deputy defense minister of the USSR. During the August Putsch of 1991, he proposed to the Soviet defense minister Dmitry Yazov that military forces be removed from Moscow. As soon as the attempted coup failed, Gorbachev appointed Shaposhnikov the defense minister of the USSR. On 21 December 1991, he was promoted to commander in chief of the Soviet Armed Forces "until such time as they are reformed." On 14 February 1992, he was appointed commander in chief of the Joint Armed Forces of the CIS. In summer 1993, he became the secretary of the Russian Security Council. Between 1994 and 1996, he represented Yeltsin at the state-owned defense company Rosvooruzhenie. From 1997 to 2004, he was the Russian presidential advisor on space and aviation.

A Foray: To Force Ratification

53. *Vasyl' Durdynets', 1994*

Vasyl' Durdynets' (b. 1937), first deputy speaker of the Parliament of Ukraine from January 1992 to April 1994. He was the coordinator of the working group preparing for the ratification of START I and the NPT during the period 1993–1994. Prior to the 1991 August Putsch, he was a member of the CPSU. In 1960, he graduated from the Law Department at Ivan Franko State University in Lviv. From 1958 to 1978, he worked in the Komsomol and Communist Party organs of Lviv, Moscow, and Kyiv. From 1978 to 1991, he was the deputy and subsequently first deputy minister of internal affairs of the Ukrainian SSR. After the Chernobyl accident in 1986, he was appointed to run the operational staff of the Ministry of Internal Affairs to provide law and order in the accident zone. From 1990 to 1996, he was a Ukrainian MP. Durdynets' belonged to the group of cautious proponents of effective disarmament. As a seasoned Communist Party apparatchik with more than 30 years of experience, he habitually sided with the official position of the president's administration, primarily that of Kravchuk, but as an MP slowly shifted to supporting all the initiatives of the working group.

In fall 1992, an immense political onslaught ensued to force Ukraine to ratify START and the NPT. It unfolded on the backdrop of Russia's fundamental change in position: Moscow at this point had abandoned the rhetoric of equality and had switched openly to establishing control over all of the former USSR's nuclear weapons. This change had become obvious after the Ukrainian MPs' official meetings with representatives of the Russian parliament and MoD to discuss nuclear disarmament issues in Moscow on 3 and 4 March. Pavel Grachev, the newly appointed Russian defense minister declared: "Russia intends to take under its control all strategic weapons of the former USSR."[89] Shortly before that, CIS Unified Armed Forces Commander Marshal Shaposhnikov also confirmed this position in his interview with Russia's Central Television.

To achieve control Russia acted bluntly, as always: it simply tried to force Ukraine to transport its entire nuclear arsenal to Russian factories as soon as possible. In the process, Russia had no regard for the fact that Parliament had not yet made any decisions on the timeline, the quantities, or the methods of arms liquidation, and, therefore, Ukraine had no legal basis for any actions with regard to its nuclear weapons.

After returning from Moscow on 7 September, I sent a written statement to Parliament's first deputy Speaker Durdynets' who was the curator of our working group. Within the document, I attempted to draw his attention to the change in Russia's political course with regard to the former USSR's nuclear arms, and suggested an official meeting concerning Ukraine's political stance with these new circumstances. Durdynets' supported my idea and the meeting was called quickly with the ministers of defense and foreign affairs, with the representatives of the MIC and president's administration in attendance. Despite working towards the initiatives accelerating the ratification of START and the NPT, the meeting resulted in an outcome contrary to its original goals.

The Ukrainian MFA was spearheading the effort to quickly ratify START and the NPT, with support from the Derzhatomnahliad, and also by this point, from the MoD. As was already discussed, these agencies were the first to present the nuclear documents for ratification by Parliament in mid-February 1992. Besides this, Kravchuk and Pliushch were externally pressured through diplomatic channels with the support of the heavy artillery of the Western and Russian media.

The tempo at which Kravchuk and Pliushch were receiving foreign delegations at that time could be evidenced by the numerous articles in *Holos Ukraïny*.[90] Every foreign visitor asked but one question and demanded a concrete answer: When would Ukraine ratify the nuclear treaties? The greatest value of the protocol publications to us now is that they allow us to reconstruct official positions of all Ukrainian leaders on the topic of nuclear disarmament.

On 17 November 1992, in Kyiv, Pliushch held a meeting with US General John Shalikashvili, the Supreme Allied Commander Europe. Pliushch told Shalikashvili that Ukraine had no intention of altering its existing course towards disarmament and neutrality. However, to guarantee the country's security, Ukraine needed to develop relationships with other countries and organizations, including NATO. Shalikashvili informed Pliushch that NATO was undergoing a conceptual change. This should have been received as an invitation to join the Alliance for security guarantees, but the implicit invitation was left without a response. On the contrary, the *Holos Ukraïny* correspondent drew a surprising conclusion from this meeting, stating that Ukraine used Shalikashvili's visit to reaffirm its neutral status, "notwithstanding any obstacles." The newspaper reported that Shalikashvili's meeting with Kravchuk was similar in content.[91]

On 19 November at a press conference in Germany, Kravchuk declared for the first time that "START ratification will commence in the Ukrainian Parliament in late December or early January next year." He stressed that Ukraine was not changing its strategic course towards eliminating nuclear weapons, but stipulated that before that could happen the country had to resolve compensation issues, as well as strengthen Ukraine's security guarantees from other nuclear states—most importantly, from Russia.[92]

On 22 November 1992, Kravchuk received two American senators, Sam Nunn and Richard Lugar; they spoke behind closed doors. According to the president, in response to Ukraine's state agency Ukrinform, the senators were interested in two issues: the first concerned economic reforms, approving of Kuchma's appointment as prime minister; the second was about the ratification of START. Nunn and Lugar also reported to Kravchuk that the US was prepared to provide Ukraine with US $175 million for nuclear disarmament. Kravchuk responded that START was already brought up to Parliament for ratification. Judging from the developments, he was pleased with the promised amount of money, evidently seeing it as a generous offer. He expressed hope that the sum would "facilitate ratification," along with his conviction that "If Russia also takes similar steps, that is, accedes to the treaties and adopts the relevant documents, I am sure that the process will proceed more expeditiously."[93]

Kravchuk reiterated Ukraine's need for security guarantees, and he even mentioned "participation in European structures," but his subsequent words made it clear that he had little idea

After the dismissal of the working group in March 1994, he did not criticize the policy of forced nuclear disarmament in his speeches, but he regularly used the arguments that had originated from the working group. In 1995–1997, he served as vice prime minister of Ukraine on matters of state security and emergencies, then as first vice prime minister and, subsequently, as acting prime minister of Ukraine. In 1997, he was appointed chairman of the Anti-Corruption and Organized Crime Coordinating Committee under the office of the president of Ukraine, and director of the National Bureau of Investigation. He was promoted to general in the Internal Affairs Service of Ukraine in 1997. In 1999–2002, he served as the minister of emergency management.

of what these guarantees should entail or against whom Ukraine would need protection. Moreover, Kravchuk seemed to sincerely believe that Ukraine's surrendering of nuclear weapons would instigate a global disarmament movement which would result ultimately in the destruction of all nuclear weapons in the world. Kravchuk announced in *Holos Ukraïny*:

> At the meeting, I articulated the following proposal: Let the Western countries declare that Ukraine and other states that are voluntarily giving up their nuclear weapons would now have guarantees, so that other countries could not use nuclear weapons against them. Moreover, these guarantees would have to remain in effect until the complete elimination of all nuclear weapons in the world.[94]

On 18 December 1992, Hans Blix, the IAEA director, arrived in Kyiv. First, he met with Pliushch. Blix's main thesis entailed not understanding why solving problems associated with nuclear disarmament was preventing Ukraine from acceding to the NPT, "because one does not preclude the other." Pliushch responded to this with one of his legendary aphorisms: "When the frosts are in the forecast, it is hard to give up the coat." He reassured that Ukraine would join the NPT, while asking "to not rush it," and instead, concentrate on the problems that prevent Ukraine from immediate accession. Blix met with Kravchuk the same day. Both the president and the Speaker reiterated that the problems with eliminating nuclear weapons were of a "technical, financial and legal" nature, and were in no way linked to the predetermined course towards disarmament, but only to the methods of implementation.[95]

On 24 December Kravchuk spoke with President Bush on the phone. START was the only topic covered by *Holos Ukraïny* that referenced the president's press office. Kravchuk acknowledged Bush's "readiness to provide assistance in the destruction of nuclear weapons." He reported on the progress of ratification preparations, confirming once again Ukraine's commitment to becoming a nonnuclear-weapon state and assuring that he would "do all that is necessary to have the treaties approved." At the same time, *Holos Ukraïny* wrote the presidents "discussed the practical issues Ukraine encountered in their implementation."[96]

On 26 January 1993, shortly after his inauguration, Bill Clinton, the new US president, called Kravchuk. The topic of conversation remained the same, as reported again by *Holos Ukraïny*:

> At the request of the US President, L. Kravchuk informed him on the developments within the START ratification preparations, emphasizing that the president and government of Ukraine would continue the course determined by the parliament. Clinton confirmed the previous administration's intentions of providing $175 million towards supporting the process of dismantling nuclear weapons.[97]

Between the two presidential calls, the head of the Parliamentary Commission on Foreign Affairs, Dmytro Pavlychko, announced in the final days of 1992 that START I would be reviewed by Parliament no earlier than February of the following year. Responding to appeals for immediate ratification, Pavlychko emphasized that the text of the agreement comprised three volumes of over 600 pages in total, and that it had been received by Parliament only two weeks earlier. It would be impossible to review the document swiftly, especially since in January Parliament would be reviewing more urgent issues related to the implementation of reforms.[98]

On 5 January 1993, a seemingly insignificant event happened, but it provided formal grounds for a further delay of Parliament's ratification vote. Parliament adopted a decree to create a working group tasked with preparations for the ratification of START, which essentially duplicated Pliushch's decree from July of the previous year. Notably, the decree altered the Group's composition: Kuchma, now prime minister, left the Group, and the number of working group members increased from 13 to 23 individuals.

In the coded political language of the day, an increase in size was an attempt at slowing down the Working Group's activities. The Group had already proven its effectiveness in gathering information about Ukraine's nuclear weapons and developing a disarmament strategy. Everyone understood this. However, the stratagem resulted in the opposite effect than was expected. The arguments we had assembled up to that point were so persuasive that they quickly delivered a nearly 100-percent consolidation in the Group's new membership. Moreover, its increased size made disseminating information collected by the Group within Parliament easier. This example had demonstrated to me the workings of Pliushch's wisdom, which he bestowed on me during one of our conversations about problems we might face with the Working Group's increased size. He said: "If you are right, they will support you. If not, then not." Yet, on 5 January the Working Group received Parliament's orders to examine the issue and to prepare a draft resolution on ratification. The orders bought us more time.

Meanwhile, realizing its plan of accelerated ratification was in peril, Russia reacted immediately to the creation of a new parliamentary working group. On 8 January 1993, as reported by *Holos Ukraïny*, Russia's first television channel *Ostankino* commented on the idea that Ukraine was dragging its feet on the START ratification, and, citing Reuters, implied that the ratification documents were not even a top priority on Parliament's agenda.[99] Concurrently, Russia moved from words to actions in spring when it stopped supplying spare parts and refused to provide warranty services on the SS-19 strategic missile carriers and warheads, until Ukraine would ratify START I and the NPT.[100]

Nevertheless, the Working Group managed to hold back parliamentary review of nuclear treaties up until 3 June 1993. By that point, however, Parliament already knew a great deal more.

Kravchuk's Disarmament Concept

On 16 December 1992, *Holos Ukraïny* published Ukrinform's interview with Kravchuk, which was entitled: "Nuclear Weapons: The Ukrainian Leadership's Perspective." It was supposed to fill in the blanks on the Ukrainian leadership's position on disarmament.[101]

The reason for this publication was the world did not really understand what Ukraine was doing, or why. This notion was expressed in one of Ukrinform's questions: "Lately, certain mass media outlets, both in the near and far abroad, have been disseminating a great deal of material suggesting that Ukraine is delaying elimination of strategic nuclear weapons located on its territory."[102] That is to say, Ukraine's position lacked clarity, and its key messages were communicated neither to domestic nor to foreign audiences. Moreover, the countries which demanded that Ukraine speed up the disarmament process, expertly manipulated the media.

Kravchuk was asked to "briefly summarize Ukraine's leadership position on nuclear weapons," and, more to the point, to answer the questions, "How was it formed? What is its essence? Has it changed as of today? And can it change in the future?"[103] The answers demonstrated that Kravchuk did not really have a clear idea, although he did use some of the language from Parliament's resolutions, which he most likely just cited. This publication is of particular value because it allows for a comparison of Kravchuk's words to his actions at certain periods in the disarmament process. The following sums up the essence of his statements.

Kravchuk appealed to Ukraine's Declaration of State Sovereignty, which documented the country's "intention to become a nonnuclear-weapon state in the future," as well as to the 24 October 1991 Statement of the Parliament in which Ukraine had reaffirmed its intensions of becoming a nonnuclear state, while doing so "in the time frame that is minimally necessary in view of the legal, technical, financial, organizational, and other capabilities, and in consideration of environmental safety."[104] He stressed that Ukraine declared its right of "exercising control over non-use of nuclear weapons located on its territory," the country's intentions of joining START and NPT, finalizing an "agreement with IAEA regarding guarantees," and on Ukraine's course towards military industry conversion. This amalgamation of information was presented as Ukraine's official position. The latter, Kravchuk underscored, "was determined by the Ukrainian parliament, and no one but Parliament has the authority to change it. The president and the Ukrainian government work to implement this position, including through negotiations with other countries."[105]

He emphasized that disarmament under the Lisbon Protocol had to take place proportionally among the four former Soviet republics, "both in terms of quantities and time tables," and with technical and financial assistance from the West, as well as "in consideration of Ukraine's right of ownership to the assets" contained in the warheads that had been transferred to Russia for disassembly. Nonetheless, a mere month after this article's publication on 15 January 1993 during negotiations with Yeltsin in Moscow, Kravchuk failed to raise the question of proportionality in reductions (which he presented as a fundamental condition for Ukraine in the above interview). The January events notwithstanding, the following two sentences in the article contradicted the president's previous statement, suggesting that he either only very vaguely understood the conditions of START or was deliberately distorting them, which led to Kravchuk's repeated public declarations of excessive commitments: "When speaking of an integrated approach, we mean the simultaneous elimination of nuclear warheads, their carriers, launch facilities, and control panels. *All* nuclear munitions remaining on Ukrainian soil will be liquidated in accordance with the previously reached agreements and the schedule established under international control."[106]

The term *"all,"* used by Kravchuk in relation to nuclear weapons, entirely contradicted his intention to achieve proportionality in reductions. Besides, no previous agreements existed that obligated Ukraine to destroy *all* nuclear weapons. Kravchuk's reference to "an integrated approach" signaled his resolve to destroy everything that had to do with nuclear weapons on Ukraine's territory, including their entire infrastructure. The all-inclusiveness in Kravchuk's approach meant the liquidation of nuclear warheads, carriers, silos and control panels. Such wastefulness, not stipulated in any of the treaties, cannot be explained by a shortage of information because, by then, Kravchuk already had ample data from the Working Group to know that START did not require the destruction of the strategic arms infrastructure, even if we liquidated nuclear warheads.

In the next paragraph, Kravchuk substantiated the need for a thorough approach to disarmament. However, his subsequent actions moved towards getting rid of nuclear weapons as quickly as possible and indicated that this passage most likely was prepared by his advisers:

> Thoughtful people understand that before consenting to something, this "something" needs to be thoroughly understood. Such an attitude is exemplified by the US's approach. Why is it illustrative? The US administration was among the coauthors of START I. Thus, it had the opportunity to learn many of these things, and to evaluate them in the process of the document's development. Even under these circumstances, it took the US Senate over a year to study the START I Treaty—and all of the consequences of its implementation for the country's security and economy—before the treaty was approved.[107]

Further on, acknowledging the fact that only the Russian Federation president had the launch controls of all the missiles on Ukraine's territory at the time, Kravchuk added somewhat illogically:

> Today, I have the capability to prevent an unsanctioned launch of nuclear weapons from our country's territory. This is to be achieved in the following manner: special signals have been installed for communication control units and strategic missile troops deployed on Ukraine's territory. They allow for blocking the launch of intercontinental ballistic missiles when such a launch has not been sanctioned by the president of Ukraine. Special means of direct communication have also been installed with command control of the strategic nuclear forces located on Ukraine's territory, with the same purpose.[108]

Here, Kravchuk demonstrated his wishful thinking, either lacking information or thinking the readers would not know better, because back in April 1992 at the aforementioned Defense Council meeting, Iakiv Aizenberh, the director of Khartron, indicated quite clearly that it would be impossible to establish control systems preventing the unauthorized launch of nuclear weapons without the consent of Russia, which had not granted it.[109]

Aside from our working group, there were two other official bodies tasked with providing Kravchuk with specific information: the National Committee on Disarmament and the State Administration Committee on Nuclear Policy. They were headed, respectively, by Deputy Foreign Minister Borys Tarasiuk and the National Academy of Sciences vice president, academician Viktor Bar'iakhtar.

The National Committee on Disarmament was formally established in June of 1992, but its activities were not officially regulated for another six months. It comprised 33 members, including nine representatives of the MoD. Despite the significant allocation of resources and the substantial authority it was given, the committee's first tangible results did not surface until February. The document it produced contained analysis of dubious quality, with its persistent advice to transfer all weapons to Russia as soon as possible.

CHAPTER 3. Breakthrough

START II: Grandstanding

The year 1993 began with the US and Russia signing their latest treaty on the reduction of strategic offensive weapons, START II. On 3 January, both countries committed themselves to reducing their nuclear arsenals by more than two thirds over the next 15 years. Nonetheless, the main point was a caveat: START II could only enter into force if all the signatories to START I ratified it.[1]

At that point in time, only Russia and the US had ratified it. During his May 1992 visit to Washington, Kazakhstan's President Nazarbayev agreed to dismantle 108 SS-18 Satan missiles deployed within his country.[2] The Belarusian parliament was also getting ready to ratify the nuclear treaties and on 2 February 1993, it unanimously approved both START I and the NPT, after which Speaker Stanislav Shushkevich signed the document. Ukraine remained the only country that postponed the debate over both treaties. As events were to show, political pressure to get Ukraine to speed up ratification was the key point of START II.

On 3 January 1993, President Kravchuk responded with a statement in which he confirmed that this signing "does not place Ukraine under any obligation and does not extend to its territory."[3] However, this statement proved unlikely to neutralize the coordinated information campaign that Russia and the US had already launched.

A press conference took place in Moscow with the US and Russian presidents, in honor of the signing of this "historic treaty." On 12 January, the Russian Foreign Ministry sent a special ambassador, Mikhail Streltsov, to Kyiv with an express purpose to inform the Ukrainian side about the content of the new Russian-American treaty. During those same days Ukrainian Minister Anatolii Zlenko after meeting with Russian Foreign Minister Kozyrev "declared his full support" for START II, according to a quote in the *Boston Globe* citing the dispatch from the Russian news agency *TASS*, and also reporting that Zlenko and Kozyrev "agreed to expedite talks on technical aspects of liquidating the nuclear weapons on the territory of Ukraine."[4]

This was quite strange since Ukraine had not yet ratified START I, which basically meant it had not agreed to an arms reduction at all. It was clearly too early to talk about the "technical aspects of liquidating the nuclear warheads," let alone about expediting talks on this subject. This meant that either TASS was reporting on an already traditional overstepping of his authorities on the part of Anatolii Zlenko, or Russian media was deliberately distorting his words. Nevertheless, the Foreign Ministry never issued any objection.

Still, the press could not help but note that not everyone agreed unanimously with the minister within Ukraine's Foreign Ministry. Volodymyr Kryzhanivs'kyi, the then ambassador to Russia, stated at a press briefing in Moscow: "Why are so many folks insisting today that Ukraine's parliament should quickly ratify the treaty and forget its obligations to its own people?" He also pointed out that given its growing financial problems, Ukraine urgently needed the assistance of the West to destroy its arsenal.[5]

Manipulating the treaty via the news to increase pressure on Ukraine was also evident in the international press. The key message was: by delaying the ratification of START I, Ukraine was interfering in the nuclear disarmament process of the two superpowers.

According to the MFA press department, more than 30 of the 50 articles reviewed in the quarter, were published in January.[6] Moreover, the surge in publications came after the US and Russia had already signed START II. Ukraine was accused of disrupting the implementation of this treaty by issuing demands for security guarantees and compensation for the cost of liquidation. Both the *Los Angeles Times* and *New York Times* wrote at the beginning of January that Ukraine constituted a threat to stability in Eastern Europe, and called on Western donors to cut off economic aid and political support, if the parliament did not ratify the nuclear treaties immediately.[7]

The MFA then wrote a rather delayed analytical note, "The issue of nuclear disarmament in Ukraine in the Western press," which it sent to the Parliament Commission on Foreign Affairs months later on 10 May 1993, noting, "The biggest wave of massive pressure on Ukraine came at the end of last year and the beginning of this one [...]. At that time, not a day went by without some article criticizing the dilatoriness of Ukraine's parliament regarding the ratification of the NPT and START I. The gist generally was to blame Ukraine for hampering the implementation of both the START I and START II treaties signed by Russia and the US, and for making excessive demands concerning national security guarantees and compensation for the liquidation of the missile complexes."[8]

On 5 April, the Russian government issued a very direct statement in which it announced: "Kyiv's position, which is in breach of commitments it has made, threatens extremely dangerous consequences [...]. [T]he effectiveness of the nuclear nonproliferation regime becomes at risk, as well as the coming into effect of both START I and START II."[9] On 24 June, the US Senate Committee on Foreign Relations met to discuss US and Russian compliance regarding START II, where CIA Director R. James Woolsey noted: "The relationship between Russia and Ukraine also threatens to stall Russia's progress in implementing the START agreements. Russia has set preconditions for exchanging START I instruments of ratification and for ratifying START II. Russia's preconditions are that Belarus, Kazakhstan, and Ukraine must ratify START I and accede to the Nonproliferation Treaty as nonnuclear states."[10]

In a similar fashion, the *Washington Post* reported: "While Administration officials say Ukraine's reluctance to dismantle its nuclear arsenal is understandable, the position has had serious costs. It has aggravated Kiev's relations with Moscow and led Russia and the United States to delay the arms cuts agreed to under the Start 1 and Start 2 treaties. Administration officials are also worried that Ukraine's stand may hurt efforts to control the spread of nuclear weapons."[11] Thus, the maneuver known as START II turned Ukraine, which had voluntarily declared its intentions to become a nonnuclear state, into the supposed main obstacle to a nuclear security regime in the world.

Immediately after the US and Russia signed the START II Treaty, the *Wall Street Journal* published an article called "Starting Over," stating that this event continued the superpowers' tendency toward ignoring the interests of the other emerging nations from the former USSR:

> START II fits within the context of the warmer relations between the US and Russia, and that is all to the good. Yet, it ignores the more precarious strategic balance among the Soviet successor states. Shutting out the other three nuclear republics on grounds of "anti-proliferation" overlooks the reality that the republics did not seek nuclear status, they inherited it. They are being treated like something less than what they claim to be and what most of the world accepts them as — independent nations.[12]

Meanwhile, START II got mixed reviews in Russia. For instance, *Pravda* published an article on 9 June 1992 called, "To Disarm, but Intelligently," by Grigorii Kisun´ko, the former general designer of the Soviet missile defense systems, in which he wrote:

> Obviously, it [START II] is being forced on us through well-planned American pressure for the purpose of destroying systems of potential deterrence while strengthening the US's nuclear offensive capabilities. Once this Treaty is carried out, any confrontation between US and Russian forces will end up with America's unbounded dictatorship, right up to the Russian Federation's complete nuclear disarmament in the US's line of sight.[13]

Further events confirmed that START II was not as important or urgent for the US, and even less so for Russia. They took three, and seven years, respectively, to ratify the treaty: the US Senate ratified START in 1996, while the Russian Duma did not ratify it until the year 2000. Ukraine was forced to ratify START in 1993.

How Ukraine's Diplomats Traveled to the US for "Security Guarantees"

Meanwhile, the camp that favored rapid ratification of START, spearheaded by Ukraine's president and Foreign Ministry, was leading the preparations for voting in the legislature from several angles. First, it was necessary to get the approval of the parliament to put the issue on its agenda. Second, requirements that were no longer even being questioned had to be at least formally met providing Ukraine with security guarantees, and financial support from the West for disarmament and compensation for nuclear materials. The first was not going to be possible without the second at that point.

However, in contrast to the Working Group, which had formulated a package of these demands, the MFA thought it was enough to have "paper" security guarantees, such as the infamous Budapest Memorandum, the US $175 million

54. Borys Tarasiuk, 2003

Borys Tarasiuk (b. 1949), Ukrainian diplomat, deputy foreign minister from March 1992 to December 1994, first deputy foreign minister from December 1994 to September 1995. He was also the chairman of the National Committee for Disarmament from June 1992 to April 1995. Tarasiuk was a career diplomat. Immediately after graduating from the Department of International Relations and International Law at Taras Shevchenko University in Kyiv in 1975, he worked at the Foreign Ministry of the Ukrainian SSR. From 1981 to 1986, he was the secretary to Ukraine's permanent representative at the UN in New York. Prior to the collapse of the USSR, from 1987 to 1990, he worked at the Central Committee of the Communist Party of Ukraine as an Instructor in the Department of Foreign Communications. Between 1990 and 1992, he was an advisor and he supervised the Political Analysis and Coordination Department and was director of the Secretariat of the MFA of Ukraine. From 1993 to 1995, he chaired the State Interagency Commission on Ukraine's accession to the Council of Europe. In 1995, he became Ukraine's ambassador to the Benelux countries and, at the same time, headed Ukraine's Mission to NATO until 1998. He was appointed foreign minister in 1998–2000 and again in 2005–2007. He has been a Ukrainian MP since 2002. Between 2003 and 2012, he headed the People's Movement.

for disarmament. It was prepared to agree to the formula for compensation that Russia had proposed. Yet, given the distribution of authorities within the country, it was the Foreign Ministry that was to make these demands real for the legislature. The MFA would negotiate both the guarantees and the compensation.

At the beginning of 1993, the Foreign Ministry began an active search for "security guarantees" on the terms that they understood. They had two options as a priority: the Russian and the American.

In the first few days of January, Deputy Foreign Minister Borys Tarasiuk led a government delegation to Washington. Tarasiuk thought, as Reuters reported on 8 January, that "for us, the government, it will be easier to convince the parliament, it will add more arguments in favour of a positive decision if this (security) declaration appears before a ratification vote."[14] The State Department responded that the guarantees the United States could name would not differ from what former secretary of state Jim Baker had mentioned the previous summer and what was being offered to other countries and was set in the United Nations Charter. In other words, nothing differed from what Ukraine already had as a UN member.[15]

The evening of 8 January, Reuters also reported that the Ukrainian diplomat who had been leading the negotiations on military issues, went home with a letter from the United States that talked about security guarantees. All the interested parties expected that this would force the parliament to ratify START I. The contents of the letter were conveyed as: "Tarasiuk was looking for a formal 'political declaration at the highest level.'" The visit had ended with a meeting between Tarasiuk, and President Bush, who normally only accepted delegations at the highest level. This was intended to indicate the weight the US was attaching to Ukraine. Bush also signed these "guarantees." Reuters also quoted some explanations from a senior US official: "My understanding is that they were given a letter [...] that describes the kinds of things that we were talking about, the kind of assurances we could make once they ratify the treaty and pledge to become a non-nuclear state [...]. [W]e put something in writing that they would take back and show their folks."[16]

The *Los Angeles Times* wrote of Tarasiuk that during this visit, "the State Department rebuffed him, but Ukrainian officials said that a compromise was then reached, allowing him to take home a document outlining US guarantees that will protect Ukraine." The article continued with: "Tarasiuk also said he reached an understanding in Washington on financial compensation but gave no details." Tarasiuk insisted that he was pleased with the results and confident that this letter would "overcome concern in Ukraine's legislature over the possibility of aggression by neighboring Russia."[17] In a telephone interview with Reuters, Oleh Bilorus, Ukraine's ambassador to the US, also expressed satisfaction "with the atmosphere of real partnership," but refused to confirm whether US officials had really given the security guarantees Ukraine was seeking.[18]

Citing the press service of the Ukrainian Embassy in Washington, *Holos Ukraïny* wrote about the "two-day visit of Ukraine's government delegation" and about a meeting between the head of the delegation, Borys Tarasiuk, and President Bush, and the handing over of the letter from President Kravchuk. The press service reported that Tarasiuk had emphasized some unresolved issues, among which were the lack of reliable international security guarantees; financial compensation; and the vagueness of the conditions, timeframes, and locations for destroying the nuclear weapons after they were transferred to Russia. In effect, the Embassy avoided the issue of "guarantees," which were being treated with skepticism in the Western press.[19]

It was around this time that Russia presented its draft "guarantees." I am unsure what this document contained, but the Foreign Ministry had decided not even to reveal its contents to the parliament. This draft was specifically discussed on 12 January 1993 at a meeting between Tarasiuk and Russia's special envoy, Mikhail Streltsov. In his notes on the meeting, Tarasiuk wrote, "I explained that we were not satisfied with the text in the Russian MFA's latest document on security guarantees for Ukraine," and said that the text was intentionally not submitted to the legislature.[20] Minister Zlenko sent these notes to Pliushch for his information, but only two weeks after the meeting. In the meantime, the notes showed, first, the Ukrainian MFA did not offer its own proposal for guarantees that might have been considered by Moscow and Washington; and second, Russia was at that given moment finally prepared to begin Ukrainian-Russian negotiations on nuclear weapons and to define the terms and timeframes during the last remaining weeks of January, which was indeed when these talks finally began.[21]

Russia's idea was that before the negotiations began, they would be preceded by two other important events with the participation of the presidents, in Moscow and Minsk, planned for January with just a week's interval in between. They were supposed to lead to a fundamental change in Ukrainian-Russian relations. Based on this bilateral reboot, which the Kremlin was hoping for, Ukraine's nuclear disarmament according to the Russian script would have already become a mere formality.

Could Ukraine Have Joined NATO in the Early 1990s?

By early 1993, for the majority of Ukraine's political establishment, security guarantees had become an incontrovertible condition for nuclear disarmament. However, who would guarantee the inviolability of Ukraine's borders and its sovereignty as a state, against whom and in what manner was anything but clear, with opinions diametrically opposed. The fault line once again ran between the new, pro-Ukrainian generation and the old, pro-Soviet one.

The idea of joining NATO as the main guarantee of national security for a nonnuclear Ukraine seemed extremely radical in the early 1990s. After decades of Cold War and the division of the world into capitalist and socialist camps which formed the antagonistic military blocs, NATO and the Warsaw Pact, a majority within the Ukrainian leadership continued to view the North Atlantic Alliance as an enemy.[22] The idea of looking for protection in NATO could not emerge in an environment dominated by yesterday's communists who had spent their entire careers propagandizing the enmity of the two camps. Nor could it emerge in an environment filled with KGB cadres whose professional duties depended on a society in which no one could be permitted to have a different opinion than what communist propaganda promoted.

The thought of "joining the enemies of communism" could only take shape in the dissident circles, where the communist system itself and its source, Russia, were seen as the enemy; particularly, among the new wave of politicians for whom communism had never been a career or way of thinking. "It is laughable to imagine that, after the unilateral disarmament of Ukraine, Russia might offer any reliable guarantees, let alone uphold them. History has shown us that our northern neighbor has more than once treated international treaties as worth less than the paper on which they were written. Our security might really be guaranteed if we joined NATO," *Holos Ukraïny* quoted Oles Shevchenko, an MP and political prisoner of many years, in March 1993.[23]

For Ukraine to accede to NATO was, of course, categorically unacceptable to Russia; it would mean the irreversible departure of the economically most powerful republic from under its geopolitical control. For those same reasons, it was of interest to the US: given the inertia of the cold war mentality, taking away an ally of Russia and weakening the enemy would be seen by the Americans as a victory.

From the very first days that our working group began preparing to ratify START I, meaning from July 1992, we proposed that only the accession of Ukraine to the European security system would offer national security guarantees. I personally insisted that the last warhead be removed from Ukraine on the day the country became a member of the European collective security system.[24]

The US at the time did more than just support the idea. On 7 December 1992, Under Secretary of State for International Security Affairs Frank G. Wisner requested a meeting with Ukraine's ambassador in Washington, Oleh Bilorus, in which he seemed to have urged Ukraine to join NATO. Arguing that Ukraine should ratify the Lisbon Protocol as quickly as possible, Wisner explained the national security guarantees that the United States was prepared to offer Ukraine at that time, if it disarmed: "We know about your problems with Russia, but nuclear weapons will not protect you against it. On the contrary, they will only make your security situation even worse. The only assurance of your security is to join transatlantic structures and cooperate with them." This quote would later appear in Bilorus's report.[25]

The opinion that the highly ranked State Department official seemed to have expressed was supported by the West. At the time, the *Economist* wrote, "a non-nuclear, western-friendly Ukraine could, like Poland, Hungary and ex-Czechoslovakia, fall under the shade of NATO's defense umbrella."[26] But from the beginning of 1993, the winds changed with the coming of a new president in the US. The Clinton Administration never saw Ukraine in NATO.

At the end of May, US Deputy Secretary of State Strobe Talbott, during his visit to Kyiv, as quoted in the *Washington Post*: "avoided the words 'security guarantee,' which convey a NATO-like obligation to defend Ukrainian territory against outside attack." Moreover, the article continued: "Talbott did not reciprocate Ukrainian interest in solving the problem with a new multilateral treaty to be modeled on the 1955 US-Soviet accord that guaranteed Austria's independence and neutrality. That treaty took nearly 10 years to negotiate."[27]

Less than a week after Talbott's visit to Kyiv, US Secretary of Defense Les Aspin showed up. Once again, the *Washington Post* reported on this event: "Aspin is not prepared to offer Ukraine the sort of bilateral security guarantee that some Ukrainian legislators have demanded as the price of giving up the nuclear weapons."[28] The next day, the newspaper wrote: "Washington does not plan to offer Ukraine membership in NATO or protection under the US nuclear umbrella or through the stationing of any US troops on Ukrainian territory, officials said. They said the United States plans instead to provide less specific security assurances [...]."[29]

In November 1993, security guarantees were established as one of the conditions for Ukraine to ratify START I, but it was unrealistic to expect the parliament at that time to concretize these in its ratification resolution with reference to joining NATO. The MFA was in charge of negotiating those security guarantees and it never raised the possibility of Ukraine's joining transatlantic security structures.[30]

Government Negotiations: Prelude

On 11 January 1993, the Russian president's press service announced that a Russian-Ukrainian meeting at the highest level would take place on 15 January. *Holos Ukraïny* reported this, citing Yeltsin's press service, and not Kravchuk's.[31] The Ukrainian legislature's bulletin quoted the head of the Russian delegation at the negotiations with Ukraine, Iurii Dubinin, who announced the items on the agenda: nuclear disarmament issues "that have become particularly urgent in connection with the signing of START II on 3 January," and "eliminating complications in bilateral relations, especially related to the Black Sea Fleet."[32]

At this time, conditions were such that the 15 January meeting could be seen in the light of plans to reboot Ukrainian-Russian relations, which had been getting more and more strained since the collapse of the USSR. Russia continued to pressure Ukraine as an independent competitor with a rapprochement that would lead to reabsorption into a new union.

On 23 January, barely a week after the Moscow summit, a meeting was set for Minsk where the ten CIS members were supposed to sign a new statute that would launch the establishment of a new union of states. The Kremlin's shift from its usual pressure to a policy of good neighborly relations would, Russia's political strategists thought, weaken resistance in Ukraine to transferring its strategic nuclear weapons to Russia and persuade it to go under Moscow's nuclear umbrella. It was also expected that the idea of Ukraine's integration with the European Union and NATO would be buried for many years to come. In the foreign policy context, friendly Ukrainian-Russian relations were supposed to foster the execution of Russia's plan, "together to Europe," whose essence was becoming increasingly obvious. After all, the CIS, which Kravchuk had

55. *Boris Yeltsin and Leonid Kravchuk at the Yalta Agreement of 1992*

called "the way to a civilized divorce among the republics," no longer satisfied Moscow's imperial ambitions. By the beginning of 1993, Russia began to work on the idea of transforming the CIS into a renewed union.

Both with the issue of nuclear disarmament and with the change of the status of the CIS, Ukraine's position was decisive. Newspaper archives testify that the then head of the SBU, Ievhen Marchuk, argued in support of two theses, Russia's "nuclear umbrella" and "together to Europe," in a policy-oriented article, printed on 12 December 1992, in *Holos Ukraïny*, a month before these events. Despite the difficult history of relations between the two states, the author argued, their further prospects lay in Ukraine and Russia "living in peace," but the nuclear weapons located on Ukraine's territory stood in the way of this and "would essentially only be a source of various tensions in Ukrainian-Russian relations." Therefore, Marchuk concluded, the basis for neighborly relations was "coordinating actions with Russia" and "keeping Ukraine under the nuclear shield," meaning the Russian shield. Marchuk also mentioned that Ukraine could move towards Europe only in tandem with Russia: "It is unlikely that we will be able to enter the European house by elbowing our partner aside."[33]

Events were to show that undertaking a "course to rapprochement," essentially restoring the USSR, was the objective of Russian diplomats during the 15 January Ukrainian-Russian talks in Moscow. Because of a blizzard, the Ukrainian delegation showed up for the meeting two hours late. As the recently assigned head of the government delegation over nuclear weapons negotiations with Russia, I was added to the group that accompanied President Kravchuk at the last minute. Even though we had violated diplomatic protocol, Yeltsin and all his team during the course of the evening radiated emphatic hospitality and paid no attention to any circumstances: the political stakes were simply too high.

Negotiations started with an hour-long meeting face-to-face between the two presidents. In time, this was expanded to include premiers, deputy-premiers, and the foreign and defense ministers. Most of the people in our delegation participated in the closing press briefing that the Russian president began with the very promising words, "We had the wisdom and the intelligence to remain friends."

In the joint communiqué summarizing the talks, the two sides documented that the two presidents had agreed "to give the necessary instructions for all matters connected to debt and asset servicing of the former USSR to be resolved."[34] The presidents also agreed jointly to appoint Vice Admiral Eduard Baltin commander of the Black Sea Fleet and tasked their state delegations "to prepare the documents for the implementation of the Yalta agreement."[35]

Still, half the communiqué concerned nuclear weapons. Kravchuk praised Russia for ratifying START I and signing START II and confirmed Ukraine's "determination" to ratify START I. Yeltsin in turn "welcomed" Ukraine's steps to ensure the nonnuclear status it had declared and announced that he was prepared to provide security guarantees and to draft the text of such guarantees as soon as possible. The governments were requested to "immediately begin negotiating to resolve all issues connected with implementing START, including the terms for dismantling, transporting and destroying nuclear warheads and components of Strategic Nuclear Forces missiles deployed in Ukraine, and the processing of nuclear components for use as fuel at Ukraine's NPPs."[36] The two presidents also agreed to identify the system for material and technical servicing and implementing the warranty and manufacturer oversight for the operations of the Strategic Nuclear Forces' missile complexes "by Russia."[37]

All these agreements between the presidents were, in fact, in violation of Article 2 of the Lisbon Protocol, which stated that, before discussing the terms for dismantling, transporting and destroying the nuclear warheads, Russia, Ukraine, Kazakhstan, and Belarus had to engage in four-way talks to agree on dividing up the reduction quotas, in other, in words, to determine who was supposed to destroy how many nuclear warheads under START I.[38] This sequence did not suit Moscow: its tactic of dividing and conquering worked much better than seeking consensus in a multilateral format. Nevertheless, Ukraine's MFA, and hence Kravchuk, had agreed to the change in the negotiating format, thereby giving Russia a substantial advantage. Agreements on the fate of the Strategic Nuclear Forces, as the presidents had agreed, were to be drawn up and signed immediately, within the month.

Whereas the Ukrainian press merely quoted the text of the communiqué, the Western press talked about the deeper implications of the Moscow talks. On 16 January 1993, the *Los Angeles Times* claimed: "President Boris N. Yeltsin swept away a major obstacle to strategic disarmament, announcing Friday that Russia has agreed to place Ukraine under the protection of its atomic umbrella."[39] The Montreal *Gazette* published an article that same day that stated: "Russian President Boris Yeltsin sweetened relations with neighboring Ukraine yesterday, promising nuclear protection, a hands-off border policy, and increased oil supplies."[40] The question of the geopolitical impact of these events was continued in the same *Los Angeles Times* issue: "To mark the turning of a new page, Yeltsin and Kravchuk held an elaborate document-signing ceremony Friday in the same lofty-domed Kremlin chamber where Bush and Yeltsin signed START II earlier this month."[41]

This turn in Ukrainian-Russian relations, according to the Kremlin's thinking, was to ensure Kyiv's agreement to substantially reduce its sovereignty. However, Moscow failed to get what it wanted. The opposition was putting pressure on Kravchuk. After the delegation returned from Moscow, the leadership of the opposition People's Council held an immediate meeting with the president. In very strong language, he was told that any shift in course towards Moscow and a renewal of the Union was categorically unacceptable for Ukraine. Moreover, signing the new statutes of the CIS would have meant effectively giving up Ukraine's sovereignty.

A sharp change in Kravchuk's mood after this meeting can be seen in the statements issued by Ukraine's MFA that appeared in the Western press just before Minsk. "This structure would lead us to the restoration of the former Soviet Union" and "We are against the formation of a new superpower on the principles of confederation or federation," Henadii Udovenko, Ukraine's Ambassador to Poland and close friend of Kravchuk's, commented on the situation for *United Press International* on 19 January.[42] Reuters quoted First Deputy Foreign Minister Mykola Makarevych on 20 January: "We don't want to put our signatures to it simply in order to stay in tune with all the rest."[43] Speaking to the *New York Times*, Radek Sikorski, who was the former deputy defense minister of Poland, predicted the further development of the situation on 21 January: "Ukraine faces a genuine security dilemma. Russia is not reconciled to the loss of its valuable territory and resources. Even Russian democrats have trouble recognizing Ukraine's right to sovereignty, let alone the hawks. The threat would take the form of pressure on the fragile Ukrainian economy, agitation among its Russian-speaking minority, and other destabilizing measures."[44]

On 23 January, in Minsk, Leonid Kravchuk refused to sign the draft new statutes of the CIS, which would open the door to restoring the Soviet Union. Having failed in its attempt to swiftly revive the USSR, Russia focused on cleaning out Ukraine's nuclear arsenal instead. An open diplomatic scuffle now broke out between Ukraine and Russia over Ukraine's nuclear inheritance.

Government Negotiations, Round One: Introduction to Methods

On 26 January 1993, at the Defense Ministry's sanatorium in Irpin outside Kyiv, the first round of talks between government delegations on nuclear weapons began. I headed the Ukrainian delegation, being Minister of the Environment at the time, but for some reason the Cabinet apparatus issued Premier Kuchma's instruction of my appointment with the note "For internal use only," in contrast to the public announcement that Iurii Dubinin, a career diplomat, had been appointed to head our Russian counterpart.

Therefore, at the Irpin, I sat facing a professional communist, a diplomat whose career had begun in 1955, when I was only four, and someone I had heard about from the time he was the Soviet ambassador to the US in the late 1980s, engaged in persuading the democratic world of the advantages of the communist system. I myself had never joined the Communist Party and had been in politics less than three years. I also had no diplomatic experience, but as a scientist, I knew very well how to work with data and I knew one thing for sure: at that moment, it was going to depend on me to ensure that Ukraine defended its right to nuclear weapons.[45]

Not only did we have completely different experience, both in life and in holding negotiations at this level, but our level of preparation differed for the upcoming discussion. Negotiations are typically prepared for by the entire staff with specialists in different areas: nuclear, military, legal and diplomatic. Materials were prepared from line ministries, agencies and special services for Dubinin. I prepared by myself. All I had was information that our working group had managed to put together. The Ukrainian MFA, which should have been providing materials and recommendations for how to negotiate, appeared to have recused itself from this work.[46]

After the Moscow agreements between Kravchuk and Yeltsin, the Russians were expecting things to be wrapped up quickly. However, I intended to clearly follow Ukrainian legislation and immediately warned that the talks would go on for as long as it took to resolve the issues. In his recollection of events, Dubinin interpreted this as follows:

56. *Iurii Dubinin, 1986*

> I was appointed to head the Russian delegation, while the Ukrainian delegation was led Yuri Kostenko, minister of environment and chair of the special working group of the Ukrainian parliament on preparations to ratify START I. Being one of the leaders of the nationalist People's Movement party, he managed to announce, even before our meeting, that the talks could go 20, 30 years. It was obvious to us, however, that an issue of such significance for strategic stability needed to be decided in the shortest time possible.[47]

In my opening speech, I first informed my Russian colleagues of the parliament's decision regarding nuclear weapons deployed on Ukraine's territory. I emphasized that, in line with international law, Ukraine passed legislation declaring these weapons its property. I went on to convey how Ukraine planned to move towards nonnuclear status, based on a political decision by the legislature, and what we proposed considering in this context at bilateral talks. Above all I proposed focusing on security issues around strategic weapons, which worried the world most of all. Separately, I raised the issue of how Russia intended to compensate the value of nuclear materials taken from the tactical weapons that were removed from Ukraine previously.

Judging by Dubinin's recollection, the Russian side heard this speech thus: "In his introductory words, Kostenko unexpectedly announced the 'right of ownership' to 'nuclear warheads.' There was no more intricate wording as in the Foreign Ministry's memorandum (the right to all the components of the nuclear warheads)."[48]

In response to my statement, instead of constructive dialogue, Iurii Dubinin started by accusing Ukraine of changing its position on nuclear weapons, arguing that, by declaring right

Iurii Dubinin (1930–2013), Russian diplomat. Dubinin was the Soviet ambassador to the US from 1984 to 1990. He headed the Russian Government's delegation for negotiations with Ukraine concerning nuclear weapons in 1993–1994. A career diplomat since 1954, he worked in the UNESCO Secretariat in Paris from 1956 to 1959. From 1963 to 1968, he was first secretary and counselor at the Soviet embassy in France. Subsequently, he worked at the First European Department of the Soviet Foreign Ministry until 1978 and was its head for seven years. From 1978 to 1991, he was Soviet ambassador to Spain, then the US, and then France. From 1994 to 1999, he was deputy foreign minister of Russia. From May 1996 to August 1999, he was Russia's ambassador to Ukraine.

of ownership, the country was claiming nuclear status, and this, he opined, was moving away from the international commitment Ukraine had made to become nonnuclear. His logic was that, in this situation, Russia could not encourage Ukraine to own nuclear weapons, as that would violate the NPT, and so he proposed that either all warheads be immediately removed to Russia or control over them handed over to Russia's Armed Forces. This mainly meant control over the missile silos in which they were being stored. Otherwise, he warned, Russian companies would no longer service the 130 liquid-propellant missiles or the nuclear warheads on Ukrainian territory. Dubinin presented the Russian view of the situation as follows:

> Proclaiming ownership over the nuclear weapons on the territory of Ukraine is already a claim of having nuclear power status. In response, I noted that the statement being made was a change in position towards nuclear weapons and was a derogation of the commitments set by the acts of the Ukrainian government, as well as through the execution of international documents, in the framework of both the CIS and Lisbon [Protocol]. We understand that it is up to Ukraine, as a sovereign and independent state, to decide what kind of policy to follow. However, as a nuclear power that is a signatory to the NPT, Russia does not have the right to transfer nuclear weapons or said control over them to anyone, or to assist nonnuclear states in producing or acquiring nuclear weapons, or to encourage or induce them to arm themselves with such weapons. Hitherto, Ukraine declared its intentions of becoming a nonnuclear state and we cannot be involved in changing this status.[49]

Despite this freewheeling interpretation of Ukraine's position and the ultimatums, I maintained equilibrium and once again argued Kyiv's position, appealing to logic. If Ukraine cannot be the owner of nuclear weapons because "it is a nonnuclear state," why then was it party to the nuclear weapons reduction treaty? The signature of Russia's Foreign Minister Kozyrev attached to the Lisbon Protocol confirmed this. Furthermore, whose nuclear weapons on its territory was Ukraine supposed to be eliminating under START I?

Once again, I turned to the Russian delegation with the proposition that negotiations start with a review of the question of security, because that was of equal interest to both Ukraine and Russia. After all, Pivdenmash, Khartron, Arsenal, and other Ukrainian enterprises were already servicing all the strategic offensive weapons, both in Ukraine and in Russia. Then I once more raised the question of compensation for tactical nuclear weapons.

My counterpart responded that the issue of tactical nuclear weapons was not on the current agenda, but the Russian delegation would pass our request along to its leadership. As to Russian companies servicing strategic nuclear weapons, Russia had no right to support nonnuclear states in either the manufacturing or acquisition of nuclear weapons, because this was prohibited by the NPT.

As the saying goes, "I talk of the chalk and you of the cheese," which sums up how the Ukrainian-Russian nuclear dialogue progressed. On the other hand, it became increasingly obvious what the Russian delegation avoided discussing during these talks: its intentions of carrying out the Arbatov Plan.

Its first phase was carried out in Lisbon: along with recognizing the former Soviet republics as heirs to the USSR with international commitments, it placed the economically ruinous commitment of the former Soviet Union to reduce strategic nuclear offensive arms squarely on the budgets of the four republics where they were deployed, while simultaneously dividing the USSR's debt, worth an estimated US $60 billion, among them. After the debt and commitments had been divided up, Russia moved on to phase two: taking over the entire Soviet nuclear arsenal. Plan A was to place all strategic offensive weapons in the hands of the Russian Armed Forces. If that did not work, Plan B was to immediately transfer all warheads to Russian territory.

The lack of legal precision on the part of the Ukrainian MFA in interpreting Ukraine's nuclear status up until July 1993 offered Russia plenty of room to maneuver, manipulating world opinion about "aggressive Ukrainians" who were reaching for the "nuclear button," and gaining unilateral advantages in the negotiation process. The Russians took full advantage of the contradictions that had accumulated in the documents signed by Kravchuk and the Foreign Ministry in conflict with decisions made by the parliament. From the onset, it was a mystery to me whether this was due to a lack of professionalism in the diplomatic corps or to a deliberate effort to allow our opponents to use such contradictions against Ukraine. Confirmation of my own conclusions, that Moscow was acting according to its own plans to turn Ukraine into a nonnuclear state, came when the Russian delegation submitted a series of draft agreements that were later signed by the premiers of Ukraine and Russia, on 3 September 1993.

Firstly, the Russians arrived with an already-prepared schedule for transferring the nuclear warheads of strategic missiles from Ukrainian territory, and it included absolutely all of Ukraine's nuclear weapons. Dubinin justified this on the basis of a 7 May 1992 letter from Kravchuk to Bush in which the Ukrainian President mentioned 100 percent disarmament. Deputy Defense Minister Ivan Bizhan, who headed the military group in our delegation, obviously rejected this draft, as the parliament had not approved a single document that would talk about the extent of disarmament. Even the then-not-yet-ratified Lisbon Protocol only mentioned reducing Soviet weapons by less than half.

For my part, I also informed the Russians that Ukraine was still considering the question of how to utilize its nuclear material for peaceful purposes and that was why we had not decided where, specifically, the dismantling of the nuclear warheads and their processing into weapons-grade uranium and plutonium would take place. Although Dubinin insisted on the joint communiqué signed by Kravchuk in Moscow on 15 January, Ukraine had not in fact recognized this in any of the parliament's decisions regarding ownership. What was more, by then the government had already received preliminary proposals from the Americans about the option of utilizing nuclear material for domestic NPPs. Therefore, I said, it was premature to talk about schedules for transferring the weapons to Russia.

Secondly, the Russians planned to subordinate to their own chain-of-command the "locations (S),"—that is, weapons storage facilities (*ob'iekty S*)— where Ukraine's missiles were being held. Here is a single sentence from the draft agreement: "Change of status, appointing and transferring the command staff of officers and ensigns at weapons storage facilities (S) shall be carried out by the Ministry of Defense of the Russian Federation with the approval of the Ministry of Defense of Ukraine."[50] This effectively meant subordinating part of Ukraine's armed forces to Russia, which we had no right to allow. Nevertheless, by summer, the new head of the Ukrainian delegation, Deputy Premier Valerii

Shmarov, initialed this agreement and in September 1993 it was signed as part of the Massandra Accords.

Thirdly, there was what the Russians called "the peaceful dividend for Ukraine," or compensation for nuclear material in the form of fuel rods for its NPPs. The quantity of fuel was to be equivalent to the quantity of fissile material extracted from the warheads. The equivalency was to be established based on data provided by the Russian side, as Ukraine did not have any official data on the specific characteristics of its nuclear weapons. The Russians proposed providing some data orally in a Q&A format. Here, too, we could not render a decision, because our working group was only at the stage of studying various propositions. Hence, we agreed to hold a separate meeting of nuclear specialists in Moscow.

Finally, there was the draft agreement providing warranty and manufacturer oversight over the operations of missile complexes of the Strategic Forces, deployed both in Ukraine and Russia. Here, the Russian side was especially insistent. Dubinin wrote: "In the Soviet Union, some of the missiles were produced in Russia, some in Ukraine. As a result, after the collapse of the USSR, Russia found itself with missiles made in Ukraine and Ukraine found itself with Russian-made missiles. Maintaining these missiles in the proper technical condition is the business of the factories that produced them."[51] Worried that Ukraine might refuse to service missiles it had produced that were now in the hands of the Russian Army, the Russians proposed that an agreement for Ukraine's military-industrial complex (MIC) to service these missiles be immediately sent for signature to the heads of government without the entire package of agreements on nuclear disarmament, as Dubinin put it, "immediately and without tying them to other issues."[52]

At this point, a break was made in the talks until the following morning. Late in the evening, I called together the members and advisors of our delegation who were providing us with the necessary specialized information and asked their thoughts about the proposals presented by the Russian side. All their responses were emotional and unambiguous: Russia was openly pressuring Ukraine to such an extent that it was hard to call the talks "negotiations between equals." From the beginning of 1992, the regular supply of parts from Russian enterprises for the regular servicing of nuclear warheads and Russian-made SS-19 strategic missile complexes had been stopped at almost all the units in the Strategic Forces deployed in Ukraine. Russia was now using this situation for blackmail. We also had no right to agree to their demand to sign just one agreement that they were interested in, on servicing Russian missiles made in Ukraine, and to postpone the resolution of all the other issues. Absolutely all the members of our delegation, the advisers and the specialists favored only a package approach to agreements.

In the morning, I informed the Russian delegation of our position: there had to be an entire package of agreements that would ensure nuclear and environmental safety of the strategic nuclear weapons deployed on both Ukrainian and Russian territory, as well as agreements that would establish the basis for destroying strategic weapons, including compensating Ukraine for the nuclear material from its tactical and strategic warheads. The Ukrainian delegation was prepared to propose that the country's leadership sign only a package that would serve the interests of both sides.

This position was not part of the Russians' plans. They made their dissatisfaction very evident and later, Dubinin was quoted in the press as saying sententiously: "Kyiv's escalating nuclear ambitions are becoming increasingly obvious. There is a feeling that Kostenko has no idea of the international implications of this gigantic issue. Ukraine is so firm, he seems to be saying, that it is prepared to confront anyone at all."[53]

When translated into Ukraine's interests, this assessment meant one thing: Ukraine had to approach every step in the most professional manner possible on this truly huge issue. Therefore, during the time we prepared for the second round, I focused on compiling information about technical issues related to the strategic nuclear forces deployed on Ukrainian territory and looked for the optimal approach to utilizing the highly enriched uranium and plutonium that would be freed up in the process of nuclear disarmament.

By 1 March 1993, the Ministry of Defense, together with the Ministry of Machine-Building, Military Industrial Complex and Conversion (Minmash-prom) of Ukraine, which included the MIC enterprises, prepared a detailed analysis of the state of the Strategic Nuclear Forces' missile complexes deployed in Ukraine, together with proposals for how to ensure that they continued to be exploited safely. With this information before us, it became clear that, despite some apparent problems with servicing and maintenance that emerged because Russia stopped supplying parts, there was no state of emergency with the Strategic Forces at that point, let alone "a new Chernobyl." However, as our military confirmed, Russia was becoming more insistent about its exclusive right to nuclear warheads and was taking various steps to bring the weapons storage facilities (S) where they were being kept under the control of the Russian Ministry of Defense.

For us, it now became important that information was being leaked in the Russian press about secret talks between Russia and the US regarding the terms for selling the 500 tons of HEU that would be released during the dismantling of the nuclear weapons. We paid attention to some important developments. For instance, despite the fact that the Russian delegation had offered as compensation for the strategic warheads to exchange HEU for fuel for Ukraine's NPPs, about 3,400 fuel rods for VVR-1000 reactors, and Russian specialists were already saying in their national press that it was not profitable to sell HEU and that Russia would gain the most: politically, militarily and economically, by storing it.[54] Another interesting leak appeared in the Japanese press.[55] Russia was busy holding secret talks not just with the US but also with the G7 countries, persuading the most developed countries of the world to give it direct financial assistance to the tune of US $5 billion to reduce the former soviet nuclear arsenal.[56]

I passed along my report on the outcome of the first round of talks to President Kravchuk and Prime Minister Kuchma, informing Kravchuk that Russia was engaging in open blackmail, refusing to service missiles until we acknowledge that Russia owned all our nuclear weapons. However, I suggested that there was a counter to this threat: using the fact that Russia was also dependent on us we could move to political counterattacks by revealing information about the state that Russian Strategic Nuclear Forces would find themselves in if Pivdenmash refused to service them. Our delegation did not even consider the option of caving in and giving up assets worth tens of billions of dollars only because someone was trying to blackmail us.

Russia Also Depended on Ukraine to Service its Nuclear Weapons

57. Stanislav Koniukhov, 2003

Stanislav Koniukhov (1937–2011), Ukrainian designer of missile and space equipment. Koniukhov earned a PhD in technological sciences in 1987 and became a full professor in 1991. He was a member of six academies, joining the Ukrainian Academy of Sciences and the Academy of Engineering Sciences of Ukraine and the International Engineering Academy in 1992, the Tsiolkovskii Academy of Space Science in 1994, the New York Academy of Sciences in 1996, and the International Academy of Space Science in 1997, where he also was vice president. He worked at the Pivdenne Design Bureau beginning in 1959, becoming its chief designer in 1991 and the general director in 1995.

In fact, Russia actively used blackmail in its "nuclear" relations with Ukraine, not just during these specific talks. In trying to get Ukraine to give up ownership of its nuclear arsenal, Moscow threatened that it would not service the nuclear warheads located in Ukraine. Its logic was simple: Russia manufactured nuclear warheads and only its specialists could service them. If they refused, the warheads would start detonating and this would lead to "new Chernobyls" on Ukrainian territory. Russian positions were based on the notion that Ukraine depended upon it to service the nuclear warheads, but Russia in turn, was not dependent on Ukraine. However, this simply was not true. Russia depended on Ukraine to an even greater degree in the servicing of its warheads.

The first time the information arose that Russia depended on Ukraine for its strategic missile complexes to be serviced, was when Pivdenmash's chief designer Stanislav Koniukhov presented this to the country's leadership on 2 April 1992, at a meeting of the National Defense Council which Kravchuk attended. He explained at the time that more than half of production of the USSR defense industry was located in Ukraine and that Russia's armed forces could not manage without it. As to strategic nuclear weapons, Ukraine and Russia each had 46 missiles with the greatest yield at that time, each of them carrying 10 nuclear warheads. These missiles had been produced at Pivdenmash and were also serviced by Ukrainian specialists.[57]

After this session, I wrote an analytical brief for Kravchuk called "Some Observations Regarding the Implementation of START," in which I proposed a specific approach to our behavior during negotiations: "As to Russian companies servicing the warheads on our missiles, given the almost equal mutual dependence on the security of both warheads and the missile carriers, the problem can be resolved through appropriate agreements. Thus, Pivdenmash will continue to carry out routine work on SS-18 missiles based in Russia and Kazakhstan, while Russia's companies will ensure the proper maintenance of the nuclear warheads on Ukraine's 45 SS-24 missiles."[58]

Nearly a year later, in March 1993, officials from the Defense Ministry and Minmashprom of Ukraine provided our working group with a report on the safety status of our warheads and their servicing by Russia. This rather comprehensive report contained a separate point entitled "Political Matters." There, in terse military style it was stated: "The Russian Federation insists that all warheads, as well as their components, whose guarantee has expired be transferred to it. The issue is being linked to ownership rights over these warheads. The supply of spare parts for Ukraine's warheads has been denied."[59] This made it clear: countermeasures that should have been applied by Ukraine's diplomats and military never did get the go-ahead. There was no question that they would be effective, as it meant Ukraine's defense industry would not only stop servicing but also supplying the carriers, equipment, and parts needed by Russia's armed forces.

This report also provided other information that confirmed Russia's dependence on Ukrainian enterprises: "Russia's rocket forces depend most for the maintenance of their weapons in battle-ready state on the Ukrainian companies VO Arsenal, VO Komunar, VO KZEA, and AP Rostok."[60] Arsenal manufactured the guidance systems for nearly all of the different strategic missiles as well as for missiles under development. Komunar made the electronic control systems used on all the missile types under development. Khartron made control systems for all the new types of missiles. Rostok made all sensor elements for the control systems. Oleksii Kryzhko and Valerii Pavliukov wrote in the report: "Without the maintenance of instruments by these companies, the stock of necessary equipment is likely to be depleted and the missile complexes could end up in a reduced battle-ready state."[61]

This offered leverage serious enough to pressure Russia and could still be used both diplomatically and informationally. Based on these data, the MFA should have launched a campaign to inform the world about the complexity of the situation. However, all the world press would hear was Russia's claims about the catastrophe the world faced if Russia refused to service Ukraine's nuclear weapons. Worse yet, the victim of all the Russian informational attacks was President Kravchuk.

Government Delegation Negotiations: Round Two

The second round of government negotiations began on 2 March 1993 in Moscow. This time, the Ukrainian delegation included the MFA's head of administration, Kostiantyn Hryshchenko, as deputy head. The delegation was also joined by Ivan Bizhan, Valerii Pavliukov, eleven advisors, including Stanislav Koniukhov, and seven experts. The Russians also had a four-member delegation, but also twenty-seven advisors: eight from the MFA, thirteen from the MoD, three from the Ministry of Atomic Energy, and another three from the Defense Industry Committee.

Not only had the position of the Russian delegation not changed, but it became outright arrogant. Refusing to review any drafts regarding the nuclear and environmental safety of strategic weapons, Dubinin argued with us that Ukraine's decisions regarding ownership of nuclear weapons was undermining the entire global security system. Therefore, Russia, which was carrying out a "peacekeeping mission" in the fight against the proliferation

58. Kostiantyn Hryshchenko, 2000

Kostiantyn Hryshchenko (b. 1953), Ukrainian diplomat. Hryshchenko graduated from the Moscow State Institute of International Relations in 1975. From 1976 to 1980, he worked at the UN Secretariat in New York. From 1981 until the end of 1991, he held diplomatic posts with the Soviet MFA. After the collapse of the Soviet Union, he returned to Kyiv to work for the Ukrainian MFA. In 1991, he became a counselor, then a supervisor within the administration of the Foreign Ministry. In 1993, he was placed in charge of the administration at the MFA that supervised armament and disarmament and became a deputy head of the Ukrainian delegation for the negotiations with Russia regarding nuclear weapons. In 1995 he was appointed deputy foreign minister and served until 1998. He then became head of Ukraine's mission to NATO and ambassador to the Benelux countries from 1998 until 2000 and, subsequently, ambassador to the US from 2000 to 2003. Between 2003 and 2005, he was Ukraine's foreign minister. He was appointed first deputy secretary of the National Security Council of Ukraine while serving as ambassador to Russia from 2008 to 2010. He returned to the post of foreign minister again from 2010 to 2012 and during 2012–2014, he was deputy prime minister under the Yanukovych administration.

of nuclear weapons, could not allow itself to be drawn into this dangerous game.

Underneath this verbal veil, Russia's strategy could be clearly seen: to force Ukraine to reject its right to its nuclear weapons in favor of Russia. All of my efforts to return the discussion to constructive decisions ended with Dubinin again and again asking whether Ukraine was a nuclear state or not, and saying that if so, the negotiations were over because the Russian delegation had no mandate to negotiate with a nuclear state.

The talks grew so heated that a break was called. Kostiantyn Hryshchenko came up to me and insisted that the Ukrainian delegation should make a move in response to such brazen behavior on the part of the Russians, which not only flew in the face of the rule of "equal parties," but was beyond the pale of all diplomatic rules on how to conduct negotiations.

Despite having no diplomatic experience of any kind, I intuitively felt that a scandalous rupture of the talks was what the Russian, not Ukraine, wanted and so I rejected the Foreign Ministry official's suggestion. After the break, the pressure shifted to unconcealed threats. Deputy Defense Minister General Boris Gromov announced that if we did not recognize Russia's right to all the strategic weapons deployed in Ukraine, his ministry would not only stop servicing Ukrainian nuclear warheads, but it would also refuse to take over from Ukraine any warheads whose guaranteed term had expired, which could lead to a nuclear disaster.

In response to this obvious blackmail, I noted calmly that, in that case, Ukraine would have to turn for assistance to third countries in order to prevent a nuclear accident. I also reminded the Russian delegation about the possible consequences if our companies refused to service Russia's Strategic Forces. The Russians clearly were not expecting such an answer. At first, silence fell on the room. Then nearly all the military started to indignantly insist that Ukraine had no right to do this because the construction of Russia's missiles was the biggest state secret in the world, and if Ukraine's MIC refused to service Russian missiles this would violate our "fraternal relations." They made other arguments in the same vein.

The Russian delegation categorically refused to discuss the issue of compensation for the tactical weapons that had been transferred from Ukraine in 1992. They stated that our demand to put this question on the agenda had been brought to the attention of the Russian leadership, but the delegation had not received any instructions in this regard and therefore had no right to discuss it. Instead, the Russians proposed yet another initiative, intended, as they put it, to "eliminate the tension around Ukraine's nuclear weapons," specifically, to cancel flight assignments for all nuclear weapons delivery systems on Ukrainian territory, to shift them to a lower level of combat readiness, and to transfer all nuclear warheads for long-range cruise missiles deployed in strategic bombers to Russian manufacturing facilities by 1 August 1993. Additionally, as part of the initiative, they proposed to transfer all the main components of ICBMs and their warheads to central manufacturing sites in Russia by 1 August 1994.

In reality, this initiative was an open provocation. The proposed pace of disarmament, one and a half years, was extremely dangerous, both technologically and for the domestic situation in Ukraine. It was the straight road to social upheaval among the servicemen in the 43rd Rocket Army, many of whom would lose their jobs in a short span of time.[62] Indeed, it was an effective means of shaking up the situation in Ukraine's armed forces, to stir up dissatisfaction with the idea of independence among the military, and to prevent them from taking the oath of service to Ukraine.[63]

As to the technical aspects of the issue, at the time Ukraine lacked 60 percent of the electronic simulators of nuclear warheads, the so-called equivalents, that were needed to replace the warheads on disarmed strategic missiles. These still needed to be produced. From this point of view alone, Russia's proposal was categorically unacceptable.

We also knew very well that Russia itself was not technically prepared for the large-scale clean-up of the nuclear weapons of the former Soviet Union. The technical capacities of Russia's enterprises were not sufficient to quickly ensure the processing and safe storage of the components of such a significant number of nuclear warheads. This was something to which Russia's nuclear specialists had admitted at a seminar in Moscow regarding the problems of nuclear safety in June 1992.[64] There were not even enough containers in Russia to safely transport warheads from Ukrainian territory: Russian orders of them had just started to go into production in the US.

Understandably, this kind of provocative face-off by the Russian delegation was not something to which we could respond positively. This is how Dubinin interpreted our reaction: "There was no answer. [...] One of my partners told me confidentially: 'Kravchuk, of course, is the president of Ukraine. However, do not think that he is in any position to achieve everything he wants or to come through on everything that he has signed. Times are changing in Ukraine [...].'"[65] Like the first round in Irpin, the talks in Moscow ended in just about nothing.

I realized gradually that Russia had moved into the active phase of subordinating the Strategic Nuclear Forces of the former USSR to its own Armed Forces. This not only fit perfectly with the Arbatov Plan but also with the Russian military doctrine, whose main purpose President Yeltsin presented at an extended session of the Russian Defense Ministry on 23 November 1992, as being: "Russia should have a modern army in line with its status as a great world power."[66] Later, *Pravda* more clearly revealed the main essence of this strategy: "After the collapse of the USSR, the Russian Federation inherited not only its nuclear missile strategic forces but also the burden of engaging in confrontation with the nuclear missile forces of the US."[67]

59. Boris Gromov, 2003

Boris Gromov (b. 1943), deputy minister of defense of the Russian Federation between 1992 and 1995. He was deputy head of the Russian delegation, which negotiated with Ukraine regarding nuclear weapons in 1993. He fought in the war with Afghanistan, where he directed the withdrawal of Soviet forces from the country. In 1989, he was promoted to colonel general and became the commander of forces in the Kyiv Military District. In 1990–1991, he was a member of the Politburo of the Central Committee of the Ukrainian Communist Party and first deputy minister of internal affairs of the USSR. In 1991, he became deputy commander in chief of the Soviet infantry; in 1992, the post was renamed to first deputy commander of the General Armed Forces of the Joint Armed Forces of the CIS. Over the period 1992–1995, he was deputy defense minister of the Russian Federation and the main military expert at the Russian MFA with the rank of deputy minister. He was elected a national deputy to the Russian Duma between 1995 and 1999. He was appointed governor of the Moscow Oblast from 2000 to 2012.

60. *Administration of the military oath to Ukraine, in Soviet uniforms, 1992*

Confirmation of such a conclusion was a statement by the Russian Federation Foreign Ministry's Department of Information, issued after the second round of negotiations had ended on 5 March, which included a comment from Russian Defense Minister Pavel Grachev about their failed effort "because of Ukraine's determination to become a member of the nuclear club."[68] Its tone was exceedingly aggressive, despite our agreement with Dubinin not to aggravate a situation that was difficult enough already.[69]

In response, the next day the Ukrainian MFA's press center merely stated, very modestly, that such statements were "one-sided and far from the truth."[70] For the first time, the fear with which the republic's diplomatic service defended itself in response to Moscow's brazen lies roused condemnation in the Ukrainian press: "No matter how fair our statement, making excuses suggests that we are not entirely in the right. Could it be that Ukraine needs to change its tactics?" wrote *Kievskie vedomosti* in response to the tone of the Ministry's statement.[71]

Commenting on the Russian MFA's statement myself, I noted that it was not a matter of Ukraine changing its policy on the nuclear issue but of Russia's desire to take over all the strategic nuclear forces of the former USSR on the territories of sovereign states and to grant itself ownership rights over all the nuclear arsenals. Forcing Ukraine to give up its natural right to own all assets located on its territory, Russia "forgot" that Ukraine was just as much a successor of the Soviet Union as the Russian Federation itself. Moreover, this effectively would cancel all agreements liquidating Soviet nuclear arsenals, including START, that were reached in Lisbon.

When I reported to the parliament on the results of the talks, I notified the legislature about Russia's intentions to ensure the presence of its military on Ukrainian soil at any cost, following one of two possible paths. The first was the indirect path of forcing Ukraine to ratify the NPT: "By ratifying the Nuclear Nonproliferation Treaty together with START I and the Lisbon Protocol, Ukraine will lose the legal right to own nuclear weapons deployed on its territory. A nonnuclear state also cannot have any access to nuclear weapons, which means that, immediately after the act of ratification, Russia will have the legal right to announce that the nuclear weapons on Ukraine's territory belong to it, and the units that service these weapons are part of the Russian Armed Forces."[72]

The second option was direct: using blackmail to force Ukraine to recognize Russia's right to ownership over its weapons: "By artificially setting up conditions prior to the talks, when for months the relevant units of the strategic nuclear forces deployed in Ukraine had not been getting the parts necessary for routine maintenance of nuclear warheads and ICBMs from Russian manufacturers, the Russian side began to use this as a way to pressure us, tying the issue of Russian supplies to Ukraine's refusal to abdicate ownership over the nuclear weapons on its soil, and insisting that the strategic nuclear forces in Ukraine be placed under the jurisdiction of the Russian Federation."[73]

At the same time, the Russians began yet another angle of attack by working on public awareness through the press. Referring to the subject of a likely nuclear accident on Ukrainian soil, Russia worked to slowly get the world to think that, in order to avoid a second Chernobyl, force would have to be used with Ukraine. The progress of negotiations during the second round of talks and the way Russia's Foreign Ministry and Minister Grachev reported on them confirmed that Russian diplomats did not shy away from this tactic for attaining their goals.

Exactly one month later, Russia launched a massive public campaign attacking Ukraine's position. The impetus for Russia's action was a breakthrough in official Kyiv's stance.

First Open Hearings on Ratifying START

At the beginning of March 1993, nuclear disarmament once again came to the forefront in the legislature: President Kravchuk and the Foreign Ministry renewed pressure for the nuclear issue to be brought up for debate in the parliament. On 3 March, Speaker Ivan Pliushch announced that the following plenary week would have a closed session in which, in addition to conversion, the agenda would include the readiness to ratify START I, joining the NPT, and the situation around the division of the Black Sea Fleet.

At this point, I approached First Deputy Speaker Vasyl' Durdynets' with the idea that, prior to such an important discussion (because there was a real danger that ratification would be pushed through immediately), the MPs, the public, and the press needed to be notified of the issues in a more professional and realistic fashion than the theories the MFA was presenting. It was time to move these closed sessions into the public arena.

The best format for this, I thought, was open parliamentary hearings on the problems with nuclear disarmament, which had not been held in Ukraine so far. They would allow for issues to be presented to a much broader audience. Our working group was meeting nearly every week at that point, but these meetings, as well as the Security Council sessions, were closed. Although I always held a briefing on our results, most deputies, the press and, thus, the public, had never heard arguments directly from specialized professionals, both military and legal. Durdynets´ liked the idea, thus, by 10 March, the Working Group was prepared for the first open hearings in the parliament.

The subject for the hearings, "Political and Legal Aspects of Ratifying START I and the NPT," was chosen because at that point the most active discussions in the halls of the parliament were over two questions: Was Ukraine a nuclear or a nonnuclear state? What about the commitment to become nonnuclear? It was important, in terms of international law, to clearly nail down the issue of the ownership of the nuclear weapons deployed on Ukrainian territory, as well as to clarify what these documents were committing Ukraine to do.

In addition to the MPs and members of our working group, the sessions hall included members of the Ukrainian Armed Forces, including the Command of the 43rd Rocket Army, two vice presidents of the Academy of Sciences, Viktor Bar'iakhtar and Valerii Kukhar, plus nine NAS (National Academy of Sciences) participants, some of whom included Nuclear Research Institute Deputy Director Viktor Havryliuk, Institute of Economics Deputy Department Head and NAS Corresponding Member Valerii Heiets´, and Institute of World Economy and International Relations Director for International Security and Doctor of Philosophy Dmytro Vydrin, among other experts, such as lawyers, international law experts, MIC specialists, and members of the Cabinet and presidential administration.

On this day, a clear political and legal analysis of nuclear weapons was presented for the first time. Speaking on behalf of the Working Group, I outlined the three key problems that made it impossible for the parliament to immediately ratify the treaty: first, the lack of security guarantees for a nonnuclear Ukraine; second, the need for financial resources to dismantle and destroy nuclear weapons; and third, legal commitments to compensate Ukraine for the nuclear material in its strategic and tactical weapons.

International law experts presented all the legal implications of the situation in which Ukraine found itself because nuclear weapons were deployed on its territory. Professor Petro Martynenko, director of the Department of Comparative Law at the Ukrainian Institute of International Relations under the Shevchenko National University in Kyiv, in his speech entitled "Main Conclusions in Assessing the Legal Aspects of Ukraine's Ratification of START I and the NPT," confirmed Ukraine's unequivocal right of ownership of the nuclear weapons in accordance with international law. He simultaneously argued that the country had no international commitments regarding nuclear weapons, as it had not ratified a single agreement or treaty related to nuclear weapons, and this meant that other countries could make no claims that Ukraine was not living up to its commitments. As he put it:

61. Viktor Bar'iakhtar, 2005

Viktor Bar'iakhtar (1930–2020), Ukrainian physicist. He was the chairman of the Commission for Nuclear Policy under the president of Ukraine (1993–1995). Renowned for his worldclass advancements in theoretical condensed matter physics, superconductivity, physical properties of solids, nonlinear phenomena in physical systems, and kinetic behavior of solitons, he authored over 500 scientific papers. Bar'iakhtar received his PhD in physical-mathematical sciences in 1965. He then joined the National Academy of Sciences of Ukraine in 1979. From 1985 to 1989, he headed the Department of Theoretical Physics and was the director of the Institute of Metal Physics of the Soviet Academy of Sciences. From 1989 to 1994, he was vice president, and from 1994 to 1998, first vice president of the National Academy of Sciences of Ukraine. He was also the founder and president of the Ukrainian Physical Society in 1990–1994.

The main basis for resolving the issue of ratification [...] needs to be not legal but political considerations [...]. The international system is such that every nation-state should care for its own security independently. [...] In line with current international and domestic law, Ukraine, as a newly independent state, does not yet have any commitments with regard to the aforementioned international agreements, in the strictly legal sense, and should make its political decisions completely freely, without any legal claims or demands on the part of other states [...]. It will have [commitments] the minute it accedes to such agreements.[74]

The various declarations and statements made by Parliament, which others often referred to as commitments made by Ukraine, he underscored, had no international legal implications, and as such could not be used as grounds for demands against Ukraine. The professor also insisted that to simultaneously ratify START and the NPT would be against Ukraine's interests. His conclusions were unambiguous: the pressure some governments were putting on Ukraine in demanding immediate nuclear disarmament had no legal basis; arguments over Ukraine's nuclear status were also baseless because Ukraine was a *de jure,* as well as *de facto* nuclear state. He summed up his arguments:

62. *Valerii Kukhar*

> Therefore, for other states or their officials to attempt to force their demands on Ukraine based on its declared principles and statements and treat these as a legal basis for such demands is legally untenable, as it is in accordance with neither the norms of international law nor international legal practices among states. Ukraine remains legally free to choose its policy regarding eliminating nuclear weapons.[75]

The final determination of the hearings was that, before ratifying any nuclear disarmament treaties, the parliament needed to have all the evidence necessary to be certain that this would not result in serious damage to Ukraine.

That day, I finally felt that we were reaching a breakthrough. These hearings launched a flood of analysis on the part of patriotically inclined staff and military. They continued to prepare reports at the meetings of the Working Group on issues that were then considered controversial or that had not been examined from the point of view of Ukraine's own interests. In effect, this entailed working up all the arguments to carry out further negotiations at any level and at the same time applying a theoretical basis to the unusual practical situation that Ukraine had found itself in. This process of self-organization by patriotic specialists looked especially remarkable against a background of almost total inertia on the part of government agencies.

Thus, by early April 1993, the director of the Department of International Law at Franko University in Lviv, Vasyl′ Repets′kyi,

Valerii Kukhar (1942–2017), vice president of the National Academy of Sciences of Ukraine in 1988–1993. He was the chairman of the Commission for Nuclear Policy and Environmental Safety under the president of Ukraine from March 1995 to July 2000, and director of the Institute of Bio-organic Chemistry and Petrochemistry under the National Academy of Sciences 1987–2013. From 1988 to 1993, he was also chair of the Scientific Council on Biospheric Issues at the National Academy of Sciences, director of the Coordinating Office for Automation, Computing Technology and Computer Sciences at the National Academy of Sciences, and a member of the Ukrainian Nuclear Society. He became deputy chair of the Council on Scientific and Technological Policy under the president of Ukraine from 1996 to 2000. From the year 2010, he was the director of the State Foundation for Basic Research.

brought out additional arguments in a brief he presented at a Working Group session regarding Ukraine's right to own its nuclear weapons:

> Ukraine has *de facto* recognized the 1983 Vienna Convention on Succession of States in Respect of State Property, Archives and Debts. Determining the question of Ukraine's succession in respect to the property of the Ukrainian SSR and the USSR revolves around the principles that can be called "the principle of territory" and "the principle of proportionality." Based on the first, all state movable property and real estate of the Ukrainian SSR and all real estate of the USSR located on the territory of Ukraine becomes Ukraine's property. The movable part of state property is transferred to successors based on this property being present with the successors. Movable property whose ownership cannot be determined should be distributed based on the principle of proportionality, which criteria are established with the agreement of all sides.[76]

Repets'kyi brought to our attention two more important details which, for some reason, were previously never mentioned by our Foreign Ministry. First, he emphasized, "when ratifying multilateral treaties, every state has the right to formulate its own reservations, if they are not expressly prohibited by the text of the agreement." Second, he explained Ukraine's rights regarding the external debts of the Soviet Union, which Russia was trying hard to shift to Ukraine: "In accordance with the Law 'On Ukraine's succession,' Ukraine had agreed to service the external debt of the Soviet Union as of 16 July 1990, to an extent that shall be designated in a separate intergovernmental agreement. Moreover, Ukraine does not have any obligation for USSR loan agreements and other contracts to which Ukraine did not agree."[77]

Scholarly circles began to look at the actions of officials in the light of national interests, a concept that was starting to take on more and more substance. Vitalii Kriukov, a senior researcher at the NAS's Institute for State and Law, focused attention on the conflict between Ukraine's interests and those of the states that were trying to make it disarm immediately:

> [O]ur policy here has proven lacking in initiative and is being carried out in a context and in directions that are largely detached from reality. They have a clearly inferior international legal basis, and are identified and promoted primarily not by Ukraine's national interests but by the largely selfish aspirations of the US and Russia. Therefore, this is happening precisely because [...] knowingly or otherwise, we are not being allowed to identify the real substance of [...] the problem. Everything constantly comes down to and is limited to the narrow framework of START I and the NPT, everything without exception being related to the problem of ratification and accession.[78]

The most active element among the military experts were those from Kharkiv Military University. The reason was obvious: it was headed by one of Ukraine's nuclear hawks, the MP Lieutenant General Volodymyr Tolubko. Three of his specialists prepared reports on "The Military and Political Aspects of Arms Reduction in Ukraine," which they presented at the Working Group session of 9 April. Major General Kovtunenko, a professor and deputy director for scientific research at Kharkiv Military University, had analyzed world trends and pointed out that nuclear weapons had been developed by nearly two dozen countries during the last few years, while not one country had destroyed them in that same period. The military specialists pragmatically noted that appeals to reduce arsenals were largely declarative rather than technical in nature. They were categorically against Ukraine abdicating the respectable status of a nuclear state in exchange for an approving "pat on the back."

Summing up the 10 March meeting, the New Jersey-based *Svoboda Ukrainian Daily* wrote: "Experts confirmed that Ukraine is a nuclear state, since the weapons deployed on its territory belong to Ukraine. MP Bohdan Horyn´ said that Russia and the United States are looking at the issue based on their own national interests, while Ukraine was looking at it based on its own interests. 'We are not talking about looking for a way to delay ratification itself; we are looking for ways to actually ratify it, but to do so taking Ukraine's interest into account first,' added Yuri Kostenko, head of the deputy's group."[79] The paper also quoted Volodymyr Butkevych, a professor at Kyiv University's International Relations Institute and UN human rights expert, as saying that Ukraine was not yet a state "that can ratify a treaty like START I. It will be ready to do so no earlier than in five years."[80]

The data and analysis, their profile raised by the preparations for the hearings, had an explosive impact.

Breakthrough

In March 1993, Ukraine's position on nuclear disarmament was declared at the highest level for the first time. The breakthrough happened on 15 March at a session of the National Defense Council of Ukraine (NDC). The meeting was presided over by President Kravchuk, with Speaker Pliushch and Premier Kuchma in attendance. The main issue being discussed was a Ukrainian strategy for nuclear disarmament in the context of Russia's ultimatum-style position at the intergovernmental talks and the threat that the Russian Federation would take control of the strategic weapons on Ukrainian territory.

As head of the government delegation at these talks, I gave a report. I presented all aspects of the negotiations process in detail and brought attention to the main divergences in positions that the delegations of Ukraine and Russia were trying to defend, which were fundamentally political in nature. They could only be set aside if the Parliament of Ukraine repealed its previous decisions regarding ownership on the territory of Ukraine, along with its nonbloc status, and the strategy for eliminating the nuclear weapons outlined in its Statement and Resolution on plans for Ukraine to gain nonnuclear status. Repealing these decisions was clearly impossible.

Given that the key problem in further negotiations remained the question of ownership, I proposed that the NDC adopt the option that was most acceptable for Ukraine: as the owner of the nuclear weapons, to temporarily transfer them to the Joint Armed Forces of the CIS for use during combat duty. This would allow for formally retaining the existing system of centralized management, while Ukraine's Defense Ministry would handle administrative command and control.

All other options would only have negative consequences. Among the main ones were the following. First, Ukraine would lose the ability to control the use of nuclear weapons located on its territory. Second, Ukraine would be unable to receive objective information about the technical condition and battle readiness of the missile forces, including the state of nuclear and environmental safety. If Russia were to control the missile systems, Ukraine would be getting information, not directly, but only through Russian sources. Third, Ukraine would lose billions of dollars' worth of costly material contained in nuclear weapons, including the physical components of the warheads. Fourth was the threat of Russian forces' deployment on Ukraine's soil.

The problem of servicing the missile complexes before they were eliminated could be resolved by maintaining their technical condition at the level of nuclear and environmental safety, but not their battle readiness. For that, even before the parliament were to ratify START I, Ukraine would propose an initiative to remove from battle duty the 130 outdated liquid-propellant SS-19 missiles and to shift the 46 SS-24 complexes manufactured at Pivdenmash to the state of reduced battle readiness. Afterwards, the warheads from the 130 missiles would gradually be moved from their starting positions and missile bases to weapons storage facilities (S). We proposed also moving all cruise missiles to weapons storage facilities (S) prior to ratification.

This would allow Ukraine to first, reduce the urgency of the problem of servicing its nuclear warheads and neutralize one of Russia's means of pressuring Ukraine during negotiations; second, raise the political repute of Ukraine and through our actions demonstrate our intention of remaining on the path towards nuclear disarmament; and, third, gain time to more broadly understand the scope of the problems connected to destroying nuclear weapons. I expressed my conviction that these measures would allow Ukraine to carry out talks with government delegations more constructively and, in the end, to sign the necessary agreements to resolve all the problems tied to the danger of strategic arms and the further elimination of nuclear weapons.

The next problem that the NDC had to resolve was ratifying the NPT. I pointed out to those present the legal fact, which Ukraine's international law experts had confirmed during the recent hearings in the parliament on the "Political and Legal Aspects of Nuclear Disarmament," that all the nuclear weapons of the former USSR that became property of independent Ukraine did so without any violations of international commitments, and entirely in line with the norms of current international law.[81] These weapons were exclusively the property of Ukraine, with all the legal implications attached to it. From the legal point of view, all other states had to recognize this fact and respect it appropriately. This meant that, under the circumstances, Ukraine could not ratify the NPT as a nonnuclear state.

Another issue to consider, which I mentioned earlier in connection with the signing of the Lisbon Protocol, was a possible legal trap for Ukraine, if it ratified START I and the NPT simultaneously with the Lisbon Protocol. This event would cancel Ukraine's legal right to own nuclear weapons deployed on its territory. Since a nonnuclear state could not also provide access to nuclear weapons, immediately after ratification, Russia would have the legal right to declare the nuclear weapons on Ukrainian territory its own property and the troops that serviced these weapons would become part of the Russian Armed

Forces. Based on this, I proposed, first, to temporally separate the ratification of START I and the NPT; second, to accede to them with certain reservations, specifically, to accede to the NPT but only when Ukraine had become a non-nuclear state, after the complete destruction of the nuclear weapons located on its territory.

The next important topic was what to do with nuclear material that would be released in the process of disarmament. Here, the NDC needed to consider a number of factors. Russia's refusal to acknowledge Ukraine's right of owner-ship over the nuclear weapons, and thus of the nuclear material in its physical form, as well as its refusal to consider the issue of compensation to Ukraine for the tactical nuclear weapons already removed from Ukraine, brought up an additional threat: if the strategic warheads were transferred to Russia for dismantling, Ukraine could forever lose their material value. After all, signed agreements meant little to Moscow if it was a matter of its own profit. In short, it would be more to Ukraine's benefit to consider other options for utilizing the nuclear warheads.

At that time, there were two realistic options, and both involved engaging a third, "civilized" party, essentially to guarantee that the Russians kept to any agreements. The first option was to sell the components of nuclear warheads, uranium, and plutonium, to other nuclear states, primarily the US. Here, the dismantling would take place at Russian enterprises, but first an agreement would have to be signed between Ukraine, the US, and Russia, which would provide Ukraine with the necessary guarantees. The second option was to dis-mantle the warheads on a commercial basis and process the highly enriched uranium into low enriched uranium in Ukraine. This would require setting up a joint venture involving specialists from both Russia and the US.

Finally, I drew attention to a fact of which most of the highly ranked offi-cials present were not aware: START I actually did not call for the destruction of nuclear warheads. This treaty only kept track of the number of destroyed missile silos and strategic carriers (as was discussed in detail in chapter 1). Given that Ukraine, in contrast to the USSR and Russia, had no industrial base to produce nuclear explosive devices, eliminating the warheads alone would automatically fulfill the requirements of START I; after all, without their warheads, the silos and carriers could not represent any nuclear threat to the world. Moreover, I pointed out, ruining the silos would be one of the most high-cost items in the process of our nuclear disarmament.[82]

The formal basis for shifting the emphasis of disarmament to the destruc-tion of nuclear warheads was several decisions by the parliament. Accordingly, Ukraine advocated eliminating nuclear weapons as such, that is, primarily the nuclear warheads. Given this, in ratifying START I, the parliament not only had a basis for amending provisions related to the destruction of the missile silos on Ukraine's territory, but was also committed to doing so while keeping its economy and the environment in mind.

In wrapping things up, I said that further talks with Russia about nuclear weapons might yield positive results only if the issues raised led to decisions that were clear and understandable to all branches of government. After a dis-cussion that lasted more than five hours, the NDC passed a resolution "On Urgent Measures to Carry Out Ukraine's Nuclear Policy," which was stamped "Secret."[83] Notably, the status of this document was very high, as the meeting in-volved top officials from every branch of government: the president, the Speaker, and the premier, while the Defense Council itself was the highest body in the state for making policy decisions regarding the country's national security.

With the new political realities confronting Ukraine in relations with Russia after Lisbon, Russia's open aspiration to take over all the nuclear weapons, to subordinate the Strategic Forces and the Black Sea Fleet, and essentially ensure complete control over Ukraine, forced the overly cautious Kravchuk to take some decisive steps. His decision dotted all the i's on all tactical questions regarding disarmament that had come up at that point. It was a clear directive to all government agencies. It was also the first document agreed to by the leadership of all branches of government which clearly confirmed that all the assets of the Strategic Forces deployed on Ukrainian territory, including the physical components of nuclear warheads, belonged to the state of Ukraine.

Having confirmed that it had no intention of gaining independent control over the strategic nuclear weapons for the purpose of attacking any other country, since operational control remained with the Joint Command of the CIS Strategic Forces, the NDC simultaneously declared that Ukraine needed to have means to ensure technical control over the non-use of nuclear weapons from its soil. The Ministry of Defense of Ukraine would be responsible for administrative management of the strategic nuclear forces on Ukrainian territory until such time as the nuclear warheads were completely destroyed, in order to establish the conditions for securely and safely maintaining nuclear weapons, and preventing their unsanctioned use.[84] In addition, the Ministry was to ensure the staff in the Strategic Nuclear Forces were exclusively citizens of Ukraine.

The NDC confirmed that the reduction of Ukraine's nuclear weapons would take place in line with the limits and restrictions designated in START I after reaching agreement with Russia, Belarus, and Kazakhstan. The terms for transferring the warheads for further dismantling outside Ukraine were to be established in a special agreement. Given that Ukraine had not participated in drafting START I between the USSR and US, the decision mentioned the need to revise the conditions of this treaty, especially regarding the use of missile silos. The Cabinet was given ten days to prepare and submit for the president's signature a new directive to the government delegation for further negotiations, reflecting this NDC decision.

There was one final decisive point. The NDC resolution obligated the executive agencies authorized to make nuclear policy (and their representatives for specific activities) to ensure that Ukraine would accede to START and the NPT on the basis laid out in the annexes to the Resolution. These were also stamped "Secret." In short, after this decision, for the first time in the entire period of disarmament, others among Ukraine's government institutions besides our parliamentary working group began to interact in a more coordinated fashion to counter the new challenges that were being generated with increasing vigor by Russia. One final outcome of the NDC resolution was that all the servicemen in the 43rd Rocket Army and the 46th Air Army now took an oath of service to Ukraine.

The Russians did not take long to respond. On 5 April 1993, the Russian government published its view of what Ukraine's nuclear policy should look like after Lisbon. Yet, behind their aggressive behavior, the Russians had another skeleton rattling in their closet.

Russia Was Not Technically Prepared to Handle All the Warheads from Ukraine

To promote the idea that the safest option for the rest of the world would be for all of Ukraine's nuclear weapons to be moved to Russian territory, Moscow developed an entire package of myths. They were based on Russia's intellectual and technical exceptionalism among other former Soviet republics. The implication was, of course, that only Russia was capable of ensuring the safe transport, dismantling, and storage of plutonium and HEU from all the former Soviet arsenals. To keep these materials in Ukraine, without the oversight of Russian specialists, would inevitably lead to a nuclear catastrophe because "backward Kyiv" did not have the necessary storage or personnel.

These myths were quite actively propagated, not only in the Russian press but also in Western media outlets. The reality was, however, that it was Russia who lacked the capacity to safely store nuclear material or even transport it in the early 1990s. The Russians themselves and the Americans talked openly about this, both in the press and in official documents.

According to Fred S. Celec of the Pentagon's Defense Nuclear Agency, Russia had a problem securely and safely transporting warheads from their deployment locations to the companies that serviced them. Indeed, American scientists had to develop a special model of ironclad car as technical assistance to Russia for disarmament. Moscow needed 115 such cars.[85]

At the aforementioned international seminar on nuclear safety issues held in Moscow in June 1992, a member of the Russian Duma's Environmental Committee, Vitalii Filonenko, noted the problem of utilizing nuclear warheads and the fact that Russia's MIC was unprepared for large-scale storage of nuclear material, primarily because it lacked containers for weapons-grade plutonium. What "large-scale" meant was that by 2000–2015 (these terms were determined either based on the guaranteed use-by date of the warhead or mandated by nuclear weapons reduction treaties), Russia was supposed to destroy approximately 15,000–20,000 warheads from tactical and strategic weapons. This was an enormous load on the companies that were supposed to reprocess the material. For this reason, Russia had asked the US to funnel US \$400 million to build a storage facility in Siberia to store plutonium, close to where plutonium elements of nuclear explosive devices were being manufactured, near Krasnoiarsk.[86]

In March 1993, the US Embassy in Kyiv officially confirmed this information. In its memorandum, the diplomatic mission stated that "the plutonium extracted from [Ukraine's] nuclear warheads [...] is supposed to be stored in structures that guarantee storage security and safety" and "the US is providing assistance to Russia in designing such structures."[87] This made it quite clear that Russia had no such structures at the time and that the US could just as easily have provided the same kind of assistance to Ukraine, had it been requested.

The most important part of the program was designing a fortified underground storage bunker to keep the dismantled warheads, together with an electronic security system, which would meet Russia's needs, as the *Wall Street Journal* wrote:

> The largest piece of the program, the design of a fortified, electronically secured underground bunker to store dismantled warheads, is under way. Russian plans call for it to be a third of a mile long and as wide as about 38 US football fields. However, US experts are unclear about what the Russians are going to put in the structure, which some have dubbed "Fort Plutonium." [...] So far, US aid is concentrated on providing emergency equipment, such as Kevlar blankets, to protect Russian warheads in shipment. The Defense Nuclear Agency is about to seek bids from US companies to construct 10,000 stainless steel drums designed to store plutonium removed from weapons.[88]

Additionally, the article went on to say, "[o]nce the canisters of plutonium are shipped in the railroad cars from bomb-disassembly sites to 'Fort Plutonium,' they will encounter another problem that US laboratories and companies will be assigned to work on: how to monitor and account for them regularly."[89]

The fact was that Russia's nuclear industry, like all its economy, was going through a massive crisis at the time. The state of Russia's nuclear centers at the time was described by a Dutch political analyst, Henk Wolsack, in an article called, "Russia: The Country's Nuclear Shield is Being Eaten Up by the Rust of Chaos and Crisis." Given that under Ukraine's large-scale denuclearization program, one of the main issues was the problem of how to eliminate nuclear warheads at the manufacturing plants in Russia, the writer asked: "What kind of plants are these? What state are they in, now that the country is moving to market relations?"[90] Wolsack summed up the situation as follows:

> In all due seriousness, it can be clearly stated that Russia's nuclear industry is on the edge of bankruptcy today. People from Russia's nuclear industry say that financial problems are making it impossible to ensure the work of the factories. [...] Academics, engineers, technicians and workers at the country's biggest nuclear centers can go half a year without pay, while exclusive equipment and machinery rust for lack of money without routine maintenance and repairs. [...] The question of closing the "secret" Arzamas-16 Nuclear Center near Nizhniy Novgorod is already on the legislature's agenda; this is where 80 percent of the workforce was once engaged in the manufacture of nuclear weapons. [...] Russian Premier Viktor Chernomyrdin announced that closing Arzamas-16 and other similar centers could lead to an unheard amount of economic, political and social misfortunes, because this would mean laying off hundreds of thousands of highly qualified specialists. Chernomyrdin called the prospect a threat to Russia's national interests.[91]

In addition, Moscow simultaneously used the threat of a brain drain from Russia to countries that the US saw as rivals as a means of blackmail. This threat was supposed to encourage the Americans to think that it might be cheaper to allocate a few million dollars to Russia, with which the country could improve its situation, than to worry later about what to do with nuclear bombs being developed all over the world thanks to Russian specialists.

Yet, this information was destroying evidence of Moscow's construct that the only way to save the world from the nuclear warheads deployed in Ukraine was to immediately move them to Russia. The Ukrainian side had serious claims that could stop the lightning-fast removal of valuable nuclear material out of the country. Nevertheless, the Ukrainian foreign ministry never once tried to use them, not in its public statements, nor at diplomatic meetings. This passivity on the part of Ukraine gave the Russians excellent openings for an attack.

Postscript to the Talks: Russia Moves to "Military Action"

"Nuclear weapons cannot belong to a nonnuclear state, the security of nuclear weapons cannot be divided," was how Russia began its public attack against Ukraine's nuclear ownership in a statement issued by its government on 5 April 1993.[92] In its statement, the Russian Federation government essentially demanded, or, in their own words, "proposed," that Ukraine immediately remove all nuclear weapons from its territory, accusing Kyiv of "laying claim to owning nuclear weapons" and violating an entire string of international agreements. Among Ukraine's "obligations" Moscow included agreements "to be ratified" by the parliament, which meant they were not in effect for Ukraine at that time. These included two decisions within the framework of the CIS, or the Lisbon Protocol, and other commitments that Ukraine had never actually taken on, such as removing nuclear weapons from its territory by the end of 1994. It was already traditional for the Russians to ignore Ukrainian legislation, according to which the nuclear weapons on Ukrainian territory had been declared property of Ukraine.[93]

On that day, a press conference took place in Moscow dedicated to Ukrainian-Russian negotiations, with the top members of the delegation, including Deputy Foreign Minister Berdennikov, Deputy Defense Minister General Boris Gromov, ambassador and head of the Russian delegation Iurii Dubinin, and officials from the Defense Ministry. Gromov clearly formulated the main reason for Russia's dissatisfaction: "For the first time an announcement has been made that it [the nuclear weapons] belong to Ukraine. The purpose of the Ukrainian position is to force Russia to recognize Ukraine's ownership of nuclear weapons." That the Ukrainian legislation had already established this as a fact in September 1991 the Russians presented as a global sensation. Then they invoked a catastrophic prospect: "During the second round of negotiations, it has become clear that Ukraine is not prepared to eliminate its nuclear weapons [...]. Ukraine aspires to become a nuclear state."[94]

These intentions became clear, stated Gromov, from President Kravchuk's Decree No. 209 and Defense Minister Morozov's Decree No. 9, as well as the taking of the oath of service by the 43rd Rocket Army. Gromov then quoted their key message: "Ukraine has gained a fundamental opportunity to use nuclear weapons."[95] Although not just the Russian but also the American military understood that only the Kremlin possessed access to the button, this sensationalist statement was disseminated in an incredibly coordinated fashion over the following month, in the Russian, American, and European press.

Official Kyiv found out about the details of the press conference from a report by the press counselor at the Ukrainian Embassy in Moscow, Vadym Dolhanov. His report was received at the Foreign Ministry the next day, 6 April. However, the Ministry only sent this highly important information to First Deputy Speaker Vasyl' Durdynets' on 22 April, while the Working Group had received it via Durdynets's resolution on 26 April, which was three weeks after the event.

In reporting on this press conference, *Holos Ukraïny* reminded readers that Colonel General Gromov had once been commander of the Kyiv military district, including in July 1990, when the parliament was passing the Declaration of State Sovereignty. Despite taking a sarcastic tone towards the general, the Ukrainian paper "attacked" him using his main argument: "The nuclear weapons are neither Russian nor Ukrainian, but Soviet."[96] This showed that even patriotically inclined writers for the parliamentary publication did not understand the key issue: ownership. Under such circumstances, informing the public was virtually impossible.[97]

In response to the Russian government's statement, the Ukrainian government printed a statement in the 7 April issue of *Holos Ukraïny*, to justify, albeit without reference to Ukrainian law, why Ukraine had the right to talk about ownership, though not of the weapons themselves.[98] The essence of the statement was that Ukraine "confirms its ownership" not of the weapons but only "of the nuclear components of the weapons.[99] This passage served as an indication that the text had been written at the Ministry of Foreign Affairs.[100] No further explanations followed in the pages of *Holos Ukraïny*. Incidentally, on the same spread in the paper was an article reporting on the press conference that clearly underscored the purpose of the Russian government's statement: to present Russia in the role of a warrior against the proliferation of nuclear weapons, which was confronting a "nuclear monster," as it characterized Ukraine.[101]

However, this was not the most important point. In that same issue, *Holos Ukraïny* quoted President Kravchuk, who claimed at his meeting with Yeltsin that Ukrainian-Russian negotiations over the fate of strategic nuclear weapons deployed on the territory of Ukraine had reached a dead end, and he proposed moving the resolution of this problem to the level of the premiers.[102]

In principle, the idea of "raising the level of the negotiations" looked legitimate. Yet, in practice, it meant that the president was prepared to change the head of the government delegation for another person who, unlike me, would not insist on the very thing to which Russia was not prepared to agree: Ukraine's right to own the nuclear weapons. The Massandra Accords signed at the beginning of September that same year showed exactly what Russia wanted from the new head of the Ukrainian government delegation.[103]

The Statement of 162 MPs

The first two rounds of talks with the Russians had shown that the parliament had to pass a resolution which stated in no uncertain terms that the nuclear weapons on Ukrainian territory belonged to Ukraine. That was how the draft statement of the parliament "On Ukraine's Right of Ownership of Nuclear Weapons" was born, which we submitted to the legislature for consideration. At that point 162 MPs were demanding that this document be passed. They had signed a "Statement on the Nuclear Status of Ukraine," which had become one of the most talked-about topics in the world press in April 1993.[104] Making use of arguments presented by international law experts who had participated in preparing propositions for our working group, the statement argued that Ukraine was the owner of the nuclear weapons on its territory and that not recognizing this fact was getting in the way of destroying the weapons:

> After Ukraine declared independence and the USSR collapsed, Ukraine became a nuclear state as one of the successor states to the former Soviet Union. This is confirmed by Article 12 of the CIS Treaty, which was ratified by all participating CIS states, by Article 1 of the Lisbon Protocol, where Belarus, Kazakhstan, Russia, and the US recognized Ukraine as an equal party to START I, and the Vienna Convention on Succession of States in Respect of Treaties, which was ratified by the Parliament of Ukraine.
>
> Unfortunately, today there is considerable speculation about Ukraine's nuclear status and Ukraine's right to own the nuclear weapons deployed on its territory, which can be seen in the positions and official statements by representatives of certain states. Because of this, we believe that the Parliament of Ukraine should confirm this right by adopting the appropriate statement.
>
> Without confirming Ukraine as the owner of nuclear weapons deployed on its territory, the parliament cannot begin the process of final debate over START I. Due to unclear information in both the Ukrainian and world communities on what exactly Ukraine's position is, if Ukraine ratifies the Treaty, it could be interpreted as Ukraine, in effect, liquidating nuclear weapons that do not belong to it. [...]
>
> Attempts by certain countries to force Ukraine to immediately ratify START I, the NPT, and the Lisbon Protocol and to refute its status of a country that *de facto* and *de jure* is the owner of nuclear weapons is unacceptable. The parliamentary committees are currently undertaking in-depth analysis of these agreements. [...]
>
> We are grateful to those states that are proposing Ukraine some funding to cover the cost of reducing nuclear weapons. At the same time, it would be a mistake on our part to agree to promises of minor amounts of funding in return for Ukraine's immediate nuclear disarmament. The questions of independence, national security, and territorial integrity cannot be subject to bargaining or any "monetary compensation."[105]

The *Wall Street Journal* reported on this in a 3 May article: "The lawmakers say in the letter that they cannot ratify any international disarmament treaties until Ukraine proclaims itself owner of all the former Soviet nuclear weapons on its territory. Without such a proclamation, Ukraine could be seen as 'liquidating. . . nuclear weapons that don't belong to it.'"[106] American diplomats reacted to the statement almost immediately. In a discussion with MFA official Vladyslav Dem'ianenko, Walter Sulzhynsky of the US Embassy openly expressed the concern of the American side "regarding the published statement of the group of deputies in the Parliament of Ukraine regarding the nuclear weapons deployed in Ukraine."[107] In his report on the meeting, the Ukrainian diplomat wrote: "He emphasized that the position of this group of deputies testifies to the fact that they either do not understand or do not want to understand the provisions of START I and the Lisbon Protocol attached to it. However, as he stated, what was even harder to understand was that a member of the government of Ukraine, Yuri Kostenko, holds this position, which is not in line with the official one."[108] The reason the MFA official did not explain the legal basis that lie beneath this statement, that is, why the nuclear weapons belonged to Ukraine, became clear within a few days.

On 23 April 1993, the next meeting of our working group took place. This time, it was closed and was joined by experts from Pivdenmash, the Kharkiv Institute of Physics and Technology, Khartron, Kharkiv Military University, the Defense Ministry, the Machine-Building Ministry, the National Academy of Sciences, and the Institute for Nuclear Research. In the evening that same day, I held a briefing, as usual, on the results of the session. When one journalist kept asking, "So, is Ukraine a nuclear or nonnuclear state?" I answered that Ukraine, *de facto* and *de jure*, was a nuclear state. News about the briefing was published. As events subsequently developed, this event was part of the plan to change the head of the government delegation, a plan in which both the Ukrainian and the Russian MFAs were involved (as will be shown further in chapter 4).

At first, a short statement from the MFA appeared in *Holos Ukraïny* on 29 April. The reason was "on 27 and 28 April, information appeared in many domestic and foreign mass media sources that cited the head of the special parliamentary working group for preparations to ratify START I, Minister of the Environment Yuri Kostenko, as saying that Ukraine is a nuclear weapons state."[109] Responding to it, the MFA stated the following: "This kind of interpretation of Ukraine's status (if the statement really was made) does not reflect the official position and should be treated as the personal opinion of Mr. Kostenko."[110] In essence, the statement coincided with one voiced by the American embassy official. However, that was not the end of it.

This brief note on the seventh page of the Ukrainian paper led to a briefing at the Russian MFA press center in Moscow the very next day, 30 April. At the briefing, the director of the Department for Information and Press, Sergey Yastrzhembsky, welcomed the Ukrainian MFA's announcement, saying that it had "disavowed Yuri Kostenko's statement that Ukraine is a nuclear state."[111]

In his report on the briefing, Vadym Dolhanov, the MFA counselor for the press at the Embassy in Moscow, wrote that the Ukrainian MFA's statement "was met positively in political and diplomatic circles of the RF" and that they simultaneously congratulated Kravchuk's initiative to "raise the level of the negotiations," that is, to change the head of the delegation at the talks.[112] Deputy Foreign Minister Borys Tarasiuk sent this note to Premier Kuchma without any commentary, as though the response had been a positive achievement for his agency.

A month remained before the question of ratifying START I and the NPT would be brought to a parliamentary session. Preparations were now ongoing on all sides: the US, Russia, and the two camps in Ukraine.

Military Doctrine: Defending from Whom and with What?

Oil was added to the fire by the fact that, earlier that year in April, the parliament was supposed to undertake the first reading of Ukraine's military doctrine. The nuclear hawks' positions were gaining strength, and this was causing their opponents, both in Ukraine and abroad, to feel worried. Might Ukraine really announce that it intended to remain a nuclear weapons state?

Obviously, nuclear disarmament had to be established in the context of three key documents, namely the national security policy, the "Key Directions in Ukraine's Foreign Policy," and the military doctrine. The first of these was supposed to establish a system of coordinates for the new state, specifically, who was friend and who potential foe. The other two had to translate this vision into the context of international relations and military policy.

Thus, two key questions had to be answered by the military doctrine: Who were the potential opponents? and How do we defend ourselves against them? For many deputies, the second question was being considered thus: We are giving away our nuclear weapons, and what will replace them? What is there in our system of defense that might be used instead of nuclear weapons? The nature of this doctrine was the only idea that was unanimous across Ukraine's political spectrum: it was supposed to be defensive.

Drafting of the military doctrine formally began right after the Law on the Armed Forces of Ukraine was passed on 6 December 1991. The parliamentary resolution that accompanied this law presented standard instructions to the government and appointed several ministries to be responsible for actually developing the doctrine, including the Defense Ministry and the MFA.

The intellectual part of this process was, as usual, concentrated in the parliament, this time in the Defense and Security Commission, chaired by Valentyn Lemish. All the drafts written by experts were submitted there. Many members of Lemish's commission were in our working group, which was developing the concept for nuclear disarmament, and so they were in command of the information we all knew, and vice versa.

The drafting process stretched to nearly two years, because they could not determine who was a potential enemy of Ukraine. The document approved on 19 October 1993 testifies that it was still not identified in the end. The rank-and-file communists who were running the country were now faced with a question to which they had no answer. No one believed any more in a military threat coming from either the US or Europe. On the other hand, Russia, for them, continued to be not only part of the same state, but even the central power.

Thus, it turned out that Ukraine had no one to defend itself against and a military doctrine was a mere formality.

"Neither the US nor other NATO countries were enemy countries in terms of their relations with independent Ukraine; same with China. Whom then should we threaten with nuclear weapons?" said Ievhen Marchuk much later, at an International Management Institute meeting in 2012. Russia was never mentioned even theoretically. Moreover, 20 years earlier, at the end of 1992, it was Marchuk who wrote a *Holos Ukraïny* article about the idea of a "Russian nuclear shield" over Ukraine:

> Sometimes we can hear statements from certain political functionaries that the very presence of nuclear weapons is a deterrent factor, a guarantee of security, ergo it can ensure predictability in Ukrainian-Russian relations. However, those who know the military and technological aspects of the nuclear missiles in Ukraine and Russia know that this is a dangerous illusion.
>
> The secrecy around this issue makes it impossible to illuminate the full depth of this iceberg. That is why emotions are roused mainly around the relatively small visible part of the iceberg and this fools many people. As long as relations with Russia are not properly established, the nuclear missiles currently deployed in Ukraine will only be a source of tension in those relations. Ukraine's leadership and the professional specialists are quite aware of this. Only coordinated actions with Russia as an equal partner can keep Ukraine under a nuclear shield, at least for a while.[113]

This was the fundamental approach to the issue by a large group of Ukraine's leadership, including the president, the MFA, the SBU, and, initially, the Defense Ministry.

Then again, the circle of people who were generating expert opinions was small, and those who were prepared to state them in public could be counted on the fingers of one hand. The most adamant among the military was the director of the Kharkiv Military Academy, Major General and MP Volodymyr Tolubko. The most vocal among MIC specialists was the chief designer of Pivdenmash, academic Stanislav Koniukhov. In the backrooms, there were plenty of opinions, but the minute we all came to a meeting with the president or the premier, most of the managers, in true Soviet tradition, began to nod their heads in agreement with whatever their higher-ups were saying.

General Tolubko was an unabashed nuclear hawk and saw giving up nuclear weapons as a betrayal of future generations. He called unilateral nuclear disarmament "an unpredictable risk, the alternative to which can only be allying with a nuclear state that would be guarantor based on economic interest." However, he did not name such a state. He carefully analyzed global trends in approaches to military doctrines and made an effort to present his thoughts in publications and analytical studies. In his article "Security and Ukraine's Military Doctrine: Formula and Philosophy," he considered a doctrine of guaranteed deterrence as one appropriate for Ukraine. As to the means by which it could actually be carried out, Tolubko gave an unambiguous answer: "The only and, unfortunately, irreplaceable instruments are the missiles and their nuclear payloads. The unequivocal fact is also that these missile systems are the most economical means, as they require a minimum of service personnel; they are also the best option environmentally, and have an established mechanism of international control."[114]

63. *Volodymyr Tolubko (left) and Yuri Kostenko (right), 1992*

Volodymyr Tolubko (b. 1948), Ukrainian military specialist, technology PhD, and professor. He was a colonel general in the Soviet military. As the most active proponent of maintaining nuclear status and the only Ukrainian serviceman in the working group, he proposed creating a Ukrainian "nuclear defense shield." "The strong are respected," he wrote, saying that it was unacceptable to sacrifice the country's defense capabilities in exchange for general words about "peace on earth." He advanced the doctrine of "guaranteed deterrence" based on owning the minimum number of nuclear arms necessary for Ukraine's defense. He completed his military education at the Kharkiv Higher Military Command and Engineering School in 1971, as well as the Dzerzhinsky Military Engineering and Artillery Academy in 1979 and the Voroshilov Military Academy of the General Staff of the Armed Forces of the USSR in 1986, both in Moscow. He served in different units of the Strategic Missile Army of the USSR. From 1986 to 1990, he was commander of the 46th Rocket Lower Dnipro Division. From 1990 to 1999, he was head of the Krylov Higher Military Command and Engineering School in Kharkiv, and subsequently of Kharkiv Military University. In 1999–2000, he was first deputy director of the Main Directorate of Intelligence of the Ministry of Defense of Ukraine and the deputy director of the General Staff of the Armed Forces of Ukraine.

However, KGB General Ievhen Marchuk, a philologist by profession, similarly to several among the then leadership, saw things differently. He interpreted nuclear deterrence as the intent "to use nuclear weapons with impunity," a strange notion for a civilized era.[115]

In addition to Tolubko, there were other officers who supported the idea of relying on rocket armies. For instance, in the Defense Ministry's report called "Cost Assessment of Select Types of Armament and Military Equipment (in 1990–1991 prices)" two numbers were compared: modernizing Ukraine's military would cost up to RUB 18 billion a year, while modernizing the Strategic Nuclear Forces only RUB 8.3 billion. The brief also stated that annual spending on operations and battle readiness of Ukraine's regular forces was 25 to 30 times higher than the cost of maintaining the Strategic Nuclear Forces.[116]

Still, Tolubko predicted that this approach "will more likely run into internal opposition on the part of highly-placed officials within the country, than complications with the outside world." He believed that "for some, this will be a matter of honoring the uniform; for others, a fear of reducing attention to other branches of the Armed Forces and losing the race for posts in agencies; and for yet others, a fear of acting decisively."[117]

The majority of those involved in the discussions argued that it was necessary to give up nuclear weapons, because maintaining them, they said, was costly and the very fact of retaining them was equated with the intent to use them. As an alternative to nuclear arms, they saw modernizing the regular army. Thus, scholar Viktor Tymoshenko reasoned that "throughout its history, Ukraine has paid a steep price for mistakes in its military policies and paid by losing its statehood." Therefore, the content of Ukraine's military and technical policy essentially meant modernizing the country's army as a priority, which would allow the Armed Forces to participate in "fourth generation" wars.[118] Colonels Vitalii Lazorkin and Valerii Kokhno of the Defense Ministry wrote on 4 February 1993 that, once Ukraine gave up its nuclear weapons, they saw a military advantage in "the achievement of higher quality parameters for the Armed Forces while reducing overall numbers," being equipped with more modern military technology, and more effectively using research and production potential to develop and make advanced weapons.[119]

Military assessments of Ukraine's possible armament needs have proven that all these ideas remained at the level of general statements without any answer as to who the potential opponents where and how Ukraine was to defend against them. In one of the briefs called "An Estimate of the Possible Quantity of Weapons," they noted that "justifying the number of forces and weapons for Ukraine and the structure of its armed forces needs to be based on [...] an assessment of the possible nature of a future war, the military capacities of potential opponents and allies, and also the military commercial potential and opportunities in the country's economy."[120] This underscored, once again, that ensuring the country's defense capabilities was only going to be possible when it became clear whom the government saw as a potential aggressor.[121]

As to the question of what to substitute for nuclear weapons, the most popular solution was the idea of precision-guided munitions (PGMs), or smart weapons. At that point the surface of the planet was already peppered with vulnerable facilities: NPPs, chemical plants, artificial reservoirs with systems of water-protective dams, among others. To render them inoperative and, for instance, leave an entire city without power, all it would take was a single precise hit, something no tanks could deter. What was needed were weapons whose potential use would intimidate potential attackers.

In 1992, Pivdenmash's chief designer Stanislav Koniukhov presented a program for developing nonnuclear deterrent systems before a session of our working group. It included two phases. The first phase was to be completed by 1998 and involved the already made expenditures to develop 46 SS-24 ICBMs and bombers. This phase entailed the following: first, building kinetic warheads, and reequipping ICBMs with them, but without remaking their control systems or flight testing of the regular carriers; and second, building air-to-surface ballistic missiles with an up to 500-kilometer range loaded with nonnuclear warheads, and reequipping Tu-22M, Tu-22M2 and Tu-22M3 bombers with them.

The second phase was to conclude in 2003. By then, Koniukhov wrote, "the following strategic arms options shall be in development: single block-warhead ballistic missiles that are equipped with precision-guided nonnuclear warheads, which contain effective countermeasures against missile defense systems, shall be based and stored in stationary silos within Ukraine's quota of allowed strategic carriers; and the An-70 transport aircraft for strategic air-defense missile systems, which shall be equipped with 2,000-km range cruise missiles and shall be loaded with both nonnuclear warheads and homing missile guidance systems."[122] Using the cumbersome ICBM complexes for nonnuclear warheads was not too realistic. Most likely Koniukhov was looking at the problem from the point of view of orders for his plant. However, any evidence of Ukraine developing its own strategic weapons would be carefully tracked by the US, as will be discussed further in more detail.

Ukrainian academics also aided their effort, discrediting the idea of equipping the Ukrainian Army with PGMs. Iurii Pakhomov, director of NAS Institute for Global Economics and International Relations, had written back in

February 1993 that precision-guided weapons "might be able to play the role of a strategic deterrent," but only "with the purpose of destroying peaceful nuclear facilities on the territory of a potential enemy." He continued: "This immediately would transform the country and political regime that employed this strategy into an inhuman state." The academic even signed his name to this bizarre opinion.[123]

The military doctrine finally was brought out for the first reading in the parliament on 24 April 1993. What interested the outside world the most was whether it would gain 226 votes in favor of nuclear status for Ukraine. The process was watched closely by some of the most influential news outlets in the world. That same day, they all reported with a lighter heart that the idea had not gained a majority. Nevertheless, the draft military doctrine did include plans to develop PGMs.

The US Embassy reacted quickly. In a 30 April memorandum, it noted that START, which Ukraine still had not ratified at that point, did not distinguish between nuclear warheads and conventional ones; therefore, Kyiv was expected not only to destroy its nuclear warheads, but also to eliminate all the carriers, the launch facilities, and the 46 SS-24 missile systems.[124] On 3 June, I addressed the parliament and noted that this kind of demand contradicted Ukraine's military doctrine, which meant we either needed to change the doctrine or change the formulations in START and the Lisbon Protocol.

On 19 October 1993, the parliament passed the military doctrine.[125] "The main purpose of the military doctrine of Ukraine is to guarantee Ukraine's national security from external military threats," the text of the document read, but it did not specify whence such a threat might come. It merely stated that "the main reasons for war and military conflict can be economic, political, territorial, ethnic, religious, or other disagreements that states are not always able to resolve without engaging in conflict." The priority forces were even more vaguely framed: "The Ukrainian Armed Forces include a variety of types of forces and other military formations and facilities that are necessary to ensure the defensive capabilities of the country and are provided for by current legislation." The only points that were clearly defined were those of Ukraine's neutral status and its intention "to become a nonnuclear state in the future."[126]

The program to develop precision-guided weapons was stopped forthwith due to two factors: pressure from the US and lack of funding.

How Experts and Nongovernmental Organizations Were Exploited

In the spring of 1993, as Ukraine began to form its own positions on the issue of nuclear weapons, nongovernmental organizations (NGOs) and experts were engaged in a campaign of pressuring the parliament. They would repeat the same claims as the MFA, the main one being that Ukraine should immediately ratify START and the NPT, otherwise it was at risk of international isolation. Their assessments and recommendations were sent to Parliament's leadership, and, from there, were forwarded on to our working group, "for consideration in your work." Among those who commissioned these recommendations were international organizations, as well as the MFA. Here are just three examples.

First, on 12 March 1993, the National Committee on Disarmament under the Foreign Ministry sent a so-called expert analysis of the problems around ratifying the START and the NPT prepared by the NAS Institute for Global Economics and International Relations at the request of the Committee to First Deputy Speaker Vasyl′ Durdynets′. One could not help but notice that not only the key claims, but also the fundamental schema of the arguments, were identical to those invariably used by the Ukrainian Foreign Ministry. "In our view, gaining nonnuclear status and undertaking disarmament is the most attractive option, one which is in Ukraine's interests and offers opportunities to increase the international weight and repute of Ukraine as a state, as the first country to reject a policy of nuclear force," was the positive spin the Institute director, academic Iurii Pakhomov, put on things. Then he went on to add the negatives: "The burden of nuclear weapons complicates the domestic economic, political, and social situation and could lead to the collapse of the state, as happened with the former Soviet Union."[127]

Although the parliament never passed a resolution to maintain Ukraine's nuclear status, but only voted on the conditions for disarmament, Pakhomov did the same switch of concepts as the MFA: he argued that Ukraine could not develop new nuclear technologies. "In a situation where there is an economic crisis that is only growing worse, for Ukraine to develop its own nuclear military technology is completely economically unjustified," he wrote, adding that "for Ukraine to preserve a nuclear arsenal, in contradiction to all its previous declarations, will bring to bear an entire slew of problems of a political, economic and technological nature."[128]

In assessing the legitimacy of this supposed analysis, two issues stand out. First, Pakhomov proposes "removing" the nuclear weapons from Ukrainian territory, instead of destroying them, "in line with the timeframes established in the Lisbon Protocol."[129] However, the Lisbon Protocol did not include any timeframes. The term of START I was seven years, which was in the official statements regarding the Lisbon Protocol issued by the Ukrainian and Russian MFAs. These latter statements, however, had no legal force. Besides, the Protocol did not discuss the means of reducing Ukraine's nuclear weapons, and the parliament resolution only mentioned destroying them. Shipping them to Russia was just one possible option and it was the one on which Moscow insisted, and which Ukraine's MFA and SBU supported. Finally, Pakhomov for some reason put an equal sign between carrying out the Lisbon Protocol and Ukraine gaining nonnuclear status, although the Protocol called for reducing not all the weapons, but only 36 percent of carriers and 42 percent of warheads, and this was with reference not to Ukraine specifically, but to the entire nuclear arsenal of the former Soviet Union. The impression was that the academic had read, not the Lisbon Protocol itself, but a Russian interpretation of it.[130]

Second, he called ratifying START "consistent actions towards fulfilling the Declaration adopted by the parliament on 24 August 1991," and warned: "This position towards nuclear weapons on the part of Ukraine has already been recognized widely in the international community. Any hesitations or turning away from this position will have negative, possibly even tragic, consequences for the people of Ukraine."[131] There is an error here in that the Declaration of State Sovereignty was passed on 16 July 1990, whereas on 24 August 1991, the Act of Proclamation of Independence was. The remaining question is: How could a mistake like that be made at a Ukrainian state scientific institution?

Finally, as was noted in the preceding section, Pakhomov expressed some fairly odd reservations about replacing nuclear weapons in the defense system with precision-guided munition, stating that the latter could only be used for destroying a potential enemy's peaceful facilities, which would instantly brand a country using these weapons as inhuman. In the end, the "expert analysis" offered not a single proposal for how Ukraine's national security might be guaranteed. Yet this was the quality of brief that the Ukrainian MFA was offering the leadership of the parliament.

Here is another example. During both 14 and 15 April 1993, there was a seminar in Kyiv called "Issues in International Environmental Safety in the Process of Converting Rocket Armies in the Context of Political and Socio-economic Changes in CIS Countries." Similar international gatherings with the participation of Ukrainians throughout 1992 and 1993 were particularly numerous because the organizers in every summary document would invariably repeat the following thesis: reducing nuclear weapons is the guarantee of world peace and whoever does not want to do this (read: Ukraine) is the enemy of all of mankind.

On 16 April, Odesa MP Iurii Romanov wrote up the recommendations of the seminar for Speaker Pliushch with a request to "personally take note of these conclusions," although no one had ever heard from Romanov before, nor would they after these anti-nuclear initiatives. The basic message was to "let go of your political ambitions and settle the ratification of START and accession to the NPT," and come to terms on manufacturer oversight of the warheads and their removal to Russian enterprises.[132]

The background against which these recommendations were presented had the standard preamble, namely, "the world community is concerned about the position of certain political actors in Ukraine, which is seen as a desire to manipulate the nuclear issue as a foreign policy instrument." As an argument for ridding Ukraine of its nuclear weapons as quickly as possible, "the violation of the rules for the technical operation of nuclear missile complexes and radiation safety norms when storing nuclear warheads, which increases the level of risk of an emergency situation arising," was mentioned. To establish an environmental angle, there was the reminder of "Ukraine's specific responsibility as a country that brought humanity the worst nuclear disaster in human history with catastrophic consequences," while the "frivolous and irresponsible approach to nuclear policy" was all but called the condition for "a negative environmental, medical and genetic impact."[133]

As a final example, on 26 May 1993, in a short notice called "For Nuclear-Free Principles," *Holos Ukraïny* printed calls to ratify START and the NPT from two organizations: "The Ukrainian Peace Council and the Ukrainian Peace Fund published a statement of full support for the efforts of the UN and the Committee of NGOs for Disarmament, whose aim is global disarmament for the sake of security throughout the whole world." In their appeal, they "called on Parliament to ratify START I and the NPT, which would lead to a better domestic political climate and an easing of tensions in international relations."[134] How the two were related to each other the Ukrainian paper did not specify.

In this way, together with NGOs and state institutions, the Ukrainian press was yet another factor of influence within the country that shaped opinion about the need to immediately disarm Ukraine in the run-up to the 3 June session.

Information War

The harshest element in the process of nuclear disarmament was the information war rolled out in the world press against Ukraine by Russia and the United States over 1992–1993. Its principle was, "See the goal, ignore the obstacles."

Despite the clichés about of a supposedly democratic press and standards of independent journalism, press clippings collected during that time make it clear what tactics Ukraine's opponents used to form a negative image of our country and force Kyiv to play according to rules that were decided between Moscow and Washington. It is easy enough to see in these articles that the Western press not only functioned as an instrument but was one of the decisive means for pressuring Ukraine and implementing the Russian-American plan of the immediate and unconditional nuclear disarmament of Ukraine.

At the same time, my archive of publications from the Ukrainian press of that same period shows that Ukraine did not offer any resistance in this war. It was unable to find its own national interests, not only in the global information space but even in its own domain. The main reason was the lack of a consolidated position in its government: unlike in the US and Russia, a kind of civil war raged around the issue of nuclear arms that was methodically stirred up from outside.

There was one more problem that triggered uneven starting conditions for the information battle. After the collapse of the Soviet Union, all the news agencies, all the embassies, and the entire Soviet intelligence system located around the world all became Russia's. Russia began to immediately use this resource against Ukraine, whose nuclear weapons became the number one topic, in Moscow's interpretation of it. In May 1993, the UK's *Independent* wrote: "Most of the Western media still report events in the republics as seen from Moscow, courtesy of the TASS news agency."[135]

The situation grew worse because Ukraine not only failed to set up its own information system but did not broadcast its clearly formulated positions on the nuclear issue. The contradictory commitments that were written into documents issued by the parliament and in statements made by Kravchuk and the MFA offered enormous opportunities for the country's opponents to manipulate conditions.

The way Ukraine's Foreign Ministry worked with information, as the one agency that should have held the media front, generally ended up reiterating what the world was saying about Ukraine's nuclear weapons. I have copies of their monitoring reports that were regularly sent to the leadership of the parliament. They are bursting with quotes from publications that clearly distorted the facts. Yet, the MFA never reported any measures it had taken to set the record straight in the face of the storm of disinformation. Its communication activities were limited to meeting with diplomats at various embassies.

In addition, the MFA generally tended to apologize for the behavior of the parliament, which, in the view of the Ministry's leadership, did not understand

64. TASS

The Telegraph Agency of the Soviet Union (TASS), a state news agency founded in 1925 by a resolution of the Presidium of the Central Executive Committee and the Council of People's Commissars of the USSR. The agency had the status of a central news agency with the exclusive right to collect and disseminate information within and outside the USSR. The news agencies of the Soviet republics were organized as part of TASS but were allowed to collect information only within their respective regions. After the collapse of the Soviet Union, Russia claimed a sole right to TASS, and was thus the only post-Soviet country to have a global news network at its disposal. On 30 January 1992, President Yeltsin issued a decree renaming the agency to ITAR-TASS, the name under which the agency operated until 1 October 2014.

the need to immediately and completely disarm. It was clear that Ukraine's Foreign Ministry was running in the wake of the Russian one. Thus, Ukraine essentially did not engage in battle on the information front.

The parliament, the government, and our working group were all systematically supplied with information about what the Western press was writing by the Pylyp Orlyk Institute for Democracy and the Consultative Advisory Council under the parliament, who issued bulletins with translations of articles on a regular basis. Fortunately, these bulletins preserved information about nearly 100 of the leading publications in the world's top papers. Here are some typical headlines in reputable Western publications: "Ukraine: Barrier to Nuclear Peace," "US presses Ukraine to ratify nuclear arms pact," "No Further Nuclear Arms Cuts in View," "Little Russia's Chip" (accusing Ukraine of "brandishing nuclear arms"), "The Bearcubs — Ukraine Must Stop Using Nuclear Weapons as Bargaining Chips," and so on.[136]

The campaign to pressure Ukraine was running full bore in all of the most highly regarded newspapers in Western Europe and North America. At different times, American papers were putting out articles on Ukraine on a daily basis, sometimes even two articles in a single issue. The pressure was unrelenting and was calculated to reach a victory, "Shock and Awe" style.

One of the widespread means was planting red herrings. In working up the idea that Ukraine was dangerous, Russia would spread anonymous commentaries about information its intelligence had supposedly uncovered, such as that the Ukrainians were working to get the codes to their nuclear weapons. In November 1992, this is how it was reprinted in the *New York Times*:

> There is no evidence that Ukraine is close to acquiring outright control over missiles on its territory — that is, the power to

launch its own nuclear strike. But there are disturbing indications that organizations in the republic, if not the government, may be trying to obtain that capability. A top-secret government research institute in the eastern city of Kharkiv is developing warhead enabling codes. According to Russian military experts, the institute even tried to enlist the help of Russians in its efforts. These experts also caution that a Ukrainian attempt to introduce counterfeit launch codes could trigger a chemical explosion, perhaps leading to the release of radiation.[137]

Needless to say, there was no "top-secret government research institute." What they were referring to was the Kharkiv Research and Development Association Khartron, which was involved in developing elements of strategic offensive weapons' control systems during Soviet times. During an April 1992 session of the National Defense Council, we did consider the "button issue," however, not for the purpose of launching, but on the contrary, for the purpose of blocking a launch (as was discussed in chapter 1). Since Moscow had strategic control over all Soviet nuclear weapons, including those in Ukraine, a constant danger remained that it could launch a missile from Ukrainian territory. The presence of such a danger was confirmed for us by Khartron director Iakiv Aizenberh. At the same time, Aizenberh rejected the technical possibilities for Ukraine to access the strategic missile launch control system to install blocking instruments without the consent of the Russians. The system was secured and built in such a way that any unsanctioned interference caused it to shut down completely; that is, the entire electronic system would stop functioning and would not work until it was relaunched from a command center located in Russia.

The Russians knew this perfectly well. Yet, a description of the event in the headlines of a news item from *Interfax* came out thus: "General Director of Khartron, the Enterprise Developing Nuclear Missiles' Control Systems: No One in Ukraine Knows Where the Missiles Deployed on Its Territory Are Aimed."[138]

Later on, it turned out that the red herring about Ukraine's aggressive intentions was not just coming from the Kremlin but was a joint Russian-American "dish." The *Washington Post* named the source of this information as someone in US government circles: "Ukraine is moving to acquire 'positive operational control' of the 176 intercontinental nuclear missiles still on Ukrainian territory and could achieve this goal within the next 12 to 18 months, US intelligence has concluded."[139] Altogether, this paper ran a very active campaign around Ukraine during May 1993, even naming the timeframes needed by Ukraine to gain "positive operational control" over its 176 ICBMs. The paper continued to escalate the situation, not even hiding the source of its information and unconcerned that it was not providing Ukraine's viewpoint in a subsequent article that stated: "The Brookings Institution's Blair estimated that independent Ukrainian targeting could take 'in the neighborhood of several months' to achieve."[140]

The Western press kept stoking fears, shaping public opinion about the need to take the "nuclear toy" away from an aggressive, unpredictable country. The *Washington Post* wrote: "The Clinton administration fears that Russia and Ukraine may be headed toward confrontation over nuclear weapons and believes Washington should move more aggressively to mediate the two nations' economic and military disputes, according to senior US officials." It went on:

"At the heart of the administration's concerns are fresh signs that Ukraine may eventually decide to hold on to nuclear arms that were deployed on its territory by the former Soviet Union and are now claimed by Russia. US officials said this week that they had received reports that Ukraine is trying to gain control of the weapons, ostensibly held under Russian lock and key."[141]

Presented in this way, the situation evolved in a very logical fashion: Russia began to prepare for a preemptive attack on Ukraine, so that the latter could not gain an opportunity to use weapons and accidentally destroy the whole world. Any sensible person could understand that this was utter nonsense, but what a news sensation! And there is even a source to cite. Reported the *Washington Post*: "The Russian comments hint at a nightmare scenario that US officials don't even want to think about: a preemptive Russian strike into Ukraine to recover or even destroy the former Soviet missiles before Ukraine moves."[142]

When reporting on the reasons why the State Department dismissed Ukraine's interest in resolving the problems associated with its nuclear disarmament, it took the line in a manner similar to the multilateral agreement between the USSR and US in 1955 that guaranteed Austria's independence and neutrality; the paper solemnly wrote: "That treaty took nearly 10 years to negotiate. Today, a more rapidly ticking clock pits Ukrainian nuclear ambitions against Russian willingness to pre-empt militarily. That is a race between nightmares that the United States must exercise all its leadership ability to prevent."[143] In perfect harmony for two voices, the US and Russia strove for a common outcome: disarming Ukraine.

The information war even included some moves straight out of the Soviet textbook. For instance, the *International Herald Tribune* published a letter from a reader in France named Peter B. Martin, who, strangely enough, was extremely knowledgeable about the subtleties of servicing nuclear weapons, and Russia's role in this process. Moreover, this ordinary reader expressed opinions that followed strictly those advanced by Russia:

> If Ukraine is left to its own devices, the missiles and nuclear warheads will slowly become ineffective through lack of maintenance. For example, the liquid fuel in the SS-18 missiles must be kept cool and under pressure, otherwise toxic chemicals escape and make the silos they are housed in dangerous. The warheads themselves tend to deteriorate. [...] The only country in the former Soviet Union that can service these missiles is Russia, and I doubt it is interested. Therefore, for once, inaction might be the best course to take to neutralize this form of nuclear blackmail.[144]

These were the assertions of the so-called reader of the *International Herald Tribune*, who somehow forgot that there were 308 strategic missiles deployed in Russia and Kazakhstan, including some aimed at France, which had been maintained by Ukraine's Pivdenmash.

The Ukrainian MFA's passivity in presenting Ukraine's actual policies regarding nuclear disarmament, as recognized by the parliament, or to be more precise, the MFA's embargo on such policies, made it possible for the US and Russia to freely interpret the reasons why Ukraine was ostensibly delaying the implementation of commitments it had undertaken. The key concept in the joint information campaign run by Moscow and Washington was: if Ukraine does not disarm immediately, it will be in violation of its commitments. The logic was simple: if it is in violation, then it is a dangerous and unpredictable partner; if it refuses to disarm, then it has plans to use these weapons against others; if it makes conditions, then it is bargaining and bidding up its price; and so on. This argument gave Ukraine's opponents the opportunity to treat Ukraine's demands for national security guarantees and funding for the disarmament process, not as a vital need but as a whim that merited being cast aside. For instance, in early May, the *Times* wrote: "Ukraine has dragged its feet over ratifying the Strategic Arms Reduction Treaty (Start I); it has refused to implement the Lisbon protocol on becoming a non-nuclear state; and it is toying with keeping its weapons indefinitely."[145]

The press did not care to learn the reasons why Ukraine needed financing for the disarmament process and guarantees for national security after the process ended, but instead emphasized its growing demands: "Ukrainian officials, meanwhile, have steadily enlarged their demands for financial compensation and security guarantees as conditions for surrendering the estimated 1,600 to 1,800 nuclear warheads."[146] The *Washington Post* further accused Ukraine of increasing its financial demands: "Some officials have suggested that Ukraine be paid as much as US $3 billion, far more than the US $175 million promised by Washington."[147]

In the terminology of the anti-Ukrainian campaign this was referred to as "bargaining," which, in the context of such a noble cause as preserving peace on the entire planet, was unbecoming. The *Washington Post* wrote in early 1993: "A senior US official said the Ukrainian delegation was told emphatically at the State Department that Washington would not engage in a bargaining process to persuade Ukraine's legislature to ratify its commitment to remove nuclear weapons from its soil."[148]

Yet another ace in the hands of Ukraine's opponents was the mutually contradictory information that kept coming from officials in different branches of the Ukrainian government. On one hand, the parliament held the decision on the force of law, which was binding. On the other, there was President Kravchuk, who did not understand that communism had ended, and without a Politburo, no center of state power existed as before; and that the parliament was not a branch of the government over which he had power. Because of this lack of understanding, he kept signing documents that promised things in his name that were not provided for, as decided by the legislature. The MFA worked in the same manner.

This disconnect was gladly exploited. For instance, some pointed to the unfulfilled promises which Kravchuk had given but could not fulfill without the approval of the parliament: "US officials are uncertain about the real intentions of Kravchuk. He has repeatedly promised to give up the former Soviet missiles and then reneged, indicating he cannot do so until the United States provides unshakable security 'guarantees' against an attack by Russia."[149]

Why was it so easy for their plan to work? Because a countering flow of information was nonexistent from Ukraine, even in the domestic press, let alone the international one. Nuclear disarmament could have been presented as the greatest gift of the Ukrainian people to the entire world, because every country was trying to get nuclear weapons into its security system to protect its own interests, while Ukraine was voluntarily giving up the third largest arsenal in the world. Alas, even in 2014, when Russia began a similarly aggressive information war against Ukraine after the Revolution of Dignity, Ukraine's Foreign Ministry worked no differently.[150] "The embassies of Ukraine in some Central European countries barely inform the press and society of that country, despite the information war on the part of the pro-Russian politicians," wrote *Ievro-peis'ka pravda* on 19 November 2014.[151]

These examples of an information war should be included in school texts, so that in the future Ukrainian journalists understand the weight of their own words, sometimes measured in the billions of dollars, and so that Ukrainian politicians can feel the potential consequences of being under the journalists' spotlight, while specialists who defend the national interests in the global informational space are capable of fighting and winning battle after battle.

The Pressure Builds Leading up to 3 June

As soon as the date of the session during which the parliament was to debate ratification of the two nuclear treaties was announced as 3 June, unseen pressure became apparent against Ukraine through both governmental and diplomatic channels. On 7 April, the White House responded negatively to a request from Ukraine about the possibility of a meeting between Premier Kuchma and President Bill Clinton or even Vice President Al Gore. The American administration's open rejection was a more than transparent hint. Washington was not pleased with Kyiv's behavior. Nevertheless, some experts thought it was a mistake to humiliate Kuchma this way, given that he was seen as a reformer.

Pressure on Ukraine was also coming from countries in the European Union, as well. The Danish ambassador, on behalf of the EU, handed the MFA a démarche in response to the developments concerning ratifying START and Ukraine's accession to the NPT. At the same time, they promised that a rapid approval of this issue in the parliament would "ease Ukraine's further integration into the international community and would foster the further development of interrelations with EU countries."[152] Nevertheless, their attitude towards Ukraine after disarmament showed that the appeals of a nonnuclear state are poorly heard, as was the case in 2008 when Ukraine requested to be placed in line for accession to the EU, and again in 2014 after Russia annexed Crimea and started the war in the Donbas, and even in 2022.

On 30 April 1993, the US Embassy in Kyiv handed a memorandum to the Ukrainian MFA in which it demanded that Ukraine destroy not part of its nuclear arsenal, as required by the Lisbon Protocol, but all of it and not even consider developing precision-guided weapons with nonnuclear warheads. "Some in Ukraine are under the illusion that if Ukraine keeps its deployed SS-24 ICBMs but arms them with ordinary warheads, this will be in line with the treaty on strategic weapons and the Lisbon Protocol. Perhaps this is a consequence of a wrong interpretation of what is expected in START and the Lisbon Protocol on ICBMs, including the SS-24," the American memorandum stated, without any diplomatic flourishes.[153]

This tone demonstrated little respect and continued throughout the entire document. "We strongly urge you to familiarize yourselves with this information, as well as those in the government and Parliament of Ukraine, who are not completely clear about commitments regarding SS-24 ICBMs." The US was effectively asking the Ukrainians to see in these documents something that was not actually there.[154] The Lisbon Protocol did not say which ICBMs had to be reduced—the SS-24, the most powerful ones with 10 nuclear warheads and their own missile defense systems, which Americans feared the most and called Scalpels because of their precision, or the SS-19. Moreover, these documents had not been ratified by the parliament, so Ukraine had no legal obligations towards them at that point in time.

On 3 May, the *Wall Street Journal* wrote about the gradual decline in Ukrainian-American relations, because of the growing pressure coming from the US, which began under the Bush Administration and intensified under President Clinton. After the refusal to receive Premier Kuchma at a high-level meeting, the paper wrote, Ukraine's Foreign Affairs Commission Chair Dmytro Pavlychko said that Kyiv had understood what the real meaning was: *Do not bother us until you decide to give up your nuclear weapons.* The thought had occurred that the Americans slowly but surely were beginning to understand that Ukraine was not Russia.[155]

During most of the month of May, the top Western papers stoked fears about Ukraine's hostile intentions. In the London *Times*, an article entitled "Little Russia's Chip," stated that Ukraine was "brandishing nuclear arms," while the *Washington Post*, *New York Times* and *Wall Street Journal* wrote about secret estimates circulating among the top levels of the American government, which reported that the Ukrainians wanted to gain operational control over their missiles.[156] The conclusions were that international pressure needed to be ramped up in order to get Ukraine's weapons moved to Russia as soon as possible. During this period the Western press, and more importantly, the Russian press, hardly made any mention about Ukraine's actual needs, those being a guarantee of security after giving up nuclear weapons.[157] Meanwhile, events in Ukraine confirmed just how urgent these needs were.[158]

At the beginning of May, nationalists in the Hungarian parliament protested against the ratification of the bilateral Basic Treaty on Foundations of Neighborhood and Cooperation with Ukraine, demanding that "Hungarians outside our borders not be abandoned," and that the document be ratified after Ukraine agreed to cede border counties, such as Berehove, and guarantee autonomy to Uzhhorod, Mukacheve, Khust, and Vynohradove, and determine the future concerning Zakarpattia by referendum.

On 18 May 1993, 203 ships, constituting 80 percent of the Black Sea Fleet, raised the Russian military flag. As reported in the *New York Times*, at the beginning of May, the Russians derailed the talks concerning the Black Sea Fleet by claiming property located on Ukrainian territory.[159]

The final blow came after pressure was put on the Ukrainian legislature, which originated from none other than Ukraine's own Foreign Ministry. On May 17, two weeks to the session, Speaker Pliushch received an analytical brief eloquently entitled, "Possible Consequences of Not Acceding to the NPT," stamped "For internal use." The document strongly recommended that the issue of START and the NPT be considered as a single package.[160] The brief had come from the Cabinet with Kuchma's signature, but the way events unfolded on 3 June made it clear who was the actual author; Kuchma's speech contradicted everything written in the brief, meaning that the arguments in this document were presented by the foreign minister and his deputy.

At this time, the Western press sometimes openly wrote about the US and Russia as a single unit on the matter of Ukraine's nuclear disarmament. In the aforementioned article, the *Times* also wrote: "Ukraine has blocked agreements preferred by both the West and Russia. [...] Ukraine has felt slighted by the West's nonchalance, its closer relations with Moscow, and the weight given to Russia's world role."[161]

I was one of the very few in Ukraine who had actually referred to them as the "union of two." Most politicians in Kyiv, especially those who were higher officials, were afraid to insult or anger both Washington and Moscow. The only Ukrainian paper that by some miracle dared to publish my opinions was *Narodna Armiia* [People's Army], which published these words on 14 April: "Iu. Kostenko thinks that the current position of the American president is nothing more than 'political pressure and blackmail.' Apparently, Moscow and Washington are 'playing' the same game."[162]

Still, at a certain point, a crack appeared in this apparent monolith, when the US began to see other benefits for itself on the horizon. "Russia thinks that Kyiv has to be relentlessly pressured, but Americans are now trying to persuade the Ukrainian government and propose cooperation on this issue," the *Washington Post* wrote at the end of May 1993.[163] This cooperation could have had a decisive impact on the geopolitical situation in the post-Soviet arena.

Ukraine Is Offered an Opportunity to Avoid International Pressure without Ratifying START

Thus, the pressure aimed at getting Ukraine to ratify the nuclear documents was intense, not only externally, but also from President Kravchuk and the Foreign Ministry. The president and the Ministry were mostly concerned about what the world would think, but they never seemed to ask themselves: Where was Ukraine to find the money for the task of disarmament, which was way beyond the means of the country's budget? In this high pressure situation, holding the parliament back from ratifying START in the version demanded by the Americans and Russians, without any preliminary agreements about compensation for nuclear material, without security guarantees, without knowing where the toxic substances might go, but taking on billion-dollar losses for destroying the weapons and the silos, had become increasingly impossible.

On the back of these developments, in April 1993, Ukraine unexpectedly received intellectual backing, from the Americans no less. Systems Planning and Analysis, Inc. (SPA), a research and consulting corporation, sent an official proposition that described what Ukraine needed to do in order to keep the warheads on its own territory, postpone ratifying START, leave its silos intact, and process the toxic liquid missile fuel into fertilizer, and at the same time put an end to international pressure.

In its letter to the Ukrainian government, the SPA specialists agreed that "reality is forcing Ukraine to carefully consider what implementing START would bring, in regard to economy and security." However, since this extra time "is causing concern about whether Ukraine intends to change its course towards nonnuclear status," they proposed an alternative possibility for Ukraine to ease those concerns and gain the maximum economic benefit without speeding up the ratification.[164]

The essence of the proposal was to use relatively inexpensive technologies that had been developed and tested by an American company that would make it impossible to launch any ICBMs while leaving their on-board and above-ground systems in working condition, and to engage international observers in oversight. In other words, the missiles would stay in their silos, where they could be maintained reliably and safely and with the warheads attached. This would eliminate the urgency to remove the nuclear warheads and would grant enough time to resolve political and technical problems with the destruction of the strategic weapons. This would allow Ukraine, with the West's blessing, not to remove everything immediately to Russia, but to develop a strategy for eliminating the third largest arsenal in the world in a way that would benefit the country the most.

At the same time, the Americans were assessing the problem realistically, in that the entity that might step forward as the "international control" would still have to be found. The SPA did not see the US or Russia in this role, but recommended an orientation towards NATO inspection, which would involve France and Germany. "If NATO were to confirm through its inspection that Ukraine is indeed ensuring the weapons are being properly maintained, arguments that it needs to be immediately transferred to Russia for the sake of reliability and safety, would be less powerful. What is more, precisely because NATO is directly familiar with the status of these weapons, it can recommend leaving them in Ukraine until such time as the political situation in Russia stabilizes," SPA's specialists wrote.[165]

As to the actual technology for "demilitarizing" ICBMs, the only option which Ukraine's own specialists could propose at the time, was detaching the warheads and replacing them with electronic warhead simulators called equivalents. The production of the simulators required considerable time and money, and there were no other options available in Ukraine at the time. What the SPA proposed was something that was fundamentally new for Ukraine. They described the basics of the technology in lay terms thus:

> The warheads on these missiles can remain in their launch silos or be moved to reliable and safe temporary storage. The phases of demilitarization include the following: remove the warhead, remove the small amount of fuel and fill it with a thoroughly controlled substance, after which the fuel can be put back into the missile, and reattach the warhead to its proper place. The controlled substance now present in the fuel will settle on the bottom of the fuel tank and plug up the channels through which fuel gets to the engine. These controlled substances are clean in the ecological sense and will not subsequently interfere with recycling the fuel into a useful chemical.[166]

Another technological problem that Ukraine could not solve in spring 1993 was repurposing the missile fuel. This process involved a highly toxic substance for which Ukraine lacked the necessary recycling and deactivating technologies, and our scientists all came to this conclusion in their analytical briefs. So, the SPA specialists proposed to not just "neutralize" the missile fuel, but to gain economically from it:

One American company has developed a process for recycling missile fuel into useful chemicals that can be used in farming and other areas. Ukraine has about 13,000 tons of liquid fuel in its strategic missiles. Liquid fuel is considered dangerous and turning it into safe material is a key aspect in the process of relieving Ukraine of nuclear weapons. The majority of the fuel can be reprocessed and sold. Environmentally, this process meets standards that are acceptable in the US. Other processes developed for eliminating missile fuel tend to involve burning it off, and they do not offer any benefits while having a negative impact on the environment.[167]

The SPA specialists continued: "The same option exists for solid-fuel missiles. Moreover, this process can be used at plants that process missile fuel for the purpose of converting them to peaceful chemical manufacturing. It is important to remember that processing is the path to economic benefits from what was earlier a significant problem in terms of value and in terms of impact on the environment." The SPA letter also noted that, since the proposed steps would move towards the results that were anticipated by the nuclear treaties, "all of this can be done without even ratifying START and the NPT."[168] For Ukraine, this arrangement could bring multiple benefits. Without assuming the financial commitments or destroying silos, Ukraine could actually keep the weapons; and, thus, the nuclear material contained within them on its territory, in addition to gaining new technologies and engaging the Europeans and Americans in the gradual elimination of the Strategic Nuclear Forces with the maximum benefit to itself. Above all, it would not be left on its own to face Russia.

The MFA sent a translation to Premier Kuchma with a cover letter, which vetted the company as "world-renowned" and adequately assessed the opportunities these propositions could open. This was confirmed by the cover letter, in which Zlenko outlined the project's three positions:

> First, taking on this project will allow us to bring the question of nuclear weapons on Ukrainian territory to a new level, since the option of using it militarily is eliminated and the threat of unsanctioned launches disappears. Second, the problem of finding a place to store nuclear warheads until they are destroyed is resolved. Third, implementing the project will either ease the negotiating process with Russia, since the proposed procedures partly fit the familiar proposal of the Russian side to remove ICBMs from combat duty, or, should Russia not accept this approach, expose its real intentions regarding the nuclear weapons located in Ukraine.[169]

However, this is where the developments really take a bizarre turn. The MFA hit the brakes on the American proposal. First, the MFA asked the Cabinet to involve specialists from the MFA, Defense Ministry, Machine-Building Ministry and Academy of Sciences to "analyze the proposals in the Project for further development, and identify the possibility and benefit of its realization."[170] Second, the timeframes were striking. The Americans suggested discussing their proposals during their 26 April to 1 May visit to Ukraine. The MFA then sent a translation of the propositions to Prime Minister Kuchma, and as the registration stamp shows, this was on 28 April. Yet, the letter was sent to me on 10 May from Kuchma's office. The only conclusion that can be drawn from this is, thanks to the actions of the Foreign Ministry, the very opportunities that would have completely broken the Russians' game were prevented from being immediately put into motion.

Finally, the positive feedback which Zlenko gave the American proposals were completely incompatible with his persistent support for an immediate ratification of START and the NPT, which he declared at a parliament session on 3 June 1993, a mere month later. Oddly enough, his main argument was that the world would use sanctions against Ukraine, if it did not comply immediately, and as though there had never been any proposals from the SPA.

CHAPTER 4. The Hawks' Victory

The Battle of June 3: The Doves' Move

On 3 June 1993, two opposing camps faced one another in the parliament. President Kravchuk and the MFA wanted both the START and the NPT treaties ratified simultaneously. Our working group wanted to block the immediate ratification and to delay it for at least awhile.

We all understood that at the end of the day there would be a vote on a draft resolution. Thus, both sides prepared their own versions, whose essence came down to contradictory statements: "approve" or "instruct to continue working." Only one of them would win.

The Working Group intended to persuade the deputies that many significant issues concerning the two treaties still remained unaddressed, as of that day; if they were not resolved, Ukraine would be unable to fulfill the commitments contained therein. We already had convinced a majority within the legislature, with 162 MPs having signed the "hawks' declaration" in April, as well as both Speaker Ivan Pliushch and Deputy Speaker Vasyl′ Durdynets′. In the Cabinet, our position was supported by the Defense Ministry and it was easy enough to present a good argument to Prime Minister Leonid Kuchma. As a technologist, he thought in numbers and facts, and had his own trusted source of information, i.e., the missile manufacturer, Pivdenmash. Nevertheless, on that morning, no one could have predicted the sensation his speech would cause.

Still, even with all these forces engaged, I understood quite well that the fate of the vote would hinge significantly on what Speaker Pliushch said, and so I had asked the day before, to have a téte-á-téte.

From his first days in Parliament, Pliushch stood out among the glowering communists as an unusual figure, because of his remarkably deep and colorful national disposition. His words spoken at the nation's highest podium seemed to later become national aphorism. Complementing Pliushch's witty nature was his ability for influential and statesmanlike thinking.

Typically, one-on-one conversations with Pliushch followed a certain pattern, which simply meant during the majority of our time together, he would do most of the talking, but after that it was anybody's guess. Having this prior knowledge, and upon entering his office, I asked Pliushch to grant me at least 20 minutes of his silent attention. Pliushch then asked his receptionist not to be disturbed. He actually listened for more than an hour without interruption, while I provided a precise and factual outline of the key aspects of the issue, which also included the dangers of a hasty ratification. I also presented options that would be in the best interest of our nation.

When I finished, there was an air of silence, after which, Pliushch raised his head and said, "Yura, I cannot trust everything you have said 100 percent, but I do not have the right not to trust you. I also want to say quite frankly that we will not be able to do everything you suggest. The one thing I can promise you for certain is that you will have my personal support for all your initiatives."

At the parliamentary session there were four speakers. Foreign Minister Anatolii Zlenko and First Deputy Speaker Durdynets' spoke during the open session, then during the closed portion, Borys Tarasiuk spoke as deputy foreign minister and chairman of the National Disarmament Committee, as did I, representing the Working Group.[1]

Zlenko went first. He attempted to scare the members of Parliament with the negative consequences of rejecting immediate ratification. In comparing his speech with Tarasiuk's interview at the beginning of January, a half-year earlier, and with the Foreign Ministry policy brief handed out to MPs at the beginning of February, it was obvious that nothing had changed in his argumentation.[2] In other words, whatever new information had come during this time, the Ministry remained remarkably stubborn and refused to even consider it.

Zlenko's main argument was based on the substitution of concepts. The logic within his speech was: not ratifying means remaining a nuclear state; remaining a nuclear state means intending to use nuclear weapons. Next, the intention to use nuclear weapons meant establishing a nuclear cycle and establishing sites for testing nuclear weapons, which in turn means spending billions out of a beggarly state budget, instead of paying salaries and pensions. In the end, his logic suggested that not ratifying meant Ukraine intended to set up its own nuclear cycle, which then would expose the country to international sanctions.

The Russian MFA seemed to use similar logic: Moscow referred to Ukraine's slow response in ratifying the nuclear treaties as Ukraine reaching for the nuclear button. Moreover, Russia's anti-Ukraine campaigns were built on the equation of these two unsubstantiated charges.

In following the Russian line, the Ukrainian Foreign Ministry prepared some domestic scare tactics on the dangers looming, should START and the NPT not be ratified, rather than looking for the best solution to nuclear disarmament. This was the logic on which the MFA based its international activities.

As someone who participated in all the consultations on nuclear disarmament held during the entire disarmament process in the parliament, the presidential administration, and the government, I never heard a theoretical discussion on how Ukraine might gain nuclear status with the possibility of independently making use of its nuclear weapons. All that was discussed was how to block unsanctioned launches of strategic missiles from Ukrainian territory, what procedure to follow for eliminating them, and how nuclear material from dismantled warheads could be used for power generation. Zlenko was also present at all these discussions.

Our response to Zlenko's speech was to present facts and legal arguments. At that point in time, a significant portion of the legislature already knew these facts, we used them during the Q&A period. The following is a fragment of the transcript of the open portion of this breakthrough session, available in the public domain. In effect, this was a dialog between Parliament and the Foreign

Ministry. Though it appears without any commentary, it provides a good idea of the fundamental difference in their positions. This is how it went:

ZLENKO, A. M.: Today is the day to make a firm decision on the question of ratifying START and acceding to the Nuclear Nonproliferation Treaty. Further delays in adopting the necessary decision will only lead to intrinsic damage of Ukraine's national interests [...]. We are creating an image of Ukraine as an unreliable member of the international community and a state with excessive military ambitions. [...]

We know that nuclear warheads are extremely complex mechanisms that teams of thousands of highly qualified scientists and technologists have worked to develop in nuclear states. The nuclear facilities where these specialists work consist of at least seven main components [...]. Even if we managed to produce some of them, through many years of extremely exhausting effort on the part of the entire Ukrainian nation and at the cost of billions, nuclear testing on Ukrainian territory is simply out of the question.

The same kind of infrastructure is necessary also to support nuclear warheads Ukraine inherited from the former USSR under safe conditions. Each of them has a guaranteed lifespan of not more than 10–12 years, after which it needs to be dismantled and completely reconstituted. [...] Our military specialists know that the majority of warheads that are deployed on the territory of Ukraine have already exhausted a good part of their guaranteed term. Russia has not sent the necessary parts to Ukraine for nearly a year now and for us to expect that this will happen in the future is pointless. At a certain point after their shelf life ends, even those enterprises that produced them in the first place will not be prepared to take on dismantling these nuclear warheads, because it is extremely dangerous.[3] [...]

A comprehensive analysis of recent developments around Ukraine's nuclear weapons testifies that if Ukraine leaves the path towards nonnuclear status, the reaction of other countries will inevitably be very negative. [...] Even if broad-based sanctions are not applied against Ukraine, we are being made to know that access to any form of cooperation and exchange in high-level technologies will be closed, the delivery of nuclear fuel to our NPPs will stop, and all standard trade and financial preferences and breaks for normal interstate relations will be cancelled.[4]

KHMARA S. I.: [...] What state do you represent? On whose behalf are you speaking? Your report was very tendentious, to put it mildly, and is clearly promoting a specific position. [...]

ZLENKO, A. M.: [...] I am prepared to provide the necessary information, and thanks to the Ministry of Foreign Affairs you now have an understanding of what nuclear weapons really are.

MOSTYS′KYI, A. B.: [...] In spring last year, the tactical nuclear weapons were taken out of Ukraine with surprising haste. Today, Russia is even refusing to talk to us about any kind of compensation. [...] Has the question of Ukraine's ownership of its strategic nuclear weapons actually been formalized under international law?

ZLENKO, A.M.: [...] I repeat. Indeed, we can talk about Ukraine's right of ownership of its nuclear weapons. But that still doesn't mean actual ownership, because having the right of ownership and exercising this right are, of course, different things. [...] As to guarantees of Ukraine's security, we're talking about the nuclear states adopting a political and legal document that would confirm their commitment to the unacceptability of any kind of use of force against Ukraine by the nuclear states. Of course, these kinds of commitments, in and of themselves, do not guarantee Ukraine's security, but they will have considerable political significance. We have already received the preliminary texts of such guarantees from the United States, Great Britain, the Russian Federation, France, and China.[5]

For Ukraine, maintaining normal, friendly relations with all its neighbors, and developing political, economic, and technological relations with the US and other Western countries are a priority. These include establishing the necessary conditions for foreign investment, undertaking deep economic reforms based on a market economy, continuing the democratization of Ukrainian society, and, above all, maintaining domestic political stability; all these are components that guarantee Ukraine's security and they will only be possible, provided that Ukraine acquires the status of a country without nuclear weapons.[6]

[...] To date, the US is prepared to allocate US $175 million for these goals, and another US $10 million to set up a Science and Technology Center in Kyiv. This sum will not cover all the costs Ukraine has to bear in connection with the implementation of START, but we believe that, in time, it will be increased through additional US contributions and the involvement of other Western countries.

ZAIETS´, I.O.: [...] I understand from your report that there is no actually signed document that would guarantee our national security. [...] Right now, we do not have a single agreement about compensation. And so, the external conditions for ratifying the treaty are really not there. [...] If START I is ratified without the external conditions being decided or clearly formulated in Ukraine's interests, then we will find ourselves facing the reality of being pressured right now, but then, after signing, we will be stripped naked.

ZLENKO, A.M.: [...] The external conditions that you just mentioned are already almost in place. Delaying any further is not working in Ukraine's favor. [...]

MOSTYS´KYI, A.B.: If Ukraine ratifies START I, are there any guarantees that it will receive appropriate compensation for the material and warheads that will be taken away from its territory?

ZLENKO, A.M.: We are going to keep working on this aspect if you give us the go-ahead. [...] I've also named the sum that is necessary for us to destroy the nuclear weapons that are on Ukrainian territory. Our specialists have calculated that we will need US $2.6 billion for this.[7] In other words, we are talking about serious money that we also don't have right now. [...] That is why we are in intense negotiations both with the US and with other advanced industrial countries that are prepared to help us. For instance, Germany has made it clear that it is prepared to help us. Sweden and other countries have, also. But only after ratification. So, we will have to do the necessary work in this area with regard to guarantees of our security after ratifying the treaty.

KENDZIOR Ia. M.: [...] You said that there are some promises from Sweden and other countries about being prepared to help us when we ratify the treaty. However, those are mere promises that could be forgotten tomorrow by those who made them today. We will be left on our own with a terrible problem. [...]

ZLENKO, A. M.: [...] We have an agreement and there are political guarantees on the part of the leaders of certain countries, at this time, from five nuclear states. Without any doubt, the main point is that these will be finalized only after we ratify START. [...]

CHORNOVIL, V. M.: [...] You submitted these two matters in a single package, meaning, the ratification of START I and the Nuclear Nonproliferation Treaty. Don't you think it would be more useful to consider these two issues separately? [...]

ZLENKO, A. M.: I think that, in order to avoid any further misadventures on this account, pardon me, it is better for START I and NPT to be considered at the same time. [...]

IAKHEIEVA, T. M.: [...] It has been a year since the Lisbon Protocol was signed. According to the Lisbon Protocol, Ukraine, Russia, Kazakhstan, and Belarus, were supposed to negotiate about the reduction quotas. Why has the Foreign Ministry not engaged in such talks so far? [...]

ZLENKO, A. M.: About talks regarding the Lisbon Protocol. Ukraine has taken the initiative more than once to have such a meeting and to get the necessary decisions regarding Article 2 of the Lisbon Protocol. However, Russia, for starters, has no interest in engaging in such talks. [...] I have to say, honestly, that we constantly keep this issue in focus, constantly raising it, but so far nothing has happened.[8]

The session was clearly not going Zlenko's way.

The Battle of 3 June: The Hawks' Move

The foreign minister's open confrontation with the legislature had failed, resulting in no security guarantees, no funding, no distribution of reduction quotas according to the Lisbon Protocol. Everything was just vague prospects that required the two treaties be ratified first, so that the international community would not suspect Ukraine of aggressive intent. It was now the hawks' move.

During the open session, First Deputy Speaker Vasyl' Durdynets', who was curator of our working group, took the podium on behalf of Parliament leadership. I was to speak during the closed session and present the Working Group's position. And so, the two of us divided up various aspects of the issue in such a way as to neither repeat ourselves, nor omit anything significant. Our reports, point for point, captured the entire set of arguments that the "hawks" had intended to convey to the members of the legislature.

Durdynets' argued that it was highly unfair to accuse Ukraine of playing a double game, of declaring its intentions to acquire nonnuclear status, while at the same time delaying the process of ratifying the nuclear treaties. He pointed out that the parliament had started working on this on 20 July 1992, just five months before Parliament's Executive Committee passed its resolution to set up a working group, while during this time, all the ministries, agencies, parliamentary commissions and independent experts were being provided with materials analyzing different aspects of the two treaties. Durdynets' categorically objected to rushing the process, especially as START was a technically complicated document that had been drafted during the course of nine years of difficult negotiations. Moreover, representatives of Ukraine, and even the Ukrainian SSR, never had access to these negotiations, although now the country's national security depended on this treaty.

Durdynets' insisted that only after a comprehensive study was completed would it be possible to discuss specific timeframes for a legislative review. Whatever the final decision was or when it would be made, he said, would have to be guided by the interests people and the state of Ukraine.

Durdynets' went on to the specifics of key issues that needed to be clarified before it could be brought up for debate in the legislature. This included proper guarantees for national security after disarmament and distinct sources of funding for the nuclear arms reduction program, commented on by Zlenko with an aside that US $175 million was completely inadequate, based on preliminary calculations that were in the billions. Another key issue was the possibility of recycling or repurposing the nuclear material removed from warheads, as the deputy speaker believed simply moving them to Russia was an unacceptable option for Ukraine. Finally, he also pointed out, the MFA had not sat down at the negotiating table with Russia, Belarus, and Kazakhstan to discuss how to divide up the reduction quotas, as the Lisbon Protocol required; without the aforementioned, there was no basis for reduction. After all, neither START nor the Lisbon Protocol "contains a formulation that directly requires Ukraine to destroy even a single missile."[9] On the other hand, reductions based on USSR's commitments could not be done at Ukraine's sole expense.

The deputy speaker clarified the tangled situation with Kravchuk's letter to Bush committing Ukraine to reduce all its nuclear weapons over the course of seven years. Durdynets' explained that over the seven years after acceding to the NPT, Ukraine was to remove from its strategic delivery mechanisms all nuclear warheads for destruction, and 36 percent of its carriers (missile silos, intercontinental ballistic missiles or ICBMs, and heavy bombers), as per the Lisbon Protocol. For a country that had no possibility to independently produce nuclear weapons, becoming nonnuclear quite literally meant getting rid of its nuclear warheads.

My report focused on the legal issues, the right to ownership of the nuclear weapons without which nothing else made sense, and the traps set up for Ukraine within the treaties that were being proposed for ratification.

I immediately made it clear that the question was not what Ukraine should call itself—nuclear or nonnuclear—given that there were nearly a dozen countries in the world that had nuclear weapons but called themselves nonnuclear, but who had the right of ownership over the nuclear weapons on Ukraine's territory. If we lost the right of ownership, then we would also lose the right to compensation. What was at stake was very costly indeed. The nuclear material in our warheads had been evaluated at nearly US $100 billion.

I pointed out some treacherous contradictions written into the Lisbon Protocol. Since it was being proposed that we accede to the NPT as a nonnuclear state, the combination would automatically take the legal right from Ukraine to own nuclear weapons. In other words, if we were a nonnuclear-weapon state, what did we actually own, and what were we preparing to reduce? This also meant changes to the status of the strategic nuclear forces on Ukraine's territory: they would become Russian, not Ukrainian, since a nonnuclear-weapon state could not have access to nuclear weapons. In this way, a ratification with the proposed wording would lead to legal conflicts, deprive Ukraine of the legal right to the valuable material within the nuclear weapons, and open possibilities to subordinate the 43rd Rocket Army to another country.

Parliament managed to avoid this particular trap. By ratifying only START I on 18 November 1993, it separated the timing of the ratification of the NPT and declared the nuclear weapons Ukraine's property. However, this came later.

One more unresolved issue remained: what would replace nuclear weapons in the country's defense system? I said that, by their very nature, nuclear weapons continued to provide a defense function, even when control over them was not held by the state where they were deployed. Eliminating nuclear weapons without a suitable replacement for deterring aggression meant losing effective components in the national security system and now opened an existential threat to the Ukrainian state.

As noted earlier, the military doctrine had already been adopted in its first reading with a special emphasis on high-powered precision-guided munition. Although there were already strong objections from the US, I nevertheless warned against a commitment to refuse this type of weapon. Since their technical specifications meant that such weapons belonged to the category of strategic weapons, Parliament needed to consider the issues in tandem: either change the formula for ratifying START and the Lisbon Protocol, or change the military doctrine.

I then laid out the latest calculations for the financial costs tied to fulfilling the conditions in START. They added up to nearly US $700 million annually and that was just for eliminating the 130 SS-19 liquid-propellant ballistic missiles. It did not include destroying another 46 SS-24 solid-fueled missiles, which the US was also demanding. The MPs understood that in the midst of a deep economic crisis, our budget simply could not assume this additional burden. Given this, the Working Group proposed that the Ministry of Foreign Affairs begin talks with the US and Russia about changing the conditions for carrying out certain provisions in START calling for the destruction of strategic missile launch facilities or silos, and keeping track of strategic arms reductions without destroying the actual silos. Without its payload, a strategic delivery platform, did not represent a threat (as discussed earlier in chapter 1).

I concluded that the Working Group could not recommend that Parliament agree to any final version of disarmament, because neither the technical, nor the financial justification was there to clarify what commitments Ukraine was in a position to accept. In order to avoid any accussations of changing Ukraine's nonnuclear course, I added: "There probably are no alternatives to disarmament, but this process needs to be undertaken in such a way that in becoming caught up in grandiose plans to save humanity from the nuclear threat, we do not end up accidentally losing our statehood."[10]

The normally garrulous deputies did not ask very many questions. Deputy Foreign Minister Borys Tarasiuk, who also chaired the National Disarmament Committee, tried to repudiate the arguments of the Working Group. The main points he made in his speech were:

> Declaring itself a nuclear state will, in practice, not bring Ukraine the real status of a country with nuclear weapons. But the appearance of a nuclear Ukraine will be a factor that will push Russia, the US, and other Western countries that might otherwise be our partners toward consolidation on anti-Ukrainian positions. As a non-participant in the Nonproliferation Treaty, Ukraine will find all forms of cooperation related to the peaceful use of nuclear energy curtailed. Yet we will be in no position to establish the infrastructure for our military's nuclear capacities quickly without outside help.

> Therefore, the Foreign Ministry has concluded that declaring Ukraine a nuclear weapons state, or that it intends to develop such weapons, is not in line with Ukraine's political, social, economic, or even military interests. For this reason, the MFA believes that ratifying START I and the Lisbon Protocol, as well as acceding to the NPT as a nonnuclear-weapon state corresponds to Ukraine's state interests, and that it will foster the further strengthening of its international reputation and its status as an influential, independent, European state.[11]

In the end, Tarasiuk was so caught up in providing examples of what might await Ukraine if it decided not to immediately ratify the nuclear treaties that Premier Kuchma could not restrain himself, and interrupted Tarasiuk's speech with the question: "Why are you trying so hard to scare us all, Mr. Tarasiuk? I think the position of the Working Group is much more reasoned and personally I support what Kostenko is proposing."[12]

Kuchma wrapped up the speeches and discussions in the closed session. For me, it was no secret that he was quite on top of nuclear issues. For one thing, we had often exchanged thoughts on this matter since the time when Kuchma was the director of Pivdenmash, and I was the chairman of the parliamentary working groups for developing a national security policy and to prepare for the ratification of START I. Secondly, everyone knew that he was still in touch with Pivdenmash, which provided him with real, and not ideological, information about the process. Thus he said:

65. Leonid Kuchma, 1999

Leonid Kuchma (b. 1938), prime minister of Ukraine (1992–1993), president of Ukraine (1994–2004). Kuchma graduated as a mechanical engineer with a specialization in aerospace technology from Dnipropetrovsk National University in 1960. From 1960 to 1982, he rose from engineer to senior engineer, lead designer, assistant to the chief designer, and technical manager for testing aerospace and rocket systems at Baikonur in Kazakhstan. From 1975 to 1981, he was secretary of the Communist Party Committee at the Pivdenne Design Bureau, and subsequently secretary of the Party Committee at the PA Pivdenmash, which at the time was the largest manufacturer of military and rocket equipment in the USSR. In 1982, he was promoted to first deputy chief designer at the Pivdenne Design Bureau, then to general director of Pivdenmash from 1986 to 1992. From 1990 to 1992, he was a member of Parliament and a member of the Commission for Defense and National Security. From October 1992 to September 1993, he was prime minister of Ukraine. In July 1994, he was elected the president of Ukraine.

66. *Kuchma speaking in Parliament. Behind him are Kravchuk (left) and Pliushch (right), 1993.*

Given the density of nuclear power plants and toxic manufacturing across the country's territory, and its geographic and geopolitical situation, Ukraine is vulnerable to any kind of aggression and the consequences would be catastrophic for us.

It is fundamentally impossible to establish an impenetrable air defense system, even using the most cutting-edge technologies. In this situation, the only real, effective, reliable system that can ensure Ukraine's security might be a doctrine of guaranteed deterrence and non-provocative defense. The technical foundation for such a doctrine could be strategic rocket systems based on the SS-24 with nonnuclear warheads.

Provided it undergoes a rational restructuring, unconditional conversion based on dual-use technology under properly planned programs, the unconditional reduction of the number of weapons, and the removal of obsolete systems, Ukraine's R&D, technological, production, and military potential would allow such a concept to work.

As the heir to USSR treaties, Ukraine is clearly obligated to ratify START I and to implement it in full. In practice, this means eliminating the 130 SS-19 missiles deployed on our territory over the next seven years. We need to engage in the necessary talks with the US about assistance in jointly carrying out the practical side of this. As to compensation from Russia for our share of the 780 nuclear warheads to SS-19 rockets,[13] that is an issue that has to be settled on a bilateral basis, mainly with Russia.

If we believe that complete nuclear disarmament will open the door to loans and assistance from the West, this is probably a profoundly mistaken notion.

As to Ukraine's accession to the Treaty on the Nonproliferation of Nuclear Weapons, we should probably wait until 1995, when the next Review Conference on extending the NPT will be taking place in [New York City], and try to resolve this issue comprehensively by considering an effective system of oversight and security guarantees that do not degrade the country or discriminate against it in any way. The MFA needs to prepare properly for the [New York City] Conference. In the meantime, it is better that we do everything we can to prevent the proliferation of nuclear weapons, including a parliamentary resolution that makes a commitment not to engage in any acts or actions that are connected to the direct export of nuclear weapons, their components, and technology.[14]

At last—and absolutely unexpected for all those present from both camps—came the final statement by the premier. The gist was this: Ukraine owns nuclear weapons; Ukraine has to be recognized as a nuclear state, at least for the time being; we need to leave 46 missiles on our territory. And the ratifications of START I and of the NPT need to be dealt with separately.

This was unconditional support for the position of the Working Group. Given the sizeable "director's corps" among the parliamentarians that was oriented by the premier's opinions, the result of the vote was no longer in doubt. Ratification would not take place that day.

The Battle of 3 June: The Outcome

What propositions were discussed at the closed session and what the formal conclusion was can now be recreated based on a policy brief labeled "For Service Use" that was in my archives. It was prepared by our Working Group for the parliamentary leadership, the government, and the president. In the end, Parliament supported propositions, and the Working Group continued to work on the problem, dividing it into two separate issues: START and the NPT.

As to the fate of Ukraine's nuclear weapons, polar opposite approaches were suggested. First, that Ukraine should gradually take on nuclear status, using its scientific, technological, manufacturing and human resource potential and use this as the main guarantee of its national security. Second, that Ukraine should completely eliminate all nuclear weapons on its territory in a short time. But that nuclear disarmament had to take place exclusively in Ukraine's national interests and security, the entire body of the legislature agreed.

An absolute majority of deputies were in favor of ratifying START to the letter. For one thing, this treaty did not call for the destruction of all nuclear weapons or all strategic weapons on the territory of Ukraine. And so, the scheme that came up during discussions was to eliminate only 130 SS-19 missiles over seven years and to share the reduction quota among the nuclear republics after four-way negotiations.

Practically all the deputies were talking about the need for serious reservations regarding the Ratification Resolution. These caveats were to cover absolutely all aspects of the problem: legal, military, political, scientific, technological, financial, economic, environmental, social, and so on, and to be written into a separate official document by Parliament.

With regard to the Lisbon Protocol, MPs noted serious contradictions between individual provisions of the Protocol and START I, as well as the illogic of considering these documents in a package with the NPT. It was precisely this combination, in the opinion of some deputies, that had led to political pressure and criticism. However, there were many strategic weapons and nuclear warheads Ukraine reduced under START, and the remaining warheads had to be reduced in line with the NPT, and the rest of the strategic weapons in line with the commitment made by the president of Ukraine in a letter to the president of the United States. The ratification bill had to reflect all of our reservations.

The participants in the session proposed revisiting the NPT after the proposed New York City Conference in May 1995, it would decide whether the Treaty would continue to be in effect. The MPs unanimously agreed that Ukraine needed a guarantee concerning national security and financial assistance for disarmament before ratifying START, especially from the nuclear states pressuring Ukraine into speeding up the process. They proposed a resolution to provide social support systems for the nuclear forces personnel and their families in the financial assistance package.

The Ukrainian diplomatic services took much criticism on the need to provide more in-depth analysis, to consider academic and military opinions, to properly apply Ukraine's legislative framework, and to considerably strengthen the negotiation process on all aspects of nuclear disarmament in Ukraine. Therefore, the idea was raised for the first time that Parliament should issue a statement that would lay out the principles and approaches of nuclear disarmament, and the world would actually hear it from the source.

The protocolary assignments by Speaker Pliushch's summary of the session included: "to take into account" the speeches by Zlenko and Durdynets'; to complete the work drafting the ratification resolutions "as soon as possible;" to submit them to the parliamentary commissions for review, who were then to "add their propositions and comments" after which the Commissions for Foreign Affairs, Security and Defense, and the Working Group, were to rework the draft resolutions and then submit them to Parliament in June of that year.[15]

On 9 June, the Working Group developed a plan of action for drafting the resolutions. The task was divided up among the members of the group, and I was to draft the text itself. It then became immediately apparent that, "by the end of June" was not enough time. The next session Parliament that was to consider the nuclear issue was nearly six months later, on 18 November 1993. Thus, the legislature had managed to prevent a catastrophe on 3 June. The days that followed showed us that this decision had other consequences, which were not anticipated.[16]

The Response to the Statements Issued on 3 June

The Western press was the first to react, and that reaction was explosive. The session had been closed, and it is the very reason that a transcript of the session cannot be found to this day. Yet, everyone was quoting Kuchma's speech. His statement became headline news in the world press.

In a 4 June article, which headlined, "Ukrainian Premier Urges Keeping Nuclear Arms," the *Washington Post* wrote: "Kuchma's remarks constitute the first public declaration by a senior Kiev official that Ukraine should join the world's nuclear-weapons club, and the speech is certain to complicate the tense and still unsuccessful efforts by the United States and Russia to persuade Ukraine to fulfill its earlier pledge to become a nonnuclear state."[17] The article continued: "Ukrainian Prime Minister Leonid Kuchma told a closed meeting of parliament in Kiev yesterday that Ukraine should become an independent nuclear power, at least temporarily...[and] also called on parliament to ratify the START I nuclear arms control treaty, which would require Ukraine to surrender some, but not all, of the 1,600 to 1,800 Soviet-era nuclear warheads on Ukrainian territory. [...] Kuchma urged that Ukraine take full control of those weapons it does not give up under the treaty."[18]

The *Washington Post* also observed: "Kuchma's speech is likely to be interpreted in Washington and Moscow as an effort by Ukraine to up the ante in its aggressive efforts to extract concessions in exchange for becoming a nonnuclear state."[19] At the same time, the paper reported that the US was already looking at the possibility of offering these very concessions, proposing that Ukraine leave all the warheads on its territory on condition of establishing international control. This was a signal that official Washington had changed its position.

The White House announced its new initiative the same day that Kuchma made his speech, a lightning-fast reaction that showed that events were being monitored in real time, together with making projections and working out various options for Washington to act, depending on how events unfolded. And here the above-mentioned article proved remarkably informed about the change in the US government's position towards Ukraine: "The new US plans reflect an administration decision to shift emphasis from 'sticks' to 'carrots' in the longstanding US drive to prevent Ukraine from becoming a nuclear power—in effect, offering to broaden US-Ukrainian ties instead of merely threatening the country with potential isolation."[20]

Russia reacted to Kuchma's statement much later and more bluntly, not bothering to consider Ukraine's interests in the situation. "If this happens, Ukraine will become a nuclear state that is more powerful than Britain, France, and China, shaking up the already shaky nuclear nonproliferation process and setting a bad example for, say, Kazakhstan. Russia doesn't need this kind of headache. Do the Americans? They don't either," opined *Izvestiia* on 26 June.[21]

The second thing that developed with Kuchma's speech was a split within the Ukrainian government over the nuclear issue, between the prime minister-Parliament faction and the president-MFA faction. What could long have been called a civil war moved from being covert to overt. The *Washington Post* wrote, "Parliament, which has ultimate authority to determine Ukraine's nuclear status, appeared today to back the prime minister's position," while President Kravchuk and Minister Zlenko were against keeping this weaponry. The article continued: "Ukrainian President Leonid Kravchuk reaffirmed today that his government was committed to its pledge to become a non-nuclear state." In addition, during a meeting with Russian Federation Foreign Minister Andrei Kozyrev, "he assured Kozyrev that the official Ukrainian position on the nuclear issue was unchanged. [...] 'You know our position. Our position is clear,' Kravchuk told Kozyrev at a meeting attended by reporters. 'As everywhere else in the world, members of parliament have varied opinions.'"[22]

The arguments presented during debates in Parliament did not alter the president's opinion. Already the next day, Kravchuk made observations during a meeting with Russian Foreign Minister Kozyrev that contradicted the premier's public call. The American press decided that Kuchma's statement reflected tendencies that were gaining momentum both in the legislature and in Ukrainian society as a whole.

Not only Kravchuk, but also the MFA, did nothing to hide the fact that there was a rift within the government. Zlenko sent Pliushch a "Review of the American press for 4–5 June 1993," based on coverage of the parliamentary session. The very presentation of the information testified to the diplomatic agency's position regarding these events: "The *Washington Post* emphasizes that [Russian] Minister A. Kozyrev expressed satisfaction about his meeting with L. Kravchuk and Ukraine's official position as stated during this meeting;" and "the *Washington Times* notes that there is a serious rift at various levels of government regarding the fate of the nuclear weapons;" while "Reuters draws out this topic by reporting on a strongly negative response in Ukraine's MFA to the statement by L. D. Kuchma regarding nuclear weapons."[23] Not only did the Foreign Ministry not conceal the fact that its position did not coincide with that of Parliament and the head of government, but the impression was that the foreign policy agency considered this an achievement that it wanted to underscore.[24]

Zlenko was clearly proud of his speech because the Ministry somewhat vaingloriously wrote in the brief: "News agencies are making much of Foreign Minister Zlenko's speech in the legislature. The *New York Times* points out in its article that the minister's presentation was persuasive and well-argued in making the case for ratifying these important international treaties as soon as possible. The local press brings up that part of the Zlenko speech that emphasizes the possible consequences if Parliament refuses to ratify START I and the NPT."[25]

The MFA also wanted to publicize the position stated in Zlenko's speech in the domestic press. On 5 June, *Holos Ukraïny* published the complete text of the minister's report. However, there was no opposing viewpoint in the issue.[26] There also were no comments on the numbers and financial indicators that the minister used to argue his position, although some MPs pointed out that they were inaccurate. Consequently, the publication of the speech without alternative opinions effectively worked as disinformation.

Kuchma's speech led to a wave of analysis in the press regarding the possibility that Ukraine could control its nuclear weapons. From the point of view of scientific and human resource potential, this option was estimated quite highly. The turning point came when the West began to hear Ukraine's arguments as well: "Ukraine says this is necessary to defend its recently won independence against Russian imperial ambitions and to prevent unauthorized nuclear launches by Russia. US officials believe Ukraine is privately taking technical steps to secure independent control over its nuclear missiles, but Kravchuk denies this."[27]

After 3 June, the rhetoric among MPs also changed. On 24 June, Deputy Speaker Durdynets´ instructed the commissions to submit their propositions regarding the ratification of the two treaties to the Cabinet. The Basic Industry Development Commission, the Agro-Industrial Complex Commission, and the Budget Planning, Finance and Pricing Commission, all supported Prime Minister Kuchma's position. The first took the understandable position that Ukraine should "accede to the NPT as a nuclear weapons state;" the second ambiguously wrote "join the NPT;" and the last argued that "We cannot blindly accede to treaties that were drawn up under completely different political conditions."[28]

The mindset changed in the US as well. As of June, the Western press took a positive line of appreciation towards Ukraine's position. The June events confirmed that this change had affected not just the American press, but official circles, too.

67. Les Aspin

Les Aspin (1938–1995), US secretary of defense under the Clinton administration from January 1993 to February 1994. After a closed session of the Parliament of Ukraine at which Prime Minister Kuchma argued it was necessary for Ukraine to preserve part of its nuclear arsenal, he flew immediately to Ukraine on 6 June 1993, and presented Washington's new approach to Ukraine. The Clinton administration was prepared to decrease pressure in exchange for a "defense partnership" and a package of economic and political incentives aimed at persuading Ukraine to undertake nuclear disarmament. These propositions included keeping the nuclear warheads that had been removed from Ukraine's strategic arsenal in Ukraine and under international control.

The US Shifts Position in Favor of Ukraine

At the beginning of summer 1993, the US position shifted fundamentally towards Ukraine. The turning point proved to be the uncompromising position of Prime Minister Kuchma at the 3 June closed hearings. Additionally, Parliament distinctly called for the unambiguous recognition of the nuclear weapons as Ukraine's property, and to bring about proper compensation for both the nuclear material and the dismantling work.

Unlike Russia, which typically blazed a trail with an ax in international politics and only ever took its own interests into account, the US generally took the other country's position into consideration. The Americans responded instantly, just as Ukraine articulated the position. The first indicator was the visit of Secretary of Defense Les Aspin to Kyiv in early June. Washington made Kyiv an offer, which would have never even been dreamt of before: to assist the separation of the nuclear warheads from the missiles, and to agree to letting them remain in Ukraine under international supervision. First, the US would not insist that Ukraine move its weapons to Russia. Second, it meant that in the technical aspects of disarmament Ukraine would cooperate not with Russia, but with the US, with fewer losses. Washington made it known that it was prepared to play on Ukraine's side, effectively against Moscow.

On 17 June, the *Wall Street Journal* published an editorial in which the American press expressed the opinion that the US should participate in regulating Ukraine-Russia relations around nuclear weapons. The paper called Defense Secretary Aspin's early June visit to Kyiv "a large step in the right direction" that should "turn the page" in relations between the US and Ukraine.[29]

An article published on 20 June in the *Washington Post* expressed a heartfelt confession of an unjust attitude towards Ukraine, and also included some sensational admissions by Western diplomats that their Russian colleagues recommended against setting up missions in Ukraine; they would be downgraded to consular sections in about 18 months.[30]

On 24 June 1993, the Senate Committee on Foreign Relations held hearings on START I, chaired by the then Senator Joe Biden. The hearings also reflected a pivotal change to Ukraine and confirmed that the publications in the press, were not just the viewpoints of individual journalists, but the expressions of a general mood among American politicians. I should say at this point that the presentation of this particular session struck me, not only by this shift in position, but also by the profound depth of the discussion of the issue. The depth of these official discussions was very far from the norm or our reality in Ukraine; it was something we could only hope for in our wildest dreams.

The main points that arose during this review, especially for allies of Russia who had recently been arguing for Ukraine's immediate disarmament, were astounding. First, the US should not insist on transferring Ukraine's nuclear weapons to Russia or pressure the country on this issue. Additionally, the US should encourage Ukraine's integration with international institutions, instead of pushing sole cooperation with Russia. The fact that Ukraine was attempting to take control of its inherited nuclear weapons was also acknowledged as protecting the world from a non-sanctioned Russian launch.

Bruce G. Blair, a Senior Fellow at the Brookings Institution, insisted on reconsidering approaches to Ukraine and its integration into international organizations:

> We should adopt an even-handed approach that rests on basic principles of international law, respecting Ukraine's sovereignty and its parliament's right to debate and decide on issues affecting their security. We should not focus exclusively on the nuclear agenda—Ukraine refuses to talk on these terms anyway. We should promote Ukrainian integration into international structures, instead of threatening to

68. *Senator Joe Biden, 1991*

Joseph "Joe" Robinette Biden, Jr.
(b. 1942). In June 1993, he chaired the Senate hearings on Ukraine's nuclear disarmament and voiced support for a change in the US strategy.

isolate it. [...] We should try to get [the Russian Federation] to back off from insisting that Ukraine turn over the weapons in two years, and sell them on the idea of multilateral storage of weapons on Ukrainian soil.[31]

The analyst emphasized the danger of the situation, should operational control be performed only by Russia, because their military command was still not under civilian control, that is, President Yeltsin, in whom the US had placed such faith. Blair also maintained that Kyiv had entirely peaceful intentions with regard to acquiring technical control. In his speech, he stated:

> Of course, to say that Russia alone exercises operational control is not completely reassuring given the constant danger of a lapse of competent civilian leadership, and given Russia's failure to thoroughly subordinate the military to civilian authority. [...] Ukraine's interest in operational control seems confined to a desire for a technical veto that would physically prevent Russia from launching missiles without Ukrainian permission. In private talks with Ukrainian officials in late 1991, Marshal Shaposhnikov pledged to devise such a veto and install it sometime in 1993. The device failed to materialize, however, although President Kravchuk was left no worse off than President Yeltsin, who also lacks a technical veto. [...] The general staff has the technical ability to instruct all the unmanned missiles to disregard any commands from any command post in Ukraine.[32]

CIA Director James Woolsey reported that the accession of Ukraine and other nuclear republics to the NPT as nonnuclear weapon states was a precondition for Russia to exchange documents ratifying START I and START II. This in turn would clarify how Russia was blackmailing the US, so that Washington would squeeze Kyiv into joining the NPT and support the idea of transferring the country's nuclear weapons to Russia. He admitted, "Ukraine appears to be the state most motivated to retain strategic forces. It is also the state most capable of maintaining them."[33]

Senator Joe Biden really seemed to understand Ukraine's concern about security guarantees, by parallel with the events in Bosnia, which had faced aggression from its once fraternal neighbor, Serbia. In summing up Ambassador Talbott's position, he noted: "The West, acting in what I consider to be pathetic abandonment of the principle of collective security and of the US Security Council resolution on Bosnia, could not offer assurances to the Ukrainian government that might give them a sense of security. The Security Council has not done itself proud thus far, and some of the same principles are at stake."[34] Unlike Senator Biden, it has to be said, the Ukrainian government in no way looked at the situation of Bosnia for comparison and drew no parallels between Serbia and Russia. And so it did not look at options for preventing such an eventuality.[35]

69. Jim Woolsey

Robert James "Jim" Woolsey, Jr. (b. 1941), director of the Central Intelligence Agency 1993–1995. From 1969 to 1970, he was an advisor to the US delegation for the American-Soviet negotiations for START in Helsinki and Vienna. From 1977 to 1979, he was the US undersecretary of the Navy. From 1983 to 1986, he was a US delegate-at-large during the START talks in Geneva, and in 1989–1991, he was appointed the US ambassador to the negotiation on Conventional Armed Forces in Europe (CFE) in Vienna. In 1993, President Clinton appointed him director of the Central Intelligence Agency. He retired in 1995, after the arrest of CIA operative Aldrich "Rick" Ames, who was a double agent for Russia.

Another parallel with Bosnia was also made by Yaroslav Bilinsky, Professor of Political Science and International Relations at the University of Delaware, during his speech at the hearings: "The long and short of my testimony here is that Ukraine does not want to become another Bosnia. Ukraine does not want tea and sympathy and a tombstone. From the United States and from the so-called world community, Ukraine expects fairness and a greater understanding of its legitimate security needs."[36]

The phenomenal aspect here was that a member of the Ukrainian diaspora addressed the US Senate with an argument, which Ukraine's MFA at the time could not offer. In contrast to most Kyiv officials, Bilinsky saw Russia as the greatest enemy of Ukrainian statehood. Professor Bilinsky stated in his presentation before the Senate Committee on Foreign Relations:

> Senator Biden, sir, two nations have the greatest moral right to possess nuclear weapons—Israel and Ukraine. The leaders of Israel have wisely decided that one Holocaust was enough. Ukraine is the victim of Stalin's genocide and of cultural genocide under Khrushchev, Brezhnev, and Gorbachev. Over five million Ukrainian peasants, over one-quarter of Ukraine's rural population, were killed in the terror-famine of 1932–33. In addition, Stalin ordered over two million Ukrainian intellectuals, workers, and peasants killed in the "Great Terror" of the 1930s. Under Khrushchev, Brezhnev, and Gorbachev, millions of Ukrainians were lost to russification—the equivalent of cultural genocide. But for the grace of God, the people of Ukraine, with their 1,000-year-old history, would have been wiped off the map. No nation in the world can deny Ukraine's moral right to keep nuclear weapons to prevent another genocide, least of all the Russians, who, albeit indirectly, have profited from the physical and cultural genocide of Ukrainians.[37]

Bilinsky's assurances that his speech did not "in any way represent the government of Ukraine," proved needless, unfortunately, as the majority of government officials in independent Ukraine had served, if not under Brezhnev, then under Gorbachev.[38] They not only did not think in terms of the Holodomor or Stalin's terrors, but in the early 1990s they actually denied them.

In discussing the situation with Ukraine, the US Senate recognized the supremacy of Parliament, Ukraine's legislature, in the matter of ratifying international treaties. Interestingly, at one point, Senator Biden called on his collocutors not to orient themselves on the government or president, but on the decision of Parliament, as in an exchange with Walter B. Slocombe:

> Mr. SLOCOMBE. My understanding of the position of the previous administration and of this is that obligations that were assumed in the Lisbon Protocol are obligations of Ukraine as a state and are not subject to whatever the process under Ukrainian law and constitutional practice for ratification—are not subject to that process. It is an international obligation of Ukraine to meet its obligations under the Lisbon Protocol.

> Senator BIDEN. That is my understanding, as well. But, as you and I both know, our Constitution preempts any international law if in fact it is not otherwise authorized. [...] That one I am certain of. The executive cannot bind the United States by treaty, absent adhering to the constitutional requirements under our system of separated powers to do so.[39]

Arguments in favor of a change in the US position were also put forward by the Center for Security Policy, a think tank linked to the Republican Party, which was not in the White House at the time.[40] The Center's views were even more radical, as they wrote: "The Center for Security Policy has long believed that it was inconsistent with America's security interests in general and the goal of a stable, peaceful European continent to demand Ukraine's nuclear disarmament."[41]

At that time, an article by University of Chicago political science professor John J. Mearsheimer had come out in *Foreign Affairs*, under the unambiguous title, "The Case for a Ukrainian Nuclear Deterrent," and the Center largely based its arguments on the case it made. Mearsheimer not only argued that the traditional view of Ukraine's nuclear weapons needed to change, but also suggested that the idea of Ukraine organizing its own nuclear forces needed to be supported: "President Clinton is wrong. The conventional wisdom about Ukraine's nuclear weapons is wrong. In fact, as soon as it declared independence, Ukraine should have been quietly encouraged to fashion its own nuclear deterrent. Even now, pressing Ukraine to become a non-nuclear state is a mistake."[42]

Mearsheimer was first in the American press to note that dogmatic approaches to the issue of nuclear disarmament were obsolete and needed to be reconsidered as soon as possible: "A nuclear Ukraine will corrode the NPT, but this damage can be limited if the United States reverses its 1991 policy of labeling Ukraine a potential proliferator and instead redefines Ukraine as a nuclear inheritor, and hence a special case."[43] In view of the situation that the world faced in the early 1990s, with covert development of nuclear weapons, he called START I and START II "vestiges of the Cold War order" and emphasized: "After all, America's ultimate goal is to create peace and stability in Europe, not ratify arms control agreements for their own sake, especially those created for another time." The best way to reach this goal—stronger stability in Europe—Mearsheimer said was "by maintaining an independent Ukraine, a goal, in turn, best achieved by a Ukrainian nuclear deterrent."[44]

He called the policy in which the US had engaged towards Ukraine's nuclear weapons "wrongheaded," and stated that it needed to be immediately revised: "The United States should have begun working immediately after the Soviet Union collapsed to quickly and smoothly make Ukraine a nuclear power. In fact, Washington rejected this approach and adopted the opposite policy, which remains firmly in place. Nevertheless, it is wrongheaded, and despite the sunk costs and the difficulty of reversing field in the policy world, the Clinton Administration should make a gradual but unmistakable about-face."[45]

At the same time as Mearsheimer's piece, *Foreign Affairs* published an article by Steven E. Miller, director of the International Security Program at Harvard University's Kennedy School of Government, who presented the opposite viewpoint.[46] Fareed Zakaria, the journal's managing editor, contacted me with a request to write an article on this same topic, introducing the topic of discussion: "Among other points, Mearsheimer and Miller debate whether Ukraine is capable of deterring a Russian attack through reliance on conventional forces, whether Kiev is technically capable of assuming control of nuclear weapons based on its territory, and whether a Ukrainian nuclear force would be exposed to preemptive attack."[47] I joined the discussion and in September 1993, *Foreign Affairs* published my article under the title: "Kiev & the Bomb: Ukrainians Reply."[48]

I focused on three key issues. First, the uniqueness of the situation with former Soviet republics: "Ukraine, as a legal heir to the former Soviet Union, has inherited the rights and obligations of its Soviet predecessor," however, "international law permits no changes to the status of a state that inherits nuclear weapons legally." And therefore, "[w]ithout the resolution of these political and legal questions, it will be difficult to regulate disarmament even if states inheriting nuclear weapons want to disarm." I offered the opinion that this situation required "a comprehensive analysis of these problems and a new international legal mechanism for their solution."[49]

Second, I stressed the extraordinary importance of assessing the role and place that nuclear weapons could have in the process of establishing the newly independent states. And here I focused on new features in the theory of nuclear deterrence:

> The authors' scenario of a nuclear conflict between Ukraine and Russia seems highly improbable. Military theory generally considers nuclear weapons a means of deterring aggression. Since there can be no winner in a nuclear war, no state will dare resort to nuclear weapons. This caution applies in equal measure to Russia and Ukraine. Besides, taking into account their industrial potential, large-scale warfare between Ukraine and Russia—even without recourse to nuclear weapons—would threaten not only Europe but the whole world.[50]

Third, I stressed the driving forces behind the proliferation of nuclear weapons. These have led to the world having not just the five legally identified nuclear states, but nearly 20 states that either *de facto* own nuclear warheads or are close to producing them. I named two factors that, in tandem, I thought had led to such a situation, concluding the article:

> Without new effective security guarantees for Soviet successor states, attempts at nuclear disarmament will grind to a halt. The unreliability of existing collective security guarantees and the inadequacy of the NPT regime have brought 20 or so states to have nuclear weapons or to the threshold of acquiring them. The NPT has not been undermined solely by Ukraine's nuclear weapons.[51]

In an article entitled "Ukraine's Nuclear Dilemma," published in *The World Today*, Serhij Tolstov looked at three possible ways that the situation with Ukraine might evolve: ratifying the Lisbon Protocol and both treaties linked to it; not joining any of the treaties until 1995; ratifying START and postponing ratification of the NPT until 1995. The author considered this last option the most advantageous one for Ukraine.[52]

The Russian press also recognized the realignment of forces around Ukraine. On 26 June 1993, *Izvestiia* announced the appearance of "an unexpected nuclear triangle," with the US joining the Ukrainian-Russian process.[53] The Russian media thought the reason for this change was that Ukraine and Russia had been unable to agree amongst themselves, and this stopped Ukraine's disarmament. As evidence of this, they cited Les Aspin's visit to Kyiv and his proposals to leave the dismantled Ukrainian weapons on Ukrainian territory until such time that Ukraine and Russia were able to share the profits from the sale of HEU to America. Indeed, this publication was an admission that Ukraine was playing an independent game and was now in charge of the process of its own nuclear disarmament. The consequence of such actions, noted *Izvestiia*, was that the US intended to move into "comprehensive political strategy" in relations with Ukraine, which would not be confined to just nuclear issues.[54]

This conclusion was echoed by the Center for Security Policy in its 24 June 1993 article titled "What Strobe Talbott Won't Tell the Senate Today: Insisting on A Nuclear-Free Ukraine Is Folly":

> The Center for Security Policy profoundly hopes that *realism*—rather than *wishful thinking* (like Amb. Talbott's about world federalism)—will determine future US policy toward Ukraine and its inherited nuclear arsenal. If so, the United States will endorse, rather than oppose, the retention of an effective Ukrainian nuclear deterrent in the context of a more balanced and visionary US approach toward all the former Soviet states.[55]

Analyzing the July 1993 events surrounding the Russian Duma's vote to proclaim the Ukrainian city of Sevastopol a Russian territory, the Center for Security Policy further stressed that:

> [...] the time has come for a fundamental reorientation of US relations with Ukraine. A secure, pro-Western and increasingly democratic and free-market-oriented Ukraine is in the long-term strategic interests of the United States.
>
> Nothing could do more to encourage such a development, to correct any lingering misapprehensions about US attitudes toward Russian aggression against Ukraine or to deter such aggression than for the Clinton Administration to endorse, rather than oppose the retention of an effective Ukrainian nuclear deterrent.[56]

Later, the *Wall Street Journal* questioned whether by supporting Moscow's demands to hand over all the former USSR's nuclear weapons to Russia, the Clinton Administration were not "unconsciously abetting the rebirth of Russian imperialism."[57] The article went on:

> Moreover, US timidity has not quieted the imperialists or discouraged the KGB—they seem to be feeling their oats. Finally, history suggests that the US fares better with the Russians with a firm line than by tiptoeing around their supposed acute sensitivities. The point of it all: Ease up on Ukraine and take a harder look at the neo-imperialism of Russia. It is there that the real threat to the stability of Europe lies.[58]

Thus, Washington decided to review its position. The White House was prepared to offer Ukraine the option of keeping its nuclear material on Ukrainian territory under international monitoring, including weapons-grade plutonium, to engage its companies in building a plant in Ukraine to process HEU into fuel for the country's NPPs, to agree to maintain the silos, to negotiate compensation for the dismantling work and social adaptation of discharged service personnel, and to start the process of Ukraine's accession to NATO. This meant that the US saw all of these as technically and politically feasible. Ukraine had to succeed in making good on these positions to which the US had agreed.

"Key Directions of Ukraine's Foreign Policy": Nuclear Weapons Belong to Ukraine

On 2 July 1993, Parliament finally passed a document which is instrumental for any state: "Key Directions of Ukraine's Foreign Policy."[59] It was passed by a vote of 226 MPs, the minimum number necessary to pass a decision. This victory by a single vote demonstrated that there were plenty of MPs in both camps who were unhappy with the text.

In the first years of independence, the foreign policy situation in Ukraine was just as confusing as when the military doctrine was being drafted, which a potential enemy and threats needed to be identified. Both documents were written in such a way that nothing concrete was stated. In the end, "Key Directions of Ukraine's Foreign Policy" gave no answers to the questions for which such documents were actually written, i.e., what Ukraine's foreign policy objectives were, in which direction the country was moving, and whom it considered strategic partners.

Stepan Khmara commented in his usual blunt manner during a Parliament session, saying "this document smells of an inferiority complex." In his opinion, Ukraine was only justifying what it would not do, although the point was to indicate what the country intended to do to defend its national interests. Khmara insisted that the document was unfinished and should not be approved. Khmara, a survivor of the GULAG, accused one of the authors of the draft, Deputy Foreign Minister Tarasiuk, of being "a shamefully underqualified specialist whose actions went against Ukraine's national interests." To the objection of Minister Zlenko, Khmara added, "Maybe he is highly qualified, but for some reason he is acting against Ukraine's best interests.[60]

The document was actually drafted by the MFA and the parliamentary Foreign Affairs Commission, which was chaired by Dmytro Pavlychko. Given the positions by the personnel in the MFA and the Communist majority in Parliament, the so-called Group of 239, there was hardly any expectation that the document would include provisions on accession to the EU and NATO, which would distance the country from Russia. Thus, Pavlychko's commission effectively battled for the sake of a single sentence that was about nuclear weapons.

The MFA version consistently avoided the question of ownership of the nuclear materials and spoke only about ownership of the contents within them. MFA's text read: "In its foreign policy activity, Ukraine will actively advocate general nuclear disarmament, will never sanction the use of the nuclear weapons that it

70. Stepan Khmara, 1996

Stepan Khmara (b. 1937), Ukrainian MP in 1990–1998 and 2002–2006, and a member of the working group on issues related to the ratification of START I. A physician by training, Khmara was also a member of the People's Council. From 1960, he participated in the dissident movement. In 1980, he was arrested by the KGB for his political and human rights activity and sentenced to seven years imprisonment and hard labor in the Gulag, and then to five years of exile. In 1987, he returned to Ukraine and became one of the leaders of the Ukrainian Helsinki Group. In 1990, he was elected vice president of the Ukrainian Republican Party. From 1992 to 2001, he headed the Ukrainian Conservative Republican Party. At the fourth convocation of the parliament, he was elected an MP, representing the Bloc of Iuliia Tymoshenko.

inherited from the former Soviet Union, and excludes the threat of using them from its foreign policy arsenal."[61] The Pavlychko commission proposed replacing this with another sentence: "Having become by historical circumstance the owner of nuclear weapons inherited from the former USSR, Ukraine will never sanction their use, and excludes the threat of using nuclear weapons from its foreign policy arsenal."[62]

The commission had approved this version on 11 May at the last of its three meetings considering "Key Directions of Ukraine's Foreign Policy," where Deputy Speaker Durdynets´ and MFA experts participated. Parliament voted on this version, meaning that it was confirming Ukraine's ownership of nuclear weapons. Moreover, at the parliamentary session, commission member Bohdan Horyn´ proposed synchronizing the "nuclear" formulation both in the "Key Directions" document and in the military doctrine. He noted that the issue had been previously and carefully reviewed in various groups. This sentence appeared also in the version of the military doctrine adopted by Parliament.[63]

After the "Key Directions" document was approved, the Cabinet, MFA and their subordinate units were charged with following the guidelines within this document in their work. The Ministry was also required to publish its basic provisions in the press, which meant the Foreign Ministry had to explain everything it had previously rejected. That Minister Zlenko was unhappy with this turn of events could be seen in his speech to the legislature, where he said that the "Key Directions in Ukraine's Foreign Policy" were not "a law in the fullest meaning of the word," but only "a general guideline for the actions of the executive in carrying out Ukraine's foreign policy."[64] The example he used was that in Russia, a similar document had been approved by President Yeltsin, without the Duma.

On 7 July 1993, and for the first time in its history, the Ukrainian Foreign Ministry sent the country's ambassadors' and permanent representatives' recommendations that justified Ukraine's ownership rights over nuclear weapons. The MFA finally recommended basing their arguments on the fact that this wording "merely establishes a situation that, from the legal standpoint, was set in the 3 August 1990 Law on the Economic Independence of the Ukrainian SSR, and the 10 September 1991 Law On Enterprises, Entities and Organizations Subordinated to the Soviet Union and Located on the Territory of Ukraine."[65]

By providing instructions to "point out that Ukraine's ownership of nuclear weapons does not contradict its aspiration to become a nonnuclear-weapon state in the future or to the related provisions of the NPT," the MFA was admitting the obvious: that the "NPT has no concept of 'ownership' written into it." It only talks about commitments not to manufacture, not to acquire or accept, and not to proliferate weapons, while the "right to use is one of the powers of the owner, equal with the right to dispose and control, that includes the right to gain benefit from property."[66] This is the very argumentation that our Working Group

71. *Bohdan Horyn´, 1996*

Bohdan Horyn´ (b. 1936), Ukrainian MP (1990–1998). Horyn´ was a dissident and political prisoner, a journalist, art historian, political analyst, philoiogist, and teacher of Ukrainian language and literature. In 1965, he was tried for "anti-Soviet propaganda," and sentenced to three years in a Mordovian Political Labor Camp in Iavas. After his release, he worked as a painter at a factory from 1968 to 1976. Between 1976 and 1990, he was the senior researcher at the Lviv Art Gallery, and a researcher at the Lviv Historical Museum. In 1988, he was one of the founders of the Ukrainian Helsinki Group (UHG) and headed the Lviv Oblast branch of the UHG in 1989–1990. Over the period 1990–1992, he headed the Lviv Oblast branch of the Ukrainian Republican Party. He was a member of the People's Movement and the Shevchenko Society for the Ukrainian Language.

developed and published back in 1992. Since the rules in the aforementioned documents from 1968, 1990 and 1991 had not been changed, the question remained: What had stopped the lawyers, analysts and leadership of Ukraine's MFA from seeing this before?

In reporting on the implementation of Parliament's decision, Zlenko submitted a copy of the disseminated recommendations to Durdynets'. In his cover letter, the minister documented his attitude towards the decision thus: "The Ministry of Foreign Affairs of Ukraine recommends using the specified argumentation to explain Ukraine's position and *reduce possible negative political fallout from the adopted decision.*"[67] This passage shows that the MFA avoided talking about Ukraine's ownership of nuclear weapons up until July 1993, and defending this right in international relations. Moreover, this was not the last passage that the agency allowed itself to publicly pass judgment on a decision of the country's highest legislative body.

But the passing of the "Key Directions of Ukraine's Foreign Policy" was to have yet another consequence. Already, on the next day, 3 July, Ukraine's Ministry of Defense added the nuclear arsenal on the country's territory (the weapons storage facilities coded "S") to the components of the 43rd Rocket Army, which was under Gen. Volodymyr Mykhtiuk. The personnel who maintained these weapons were to swear an oath of allegiance to Ukraine. More than a year earlier, in May 1992, the personnel in the two nuclear-technical units of the 46th Air Force had already done so. Since the strategic bomber pilots had also already taken the oath, Russia spread information that Kyiv had gained an opportunity to use nuclear weapons.

The Russian parliament responded with a majority vote carrying a resolution that declared the Ukrainian city of Sevastopol as Russian territory on 9 July 1993. Then, on 5 August, the Russian government issued another statement, which described the "serious consequences for international stability and security" of the steps taken by Kyiv. The Kremlin also hinted, "a dangerous precedent had been set, which can be used by countries on the threshold of owning nuclear weapons."[68] Consequently, the negotiations in Kyiv continued between the Ukrainian and Russian governments on nuclear disarmament.

Government Negotiations, Round Three: The Massandra Accords

Thus, Parliament's decision not to ratify START and the NPT at the beginning of June without setting prior conditions, along with Kuchma's sensational statement, revealed that Ukraine—that is, the parliament and the head of the government—had formed a position that was different from what President Kravchuk and the Foreign Ministry had proposed. The sudden arrival of the US defense secretary in Kyiv, and the hearings in the American Senate, which also took place in June 1993, showed that Washington's position had shifted in response to the political events in Ukraine. The Americans were prepared to change from being forceful to advocating broad-based cooperation in various spheres.

Accordingly, by the beginning of July, Clinton proposed to Yeltsin at the G7 summit in Tokyo that negotiations over the terms to ratify START by the Ukrainian legislature should move to a trilateral format. The start of such talks, which were planned to take place in London, would indicate that Ukraine had stopped being the object of its own disarmament and become a full-fledged subject in this process.

Russia clearly understood that moving negotiations to a trilateral format would completely change the alignment of forces to Ukraine's benefit, and this would not only affect nuclear issues, but geopolitical ones. The independent entry on the international arena by Ukraine along with its partnership with the US seriously undermined the Russian monopoly on Ukraine's politics, and had opened the path to European prospects for Kyiv. Russia's leadership could not allow this, as it would put an end to its plans to restore the USSR.

Given this new situation, Russia decided to change tactics. With no way to steer the Ukrainian legislature, the latest frontal attack on Ukraine through the world press did not have the desired effect, and European and American diplomatic channels were already prepared to make greater concessions on the Ukraine issue. They would have to bet on Kravchuk and the MFA, who both were leaning towards Russia's approaches. However, in the end, even this was inadequate. To get the desired results, Russia would have to establish control of government entities engaged in the nuclear disarmament process. One of these key elements was the head of the Ukrainian delegation currently at the negotiations with Russia.

The need for a replacement became inevitable after the second round of talks in March 1993. Kravchuk took the first step. In early April, while an American delegation visited Kyiv, he said, "Ukrainian-Russian negotiations on the fate of strategic nuclear weapons located on Ukrainian territory have run into a dead end." In addition, he proposed Yeltsin to raise the talks to a higher level, that of the premiers.[69] Then, on April 20, the next step was taken, which set up a combination including a journalist from a Ukrainian paper, and both Ukrainian and Russian MFAs. The plan was for the journalist to insist on asking me if Ukraine was a nuclear weapons state. I answered that it was, indeed, both *de jure* and *de facto*, although this was not the first time I had said something like that, or that the press had written substantially about it. It was only after the 23 April publication that the Russian MFA finally issued a statement, which accused Ukraine of changing its position and harboring aggressive intentions. The Ukrainian MFA immediately responded that this was not an official position, but just my private opinion. Then, they immediately informed Prime Minister Kuchma, who had appointed me the head of the delegation, of my ostensibly aggressive commentary, which had negatively affected Ukraine's peace-loving image.

I really do not know what happened after this, but I do know that the MFA did put some intensive pressure on President Kravchuk, and he was unable to withstand it. Kuchma then immediately after his resignation as PM in June 1993 told me that I had been removed from my position leading the government delegation at the insistence of the president. Nevertheless, the negotiations did not move to the prime ministerial level. The Ukrainian delegation was now headed by the newly appointed deputy premier Valerii Shmarov.

Shmarov was appointed to the government from his position as director of one of Ukraine's MIC enterprises. Not knowing (and not wanting to know, as I was later convinced) the entire pre-history of nuclear disarmament and the specifics of Ukrainian-Russian negotiations, Shmarov was eager to cut a deal. From my understanding, the idea of his deal derived from his belief that the Russian delivery of parts (i.e., from Russian companies) for Ukraine's Strategic Nuclear Forces, and all maintenance work on the nuclear warheads currently had been suspended, because of my personal position regarding nuclear arms. All, of course, threatened Ukraine with the prospect of new Chernobyls. I sensed this as Shmarov's position during our conversation, which was right before the next round of Ukrainian-Russian talks.

When I came to Shmarov to hand him the materials and to bring him up to date on the progress of the talks, I was certain that it was important for him to get all the information put together by our Working Group, which he would now have to use to defend the country's interests. Shmarov responded to me, with an air of perfunctory courtesy, in Russian: "Yes, yes, put everything over there, Yuri Ivanovych." He had clearly told me that, although my information was important, he planned to handle the talks by following directives and objectives that were given to him by President Kravchuk. As a director, he completely understood what it meant to be left without supplies from Russia. Thus, he said, he was prepared to look for a compromise in order to maintain the security of the Strategic Nuclear Forces.

At the very first nuclear arms talks in August 1993, when Ukraine was already represented by its new leader, Russia brought along three draft treaties. The Kremlin hoped to use this bypass, acting on the principle, "When God closes a door, he opens a window," to achieve the results it had not been able to achieve through Parliament. Needless to say, the three drafts only covered key issues that mattered to Russia.

The first point without exception was to move all of Ukraine's nuclear warheads to Russia for utilization according to the agreement. First, agreeing to this meant that the executive branch was overreaching its authority, because Parliament had not yet ratified START. Second, this negated any further cooperation with the United States in the utilization of nuclear materials to develop Ukraine's own nuclear cycle and to reduce its energy dependence on Russia.

The second point was the control over weapons storage facilities (S). According to these treaties, Ukraine would merely guard them, and all specialized functions in handling nuclear warheads would be performed by the Russian military. In effect, under this arrangement Ukraine would lose all rights, as the owner, to control the use of its property by a neighboring state.[70]

The third point was the mutual guarantee of supervision concerning the exploitation of strategic nuclear weapons. With this package, Russia resolved all strategic issues that were important for its version of Ukraine's nuclear disarmament to take

72. *Valerii Shmarov*

Valerii Shmarov (1945–2018), Ukrainian politician. Shmarov was the vice prime minister of Ukraine for the military industrial complex from June 1993 to July 1996, as well as the defense minister from August 1994 to July 1996. He was later a Ukrainian MP from 1998 to 2002. As the head of the Ukrainian delegation, his actions played a decisive role in blocking Ukraine's cooperation with the US on the matter of destroying nuclear warheads and moving negotiations to an exclusively bilateral format with Russia. His signing of the agreements presented by Moscow during the third round of negotiations between the Ukrainian and Russian delegations in August 1993, known as the Massandra summit, resulting in the Massandra Accords, was the reason the Ukrainian delegation was subsequently absent in London during the trilateral negotiations.

The signing of this package of "nuclear documents" led to a change in Washington's approach to Ukraine's nuclear disarmament and to the signing of the Trilateral Statement and the Budapest Memorandum. These two agreements lacked mechanisms for Ukraine's national security, an omission that would prove costly in 2014, when Russia first attacked Ukraine, and then in 2022, when the full-scale invasion and the shelling of civilians by Russian forces began.

73. Boris Yeltsin and Leonid Kravchuk in Masandra, 3 September 1993

place. Bearing this in mind, the first point of the three above, was the most important.

Shmarov signed off on all three treaties. As the head of the Russian delegation, Dubinin later wrote: "By the end of the second day, all issues had been decided but one, the withdrawal schedule. We agreed on the liquidation of all nuclear warheads located in Ukraine, their disposal in Russia, and the payment procedure for all operations [...]. All talk ceased on Ukraine's right of ownership of nuclear weapons."[71]

In signing off on these documents, Shmarov was not only completely wiping out all the previous work with American companies to reduce Ukraine's energy dependence on Russia and to get Ukraine into the contemporary global system of nuclear power use, which was providing 50 percent of Ukrainian economy's power needs. He was also guaranteeing that Ukraine would not get out from under Russia's political guardianship or be able to integrate into the rest of Europe up to the Revolution of Dignity in 2014. The main thing that Shmarov cared about was that a deal had been cut. "Actually, there was nothing left to negotiate," wrote Dubinin.[72] He described quite well what the political consequences of the initialed draft agreements would be for Ukraine:

> The draft agreements that I brought to London unraveled all the main knots in the most urgent world issues. One would have thought that this would have brought the American representatives, if not joy, at least satisfaction. But that was not the case. Talbott's reaction to our announcement of the good news came down to a request to discuss our draft documents with American experts. The latter behaved more openly and did not disguise the fact that they had not expected us to come to an arrangement with the Ukrainians without the mediation

of the US. Our colleagues congratulated us and noted that there was no longer anything to discuss in a trilateral format in London. Maybe that was why Talbott was disappointed.[73]

To this day, I do not understand why only Dubinin went to London, and Ukraine did not have any designated representation there. Who made this decision on behalf of Ukraine?

One interesting fact is that neither Speaker Pliushch, nor the Working Group preparing for the ratification of START, were provided with any information by the Ukrainian MFA regarding the option of moving Ukrainian-Russian negotiations to a trilateral format. In the end, this format arose in January 1994 in Moscow. However, its purpose was already different, not actually helping with Ukrainian disarmament and its European prospects, but strengthening Russia in an effort to become a nuclear monopolist in the post-Soviet arena and to encourage further imperialism, which led to the war in 2014.

Then, on 3 September, the presidents of Ukraine and Russia quickly agreed to adopt all three draft treaties in Masandra. The treaties were supposed to be signed by Prime Ministers Kuchma and Viktor Chernomyrdin, but there was the final, open question: What was the timeframe for removing strategic nuclear weapons from Ukraine's territory? In the draft agreements, they would have to be removed no later than 24 months after Ukraine ratified START I.[74]

The agreements clearly conflicted with Ukrainian legislation. For one thing, START had not been ratified by Parliament, which meant that Ukraine had not established the timeframes or the quantities, or the conditions in which the nuclear weapons were supposed to be reduced. The overstepping of authority by Ukrainian officials who initialed and then signed these documents was so glaring according to Dubinin that they decided to put them into a "special protocol that was not for publication."[75]

Defense Minister Kostiantyn Morozov publicly spoke against the agreements, criticizing Kravchuk's position. At that time, it was an exceptional move for a military man. The Ukrainian delegation still had the strength to bring the documents in line with legislation during the final draft. In Dubinin's interpretation, these were efforts "to distort the essence of what had been agreed."[76] However, Kravchuk won in the end, and the protocol was signed just the way Russia insisted.

Diplomatic Heroics

The defeat was obvious and seemingly irreversible. However, Ukraine's diplomatic team still had a single individual who was not prepared to accept capitulation, the advisor to the president on foreign affairs, Anton Buteiko. He took the text that had already been signed by the premiers of Ukraine and Russia, and by hand crossed out the word "all," i.e., meaning warheads, and wrote in "as provided for in the Treaty." This revision changed the essence of the text fundamentally.

START was the only document to regulate the reduction of Ukraine's nuclear warheads. That number was not "all," as Russia desired, but only 42 percent of the warheads. Moreover, even this level of commitment was supposed to be reached only after quadrilateral negotiations among Ukraine, Russia, Belarus and Kazakhstan. Russia blocked them after Parliament ratified the treaty. Essentially, in signing the Massandra Accords, the executive branch attempted to overrule the decision of the legislature, which was in fact in violation of Ukrainian law. These violations by highly placed officials were precisely what Buteiko wanted to prevent.

Although Kuchma and Chernomyrdin agreed to present the agreement as "a special protocol not subject to publication" during their talks, the *Kievskie vedomosti* managed to publish the text with Buteiko's amendment on 9 August.[77] The parties who were to gain from this became evident on 21 September, when the Russian government issued a notice referencing this very publication, accusing "a certain part of the government administration" of preventing Ukraine from executing "the commitments it has undertaken."[78] The significance of Buteiko's action and the price of his patriotism became evident after the Masandra summit; Chernomyrdin approached him afterwards and promised that this would not be forgotten.

Despite enormous pressure, Ukraine refused to return to the previous text, as was made clear by the publication of this secret information. Russia then declared the protocol null and void, which, in effect, meant that it did not affect the essence of the agreement reached between the heads of both states and government officials of both nations. In short, you can amend the protocol as much as you want, but the agreements will still be executed. Indeed, Russia's main victory was that these agreements did, in fact, guarantee that Ukrainian companies would keep servicing the Russian Strategic Nuclear Forces; they also guaranteed the removal of all of Ukraine's nuclear warheads to Russia's factories, sooner or later. Russian diplomats achieved this victory, too, despite the desperate move by Anton Buteiko.

Even so, one more frontline remained, which was the ratification of START by Parliament. The Working Group continued to draft a text that would grant Ukraine the right to take advantage of all the benefits of ownership. We continued to work with MPs, including the People's Movement and Communist Party members. We needed at least 226 votes, and, in the end, Parliament held down the fort.

START Ratification: The Night Before

The Ukrainian legislature only returned to the nuclear issue in November 1993. Aggravated by the lack of systemic reforms, the already severe economic crisis turned into a major political one. In early June, the Donbas miners went on strike, demanding a referendum of confidence for both the president and the parliament.

In response, on 17 June Parliament scheduled an advisory referendum for 26 September 1993. However, after the 21 September resignation of Prime Minister Kuchma, the parliament adopted another measure that not only cancelled the referendum but scheduled a snap election for 27 March 1994, which followed the early presidential election on June 26 of that same year. In the course of these two events, the Working Group was also preparing its final proposal regarding START and the NPT.[79]

That summer, 9 and 25 June, and 9 July, our Working Group reviewed the key issues concerning draft preparation for parliamentary resolutions. The protocols listed the following documents in development: "Comprehensive Program to Eliminate Strategic Arms," "Setting the Limits and Restrictions on Strategic Arms in Relation to the Ratification of START I," "Financial and Economic Provision for Carrying out the Conditions of the Treaty," "Environmental Safety during Disposal Works," "The Date When Ratification of START Should Be Entered on the Agenda of the Parliamentary Session and the Presenter of this Issue."[80] Additionally, the group heard propositions again from specialists regarding nuclear material made available through the process of nuclear disarmament, the options for long-term storage of nuclear warheads on Ukraine's weapons storage facilities (S), and preparations for Ukrainian companies to work in the elimination process.

Based on these hearings, the Working Group developed a package of propositions for the legislature and executive. We insisted that these measures should be carried out prior to considering the ratification of START. Because Ukraine had announced its intentions to become nonnuclear "in the future," while START required only a partial reduction, we proposed the following two steps. First, we needed a resolution from Parliament, "On the ownership of nuclear weapons located on the territory of Ukraine." Second, we had to divide the issue of eliminating Ukraine's nuclear weapons into two phases: to first meet the requirements of START concerning Ukrainian territory, then to reduce the remaining nuclear arms, which was not provided for in START, according to procedures to be established in other treaties.[81]

Given that the main reason for the US and Europe to pressure Ukraine to disarm was the threat of a nuclear strike from its territory, we proposed to Parliament that the president take a few steps tha would remove any concerns regarding this possibility prior to the ratification of START. First, remove the SS-19 missiles that had reached their guaranteed lifespans from combat duty. Second, subordinate the weapons storage facilities (S) where nuclear warheads were being stored to the Ministry of Defense of Ukraine, start negotiating to establish an international control system over their storage, and sign an agreement with Russia about servicing the main parts that were kept in the weapons storage facilities (S). Finally, we recommended that the president (read the MFA) finally start negotiations with Belarus, Kazakhstan, and Russia about allocating the limits and quotas in Article II of the Lisbon Protocol and change the implementation procedures for destroying ICBMs.

All this was to ensure a high quality of debate in Parliament during ratification and to have the legislature make a decision in line with our national interests. The US was already prepared to support this position, which was evident with the Ukrainian delegation to Washington on 16–19 September 1993.

Despite the fact that the Ukrainian representation was not very high level—it was headed by former dissident and Foreign Affairs Commission Deputy Chair Bohdan Horyn´ and also included eight deputies from various other commissions—the group had many meetings at the level that was in fact shaping the position of America's top leaders. During various talks at the Pentagon with two deputy secretaries of defense, at the National Security Council with the chair and a senior director, and at the State Department with the head of the US delegation for strategic nuclear arms and ambassadors who were deputies to Talbott, the Ukrainian MPs were given a very clear message that the Americans were prepared to make a radical shift in their relations with Ukraine. Horyn´ wrote about this later in his summary report of the meetings.

First, the US wanted the tightest possible contacts with Ukraine, including confidential ones, especially on the subject of nuclear disarmament. Second, an independent Ukraine was in line with the national interests of the US, and Sevastopol and Crimea were Ukrainian territory (this was stated in the context of the Russian Duma's July decision regarding Sevastopol). Third, it would be a mistake for Ukraine to limit its relations within to CIS. As a major European state, Ukraine needed to think globally, and should become a self-sustaining state outside Russia's "oversight." Fourth, the US was prepared to provide Ukraine with a variety of assistance if systematic democratic reforms were launched, which also included the military. Fifth, the US could become an intermediary in relations between Ukraine and Russia and offer flexibility in its strategy towards Ukraine regarding the liquidation of strategic nuclear arms. The key here was for Ukraine to eventually become a nonnuclear-weapon state, and unlike the Russians, the Americans did not set any ultimatums. Sixth, the US also had expressed its willingness to set up a nuclear disarmament fund for Ukraine. However, the indifference of Ukraine's MFA with regards to this fund and its content made it impossible to realize as a workable financial mechanism for nuclear disarmament.

Such were the circumstances as we approached the latest review of START I and the NPT in Parliament. It was planned for 18 November and as previously, the two nuclear issues were to be considered together in a similar fashion.

Once again, MPs were presented with two antagonistic draft resolutions, although both were called "On ratifying the Treaty between the USSR and the USA on the reduction and limitation of strategic arms, signed in Moscow on 31 July 1991, and the attached protocol, signed in Lisbon on behalf of Ukraine on 23 May 1992." One was from President Kravchuk, the other from the Working Group.

Our working group once again took the categorical position that these two treaties, START and the NPT, needed to be separated over time. In contrast to the situation on 3 June, we had already formulated a clear list of reservations regarding the ratification of START, which we presented to Parliament. These propositions were congruous with the concept of effective disarmament and we handed them out to the deputies to familiarize themselves. They had taken into account all possible threats and potential objections, which was why our draft, in addition to the "yes" that the world was waiting for, contained more than a dozen points setting fairly stringent conditions. We were proposed that Ukraine fulfill the Lisbon Protocol only provided that:

1. All assets of the strategic and tactical nuclear forces located on the territory of Ukraine, including their nuclear warheads, are property of Ukraine.
2. Ukraine is not bound by Article V of the Lisbon Protocol about joining the NPT as a nonnuclear-weapon state.

3. Ukraine has administrative command of nuclear forces on its territory. [...]

4. Ukraine can gradually eliminate its nuclear weapons on condition that it obtains security guarantees and guarantees that other countries will refrain from political or economic pressure in order to resolve any disputes.

5. Ukraine reduces 36 percent of its carriers and 42 percent of its warheads, in line with Art. II of the LProtocol, but retains the option of destroying additional carriers and warheads following procedures that Ukraine will determine itself.

6. Ukraine fulfills its commitments under the Treaty, but based on its legal, technical, financial, organizational and other capacities, providing the necessary nuclear and environmental safety. Given the economic crisis in the country, doing so will only be possible if enough international financial and technical assistance is provided. [...]

7. Should the disassembly of warheads occur outside of Ukraine, it directly monitors this process to prevent any utilization of nuclear material to create new nuclear weapons.

8. Disarmament takes place on condition that the components of the nuclear arms are returned to Ukraine for peaceful utilization or compensation for their value given to Ukraine, as foreseen in separate agreements. The conditions for compensation extend to the tactical nuclear weapons that were moved to Russia in 1992.

9. Given that Ukraine was not involved in the negotiations to draft the Treaty, the president and government engage in negotiations with other states and international organizations regarding changes to certain provisions in the Treaty. Seven positions are noted, among them security guarantees, the use of missile silos for peaceful purposes, guarantees of compensation for nuclear materials and the conditions for their utilization, and the provision of financial and other assistance for disarmament.

10. The president approves a schedule for the destruction of the strategic nuclear arms covered in this resolution to ensure proper control over its implementation.

11. The Cabinet of Ministers should be obligated, when planning the 1994 budget, to anticipate a separate expenditure line for Ukraine to carry out its commitments under this Treaty.[82]

At the very end, we proposed a rule that "Ukraine will exchange ratification documents only after the conditions in points 5, 6, 7, 9, 10, and 11" are met. As to the NPT, only the coming into effect of START would "open the path to a decision by Parliament on joining" this Treaty.[83]

Kravchuk submitted his proposition as well. His main point was to focus solely on the number by which nuclear arms had to be reduced. Together with the MFA, he lobbied for the idea of reducing all of Ukraine's strategic nuclear weapons. He insisted on completely removing paragraphs that committed Ukraine to specific reductions or or to reductions of 36 percent and 42 percent, and Kravchuk's proposal left out mention of any specific numbers of carriers and warheads anywhere. The evening session of Parliament on 18 November 1993 was to give an unambiguous answer to the question of what the position of the Ukrainian legislature was.

Victory, 18 November

On 18 November, Parliament considered the ratification of START I and the NPT in a closed session. With two draft resolutions in hand, Pliushch started with Kravchuk's, although many of the MPs wanted the Working Group and Commissions draft to be considered first.

The President's proposition received only 162 votes. Although at the press briefing afterwards, Pavlychko claimed that Kravchuk was not offended—after all, that was democracy in action—in actuality, the president was very worried and anxious, and the press picked up on it.[84]

After the president's resolution, Pliushch immediately put the Commissions' version to the vote. An absolute majority of 254 voted in favor of setting aside accession to the NPT, something that Kravchuk later referred to as our biggest political mistake. Parliament then voted in favor of our version of ratifying START with the aforementioned reservations. These reservations meant that neither the weapons nor their components could be taken out of Ukraine without specific agreements about compensation for their materials or financing for the dismantling of the missile silos. Ukraine was the last of the "nuclear republics" to ratify START, but it was not prepared to exchange ratification documents until all its conditions had been met.

This resolution decreased the dangers Ukraine would face if it were to unilaterally disarm. The document put an end to the open theft that had begun when the Russian Federation started removing tactical weapons from Ukraine without any obligations to compensate the country. This not only protected the country from bankruptcy, as it would not have to undertake the financing of the disarmament, but it opened the door to possible billions in revenues to the state budget, along with economic, scientific, technological and security integration into the European community. Besides, Ukraine was clearly acting within the framework of international treaties and national legislation. Above all, it was acting in the best interests of the nation.

Immediately after the resolution was passed, government-wide public statements were issued. The president expressed his opinion in *Holos Ukraïny* in a strange format that did not involve direct communication with the press but rather in a conversation between President Leonid Kravchuk and his press secretary. He did not criticize Parliament's decision. On the contrary, he even tried to argue in its favor, but the message was clear that he had wanted a different outcome. During this conversation with his press secretary, Kravchuk noted:

> The position of the president on the matter of the ratification of START and in particular about acceding to the Nuclear Nonproliferation Treaty has broad support both in Ukraine and beyond its borders. Still, Parliament decided that it was appropriate to ratify only START and the Lisbon Protocol with certain reservations. There are many reasons for such an approach, among

them the main one being that the international community had not considered the complexity and specific nature of Ukraine's situation and did not demonstrate initiative and a constructive desire to work with Ukraine during the preparation for ratifying START.[85]

Obviously, together with the blaming of the "international community" there should have been an assessment of the efforts of the executive branch to make sure that "the complexity and specific nature of Ukraine's situation" were made understandable outside the country's borders, but it was not there. Nevertheless, the president showed a willingness to implement the document that Parliament had approved.

Representatives of the two other branches of power, the legislature and the executive, attended the press briefing after the ratification *en masse*, including First Deputy Speaker Durdynets', Parliamentary Standing Commission for Defense and Security Chair Lemish, Head of the parliamentary Standing Commission on Foreign Affairs Pavlychko, Deputy Premier Shmarov, Foreign Minister Zlenko, and Defense Minister Vitalii Radets'kyi. From the publication in *Holos Ukraïny*, it is evident that none of the initiatives or observations regarding the resolution was mentioned by any of them. On the contrary, their silence proved the necessity of each reservation listed in the resolution. The parliamentary newspaper provided Durdynets's arguments in favor of the positions in the resolution regarding the need for security guarantees:

> In meeting the world community halfway, our country also would like its own security not to suffer as a result. That is why, in ratifying START I, Parliament expects from countries in the nuclear club, first and foremost Russia and the US, guarantees of its national security [...]. Such guarantees cannot be just promises made at a cocktail reception by Bill Clinton or Boris Yeltsin. There is a long-standing practice of interstate agreements and treaties that are legal documents guaranteeing security. The most acceptable today would be a trilateral agreement between Ukraine, Russia, and the US, on mutual security guarantees.[86]

There were no objections to the need to finance the disarmament process, either, as the paper reported what was said at the briefing:

> Most of the talk at the press briefing was about the financial side of denuclearization issues. Vasyl' Durdynets', Vitalii Shmarov and Anatolii Zlenko all pointed out that the parliamentary resolution did not establish any specific numbers. But it was understood that Ukraine was in no position to carry out all the requirements of START I by itself. After all, estimates are that it will cost US $1.6–1.7 billion. And destroying the entire nuclear arsenal located on Ukrainian soil would cost nearly US $4 billion. Washington's promised US $175 million also does not solve the problem.[87]

Shmarov mentioned the proportions and timeframes for reduction: "The deputy premier announced that the destruction of 36 percent of carriers and 42 percent of warheads would take seven years."[88]

Zlenko effectively confirmed the inaction of his agency on the issue of compensation for the tactical weapons. *Holos Ukraïny* published his response to a question in this context: "Analysts say that it is pointless to expect compensation for the theatre ballistic nuclear missiles any time. However, at the press briefing, Zlenko noted that during a recent meeting with his Russian counterpart, A. Kozyrev, this issue was raised, and the head of Russia's diplomatic mission reassured him: 'We will return to this question, too, at some point.'"[89]

Ultimately, the publication quite realistically assessed the content of the resolution: "In ratifying START, Ukraine has shown its good will and desire to completely get rid of 'nuclear death' on its soil in the future. But bitter experience with Russia's removal of its tactical nuclear weapons with no compensation and with its northern neighbor's claims to Ukrainian lead our state to demand reliable guarantees of its security and compensation for the rockets already taken to Russia and for the ICBMs and planes that will now be taken away as well."[90] Indeed, this was one of the few Ukrainian publications that explained what was really going on.

Russia's reaction to the resolution was instantaneous. Its government issued a statement that, in typical unceremonious manner, called the ratification an "insult to important international agreements whose basic provisions were effectively negated by Ukraine's lawmakers." Moscow announced that "the decision of the Ukrainian parliament in relation to START I cannot be recognized on the basis that the provisos contained in it misrepresent the purpose of the treaty."[91]

In addition to this, the Russians once more threatened Ukraine with the refusal to service their warheads. The phrasing is worth noting, for it demonstrated Russia's brazenness, as the Russian diplomats used it without fearing that they might be condemned for using the nuclear security issue for blackmail: "The November 18 parliamentary resolution has created a situation in which Russia is not a position to service the strategic nuclear arms on Ukrainian territory, as it has done until now. This kind of servicing under the current circumstances would mean that Russia was acting in violation of the Nuclear Weapons Nonproliferation Treaty."[92]

In 2004, Iurii Dubinin wrote, "In effect, a new document has been made that has nothing in common with START I, but is convenient for certain political forces in Kyiv. Based on this, the Russian government has announced that the decision of the Parliament of Ukraine regarding START I cannot be recognized." However, this was all the same dog-whistle politics, implying that only Russian national interests mattered, not the right of another country to self-determination or the rule of law.[93]

Yet when Russia itself ratified START II, it included almost the same reservations, content-wise, as Ukraine had in ratifying START I. Thus, on 4 May 2000, Russian Federation Acting President Vladimir Putin signed into law the ratification of START II. The document contained an entire package of provisos. The key points here, too, were "ensuring the national security of the Russian Federation," "economically optimized utilization of existing infrastructure for the Russian Federation's strategic nuclear forces," "significant reduction of the cost of disposing of and utilizing strategic nuclear arms," "expanding opportunities for using the components of Russian Federation's decommissioned ICBMs and their infrastructure in the interests of developing the domestic economy," "leaving the Russian Federation the option of a variety of approaches to the building of its own strategic nuclear forces, including their structures and stockpile of weapons."[94]

Russia also left itself the option of withdrawing from the treaty for a host of reasons. Some of them could be seen as subjective because no criteria for assessing those cases were provided. These include "the emergence of threats to national security" through the actions not only of the US in the framework of START II or other treaties concerning missile defense, but also on the part of other states that had not joined the treaty. Russia found that ceding authority for "approving and implementing" decisions involving military construction "constitute a threat to the national security of the Russian Federation." This included "the response to emergency situations, including those economic and man-made."[95]

Article 5 described the obligations that the other side was supposed to fulfill, while Articles 6, 7, and 8 spelled out the requirements of the president, legislature, and government in the process of implementing the treaty. Among other things, the president was to approve a program to develop the Russian Federation's Strategic Nuclear Forces within two months of the coming into effect of this law, while the government had three months to draft a targeted program for disposing of and utilizing warheads "that provides the opportunity to use components and infrastructure in the interests of developing the domestic economy." The government was also to submit a report on the state of the Russian Federation's Strategic Nuclear Forces to the Duma and advance the implementation of START I, START II and the missile defense program, and "take measures to utilize the economically optimal methods and means to dispose of and utilize strategic nuclear weapons."[96] It took Russia more than seven years to ratify START II; it was signed by Yeltsin and Bush on 3 January 1993.[97]

After Ukraine ratified START I, together with the Russian government's statement—it was announced under the headline "Russian government considers Ukraine's ratification of START I with reservations a gross violation of international commitments regarding nuclear arms," *Interfax Ukraine* published another headline "General Manager of Khartron R&DE,[98] maker of nuclear missile control systems: 'No one in Ukraine knows where the missiles located on its territory are aimed'" in its *News from Ukraine* section.[99] In the actual article, the point was that Ukrainian missiles were controlled from the territory of Russia, but the headline made it sound like the Ukrainian side was the unpredictable and unprofessional one. As always, Moscow was very much on top of the application of information warfare.[100]

An analysis of publications in the Ukrainian press over the end of November and beginning of December in 1993 shows that I was the only government official who explained the economic and political significance of the passing of the ratification resolution. In the *Ukraïns'ka hazeta* of 2 December 1993, I explained in detail the conflicts of interest between Ukraine and Russia and argued why the text of the final resolution was the only possible one at that point.[101] My voice was not enough to prove that this was not politics as high-level abstractions, but that it all came down to a single question: Will there be money in the budget to pay salaries, pensions and healthcare, or will all of it be spent instead on disarmament, which is not needed by us but by the Russians and Americans?

The lack of interest in this subject among most Ukrainian press and government officials led to a situation where Ukrainians themselves did not really understand that they were being asked to disarm at their own cost. Therefore, when Kravchuk caved in, Ukrainian society had no idea what had happened. It still does not understand it today, more than a quarter-century later.

Nevertheless, it was not the position of the press that proved decisive for the way events developed. After Parliament passed the resolution, the executive branch was supposed to take the decision and convert it into action: develop horizontal agreements, convey the country's position to foreign diplomats, and so on.

One quote that came up in several articles and monographs in almost identical wording was: "At the time, Ukraine's diplomats were faced with the task of avoiding possible political and economic isolation for Ukraine because of a negative reaction in the world community over Parliament's 18 November 1993 Resolution, especially its reservations in ratifying START and the Lisbon Protocol and not joining the Nuclear Weapons Nonproliferation Treaty."[102] Not one author even commented on the fact that the executive branch was not implementing Parliament's decision but was adjusting it.

As it turned out, even those opportunities that the ratification offered could be buried thanks to the actions of the executive branch. Ukraine did not have a clear executive chain of command, and executive bodies were not used to acting in accordance with the law. The impunity of officials who broke the law and damaged the country to the tune of multiple billions demonstrated that Ukraine still lacked the rule of law decades after these events.

Instead of carrying out the decision of the legislature, the Foreign Ministry decided to take up negotiations in preparation for the "Trilateral Statement of the Presidents of Ukraine, the US and Russia," initiated by Moscow and supported by Washington. The active phase of these talks began in mid-December 1993 and ended on 14 January 1994 with the signing of one of the costliest documents in Ukraine's history.[103]

What Must We Destroy? The Letter and the Spirit of START

The question of what exactly Ukraine was supposed to destroy, in line with the letter and the spirit of START was always speculative. For this reason, it became subject to a fair bit of mythologizing. The ultimate answer to this was not known even among the country's leadership. The Lisbon Protocol, which brought four nuclear republics into the circle of START executors instead of the USSR only complicated the situation further. In terms of Ukraine's commitments, there were several interpretations.

The first version, which had no documentary basis, was that Ukraine was supposed to get rid of its entire nuclear arsenal, handing it over to Russia, and destroy all its launch facilities.[104] In addition, the option of rearming the Ukrainian army with high-powered precision-guided weapons was out of the question, as START supposedly did not differentiate nuclear and nonnuclear warheads. This interpretation was constantly used by the US, Russia, and Ukraine's own MFA. They cited the Declaration of State Sovereignty and the Statement of Nonnuclear Status issued by Parliament, both of which only mentioned intentions in an undefined future, as well as on Kravchuk's letters to

Bush, which had no legal force without a supporting decision by the legislature, as it also only spoke of intentions.

The second version was that Ukraine had to reduce its share of the 42 percent of warheads and 36 percent carriers, including launch facilities. This matched the letter of the Lisbon Protocol the most, but in order to implement it, it would be necessary to hold four-way talks between Ukraine, Russia, Belarus and Kazakhstan.

A more in-depth analysis of this approach was made by specialists at the Ukrainian MFA on 13 July 1993. A document called "On Possible Options to Reducing and Limiting Strategic Arms in Line with START I," which was accompanied by a letter signed by Deputy Foreign Minister Tarasiuk, was submitted to the parliamentary leadership.[105] The document showed that even the MFA, which had consistently insisted that Ukraine should immediately and completely disarm, was well aware of other options that were legally correct from the point of the international treaties.

Although certain numbers raise questions, the analysis itself contains significant legal conclusions, as illustrated by these statements:

> Generally speaking, the treaty requires that, over seven years, reductions take place [...] in Ukraine: carriers (ICBMs and heavy bombers) and launch facilities from 212 to 135–136 units or by 36 percent; warheads from 1,514 to 884 units, or by 41.6 percent, or to 0, meaning by 100 percent, depending on Parliament's decision. [Parliament approved the 42 percent target on November 18, 1993]
>
> The treaty itself does not specify what kind and how many types of strategic nuclear arms (ICBMs, submarine-launched ballistic missiles [SLBM], heavy bombers) each of the parties is expected to destroy out of the overall total of subject carriers, silos and warheads. Because of this, Ukraine could carry out a variety of reductions, such as: destroying 36% carriers, or 76–77 ICBMs of 176 ICBMs (130 liquid- and 46 hard-fuel missiles), while heavy bombers would not fall within the reduction quota; or destroying all 36 heavy bombers and 40–41 ICBMs, which would also amount to 36 percent of the total number of nuclear weapons carriers; or any combination thereof. TU-95 and TU-160 heavy bombers equipped with air-launched cruise missiles (ALCM) are to either be destroyed or re-equipped into heavy bombers with nonnuclear weapons, surveillance aircraft, aerial refueling aircraft, interceptors, and so on, in line with Section II of the treaty on procedures for repurposing. [...]
>
> In conclusion, under the terms for reducing and limiting under START I, Ukraine will have completely met its obligations under the treaty by reducing around 77 of any type of strategic nuclear weapons carriers or 77 SS-19 liquid-propellant ICBMs. Moreover, the categories of carriers and quantities are supposed to be determined primarily based on the conclusions of military specialists, taking into account the expected plans in building up the Ukrainian Armed Forces.[106]

The third version was that the reductions under START were to follow the percentages indicated and everything else belonged in separate agreements, and launch facilities did not have to be destroyed as there would be nothing to launch from them without the warheads, which Ukraine did not make anyway.

74. *Inside a missile silo*
View of the missile silo without the lid that hermetically seals the launch container. One of the stages of destroying the silo.

Our Working Group insisted on this very approach and this was established in Parliament's ratification resolution of 18 November 1993.

In an interview with *Ukraïns'ka hazeta* on 20 January 1994, I argued for this approach as: "Given that Ukraine does not manufacture its own nuclear warheads, removing the ones we have on our territory automatically means our strategic nuclear arms have been eliminated. Even if entire strategic missiles remain in the silos, at best we might be able to launch missiles with nuclear waste from the Chernobyl sarcophagus, but not nuclear warheads."[107]

By then, the first attempt to destroy a silo had already taken place in Ukraine. *Ukraïns'ka hazeta* mentioned it as well: "The dismantling of a training launch facility in Pervomaisk showed that the real economic and ecological costs were higher than even the most pessimistic predictions. This was also visible to all those who saw the photo reportage of this barbaric act in *Izvestiia* on 11 January 1994.[108]

The Americans were very cautious about the destruction of silos at home, understanding the danger to the environment that this process represented. In September 1992, when Russia's *Izvestiia* reported that the US Senate was prepared to ratify START, they noted an "interesting amendment by Sen. Larry Pressler," who insisted on the "missile silos, many of which were in his home state of South Dakota, being destroyed as safely as possible, so that underground water sources and artesian wells were not damaged in any way."[109]

The fourth reading was that START called for the destruction of silos but not warheads. This was postulated by the American science journalist William J. Broad in an article called "Nuclear Accords Bring New Fears on Arms Disposal," published in the *New York Times*:

The new risks are emerging because arms reductions call for weapons systems to be disabled—by demolishing missile silos, for instance—and do not, as is often supposed, require nuclear warheads to be destroyed. In practice, warheads and their firing mechanisms will be dismantled, and their nuclear cores either stored, pending future disposition, or recycled into new weapons. Even Russia's fearsome SS-18 missiles, whose retirement crowned the recent accords between President Bush and President Boris N. Yeltsin, are exempt from actual ruin. The 12-story behemoths are simply to be taken out of military service and stored, cannibalized, or used for firing payloads into space.[110]

These wildly divergent interpretations only confirm that Ukraine had a broad field for diplomatic activity to carry out negotiations to advance the Ukraine's national interests—political, military and economic—and wrest the kind of disarmament that would both benefit it and be safe.

The brief that came from within the MFA proves that this agency was more than adequately informed about the letter of the nuclear treaties. This once more raises the question, why, being in possession of this information, the MFA did everything in its power to push Parliament not to destroy its nuclear weapons, but to transfer them to Russia and then instead destroy nonnuclear carriers with Ukrainian budget money with the most costly and damaging procedures possible?

What Ukraine Really Committed to Do: The Five Documents

Yet another item for speculation was Ukraine's commitments in terms of nuclear disarmament. Despite the public nature of all the documents that established them, some still believe to this day that in the early 1990s Ukraine promised to give away all its nuclear weapons during a seven-year period.[111] This is absolutely untrue. To avoid loose interpretations, I will name and cite the documents approved by the highest lawmaking body in Ukraine, the parliament, which established Ukraine's policy regarding nuclear weapons. This is what was supposed to guide the president and the executive branch. There are only five documents:

1. "Declaration of State Sovereignty," adopted by the Parliament of the Ukrainian SSR on 16 July 1990. Declared the "intention of becoming a nonnuclear state in the future." There were no timeframes or other specifics in this document.[112]
2. Law of Ukraine "On Entrepreneurship" dated 10 September 1991.[113] Recognizes as property of Ukraine everything that is located on Ukrainian territory at the moment of the declaration of independence, in line with the 1983 Vienna Convention on succession of states.[114] This extends fully to nuclear weapons and the Black Sea Fleet, which was based in the territorial waters of Ukraine.

3. Parliamentary Statement "On the Nonnuclear Status of Ukraine" dated 24 October 1991. For the first time as a state, Ukraine testified to its intentions to become "nonnuclear." The timeframe for fulfilling these intentions was only stated as "in the future" and "in a minimal term," nothing more specific, but the conditions for this to take place were stated: "depending on the legal, technical, financial, organizational and other capacities, with the necessary assurance of environmental safety."[115]

4. Parliamentary Resolution "On Additional Measures to Ensure that Ukraine Gains Nonnuclear Status," dated 9 April 1992. Contains no quantitative or temporal commitments on Ukraine's part in terms of nuclear disarmament. On the contrary, Parliament forbade the removal of tactical nuclear weapons from Ukrainian territory until a mechanism for international oversight of their destruction has been developed and instituted, which should ensure that the nuclear components in these warheads are not reused to make weapons and their export to other countries banned. It was recommended that the president of Ukraine enter into negotiations with the leadership of nuclear states to find a comprehensive solution to the problems connected with the elimination of nuclear arms. The parliamentary commissions were instructed to engage specialists from the ministries, agencies and the Academy of Sciences of Ukraine, and independent experts, and to review the entire complex of issues related to nuclear disarmament from the point of view of security guarantees and Ukraine's foreign policy interests, especially economic, financial, environmental, organizational and so on, including the utilization of nuclear materials for peaceful purposes.[116]

5. Parliamentary Resolution "On Ratifying the Treaty Between the USSR and USA on the Reduction and Limitation of Strategic Nuclear Arms Signed in Moscow on 31 July 1991 and the Attached Protocol to it Signed in Lisbon on Behalf of Ukraine on 23 May 1992," dated 18 November 1993. It once again declares that all assets of strategic and tactical forces located on the territory of Ukraine, including their nuclear warheads, are property of Ukraine. It also gives an extremely clear answer to the question of which quantitative and temporal commitments Ukraine took upon itself: to reduce 36 percent of carriers and 42 percent of nuclear warheads over the course of seven years, but only on condition that sufficient international financial and technical assistance is provided for this purpose, reliable national security guarantees, and based on the available legal, technical, financial, organizational, and other capacities, and the necessary nuclear and environmental safety. In addition to this, the conditions and sequence of transferring nuclear arms from the territory of Ukraine for refitting should include returning the components of these nuclear arms to Ukraine for further utilization for peaceful purposes or compensating Ukraine for their value. The requirement regarding compensation is extended by this resolution to cover the tactical nuclear weapons that were removed to Russia in 1992.[117]

The latter resolution does not exclude the option of reducing nuclear warheads beyond the indicated proportions, that is, more than 36 percent and 42 percent, but this would occur only based on a separate procedure that

Ukraine might propose. Moreover, the resolution states, "Ukraine will exercise administrative management of the nuclear forces on its own territory" and "does not consider itself obligated under Article V of the Lisbon Protocol," which mandates joining the NPT as a state that does not have nuclear weapons. An important point in the resolution is the fact that Ukraine will eliminate its nuclear weapons gradually, which will allow it to monitor both the reliability of the security guarantees it is given and the guarantees that there will be no economic pressure on Ukraine to resolve any possible disputes.[118]

It is also worth noting that the greater portion of the events that are described further took place prior to the passing of this last resolution. In other words, prior to November 1993, ratification commitments did not yet exist for Ukraine.

In this way, Ukrainian law only, first, confirmed the country's right of ownership over nuclear weapons; second, expressed the intent to become nonnuclear at some undefined future point; and third, called for undertaking comprehensive measures to disarm in such a manner that would benefit the country's economy and national interests. Based on this, Parliament had previously prohibited the removal of tactical weapons to Russia. Furthermore, the 1993 Ratification Resolution documented the fact that all that was being discussed at this point was reducing less than half of Ukraine's nuclear arsenal.

75. Missile silo being decommissioned

CHAPTER 5. **Going into Reverse**

Clinton's Detour

By early January 1994, it was clear that the US plans for Ukraine had changed radically. In the summer of 1993, Washington was ready to put its geopolitical bets among all the post-Soviet states on Ukraine. The evidence came from the positive propositions, which were discussed earlier, to develop a strategic partnership. It would be a mistake for Ukraine to limit its relations to the countries of the CIS or even Eastern Europe. Therefore, as a major European state, Ukraine needed to start thinking globally. This was how the issue was presented in Washington after the 3 June 1993 session of Parliament.

In terms of Ukraine's long-term prospects, this meant a complete detachment from Russia via economic reforms and democratization. In the short term, Ukraine needed to accept US support in its nuclear disarmament efforts, while maximizing Ukraine's national interests and, considerably reducing its energy dependence on Russia by using the nuclear material for peaceful power generation.

However, by signing the bilateral Massandra Accords with Russia, Ukraine effectively rejected cooperation with the US, both in terms of dismantling its nuclear warheads and in further utilization of highly enriched uranium and plutonium in a peaceful energy industry. The final straw was the Ukrainian government's failure to act after the passage of Parliament's ratification Resolution of 18 November 1993. By doing this Ukraine demonstrated no intentions of being an independent political player, and thus, separate from Russia. Therefore, having received no answer from Kyiv to its propositions, the US put an end to its plans for a partnership with Ukraine and returned to what it had started in 1991, which was the division of spheres of influence on the post-communist territory between Washington and Moscow.

In early 1994, the US and Russia found a geopolitical compromise in their spheres of influence on European security. Without drawing new military borders in Europe, Clinton and Yeltsin agreed some of the former Warsaw Pact countries would move to NATO's sphere of influence (such as, Poland, the Czech Republic and Hungary), while other post-Soviet countries would remain under the Kremlin's nuclear umbrella. It was in this context that Moscow and Washington formed their joint position on Ukraine's immediate nuclear disarmament.

This compromise between Russia and NATO became the Partnership for Peace, which was announced at the NATO summit in Brussels on 10–11 January.[1] The program filled the vacuum created by the dissolution of the USSR and the demise the Warsaw Pact. In a post-summit joint press conference with the two presidents, Boris Yeltsin emphasized the conception of the new partnership held "an underlying element of Russian-American cooperation."[2]

Ukrainian positions were also critically weakened by a rolling political crisis ensuing in September 1993 with the resignation of Prime Minister Kuchma and the declaration of early parliamentary and presidential elections, which then grew into a severe economic and social crisis. *Forbes* weighed in on the situation on 3 January 1994: "Ukrainian economic mismanagement has been so severe that the economy is on the verge of collapse. [...] As collapse draws closer, more and more Ukrainians will look toward reunification with Russia as the solution."[3] The majority of Western experts shared this assessment on Ukraine's state of affairs on the eve of the US presidential visit to Europe.

Under these circumstances it was easy to corner Ukraine with its back against the wall. They were ideal for pushing Kravchuk to sign a document that would completely override the 18 November 1993 decision of Parliament. By then, the decision was unpalatable not only for Moscow, but also for Washington. Thus, Clinton started the year with an array of momentous events: the NATO security summit in Brussels where Russia was invited to join the Partnership for Peace, and numerous official visits to Eastern European countries, including those of the former USSR. He planned on a decisive breakthrough in nuclear disarmament of Belarus, Kazakhstan and of course, Ukraine, intending to put this issue to rest once and for all.

Clinton was personally interested in the issue of nuclear disarmament. At the time, this topic greatly affected his administration's reputation, as reported by the *Wall Street Journal* on 10 January 1994: "Clinton administration non-proliferation policy is being confounded by North Korean defiance, Iraqi sleight of hand and Iranian dissemblance. The President clearly hopes that he and Boris Yeltsin together can win at least one battle. The assault on Ukraine even has a certain surface plausibility."[4] However, the American press also remarked on the flip side of this coin. While siding with Yeltsin, Clinton was also clearly playing into Russia's imperial ambitions that the latter still had for the post-Soviet space.

Imperialism was still very much alive in the early 1990s. This was a period during the former-Yugoslav wars when a peaceful dissolution of Yugoslavia into independent states was prevented by Serbia's imperial ambitions. Russia, with the help of the KGB, reinstated its influence in Georgia; for example, just two weeks before the events discussed here, the alleged suicide of Georgian President Zviad Gamsakhurdia was announced.[5] Russia's special parliamentary elections in December 1993 secured victory for the nationalist, anti-Western, Vladimir Zhirinovsky's Liberal Democratic Party with nearly 23 percent of the popular vote, which placed it well ahead of the pro-incumbent party, Russia's Choice (15.5 percent), and the Communist Party (12.4 percent).

76. Vladimir Zhirinovsky

Vladimir Zhirinovsky (1946–2022), a neo-imperialist Russian politician known for his xenophobic and anti-Ukrainian statements. The founder of the Liberal-Democratic Party of Russia (LDPR) in 1989, Zhirinovsky served as its president and leader. He supported the August Putsch of 1991. From 2000 to 2011, he served as the deputy speaker of the Russian Duma. He is currently considered *persona non grata* in Ukraine.

Because Russia's imperialistic ambitions did not subside after the break-up of the Soviet Union, the US press criticized Clinton for his inconsistent approach to NATO's eastward expansion. NATO's policy did not include the former socialist republics' accession to the Alliance but allowed for joining in the Partnership for Peace program. The *Wall Street Journal* asserted: "It is there [in Russia's neo-imperialism] that the real threat to the stability of Europe lies."[6] Criticizing Clinton, he posed a rhetorical question:

> [W]ith his timid approach toward the eastward expansion of NATO, Mr. Clinton already is being called overly sensitive to the security concerns of Boris Yeltsin and not sensitive enough to those of former Soviet dominions. His position on the Ukrainian nukes, even though inherited from the Bush administration, exposes him to a key question: Is the president on his European tour resisting or unconsciously abetting the rebirth of Russian imperialism?[7]

Clinton's planned visit to Minsk, the CIS headquarters, was being viewed from these perspectives. According to the original plan, Clinton was scheduled to travel from Minsk to Moscow. A Kyiv visit was not arranged because of the White House's dissatisfaction with Ukrainian state leadership's stance on nuclear disarmament. Kravchuk was to be summoned to Moscow for negotiations. However, the US president had abruptly changed his mind and on the evening of 11 January, he landed in Kyiv's Boryspil Airport. This was the first stop on Clinton's official tour of the former USSR.

As it turned out later, this event was preceded by intense trilateral negotiations in Washington which Ukraine's MFA kept a secret. Even US Department of State spokesman Mike McCurry refused to reveal its outcomes to the press. He only emphasized that "a meeting with Clinton [during his tour of the post-Soviet countries] will not take place, if negotiations are not successful."[8]

The fact that US president's plane changed its route by landing in Boryspil, could mean only that the negotiations were successful, but not for Ukraine. This was apparent, not only from the documents signed three days later in Moscow, but from US president's attitude during his stop in Boryspil. As one of the meeting participants described to me, President Kravchuk, who was accompanied by the Presidential Honor Guard and a group of young women and men in traditional embroidered shirts carrying bread and salt, were stomping about in the freezing cold for nearly an hour in front of the Air Force One's closed doors. This diplomatic slap on the face had set up the humiliating meeting for the Ukrainian officials that followed.

The negotiations commenced right at the airport with one-on-one talks between the two presidents in the Official Delegations Hall. The two were subsequently joined by the acting prime minister, Iefym Zviahils'kyi, Ukrainian ministers Anatolii Zlenko, Ievhen Marchuk and Vitalii Radets'kyi, as well as by US Secretary of State Warren Christopher and Deputy Secretary of State Strobe Talbott, among other officials. Immediately after the meeting, Kravchuk and Clinton held a joint press conference, which essentially confirmed that Kravchuk had surrendered without a fight.

Holos Ukraïny two days later reported: "both presidents, not concealing their feeling of contentment, declared: indeed, the Treaty, stipulating to the elimination of 176 intercontinental missiles and of 1,800 nuclear warheads, will be signed by them and Boris Yeltsin in Moscow. Moreover, both presidents

77. Clinton and Kravchuk, 1993

believe that this would begin a new era in Ukrainian-American relations, and serve as an impetus for economic, political, and security cooperation."[9]

However, despite the optimistic opening of the meeting, President Clinton remained terse and laconic, as he replied to the journalists and their barrage of questions. Particularly restrained were his responses about security guarantees and financial assistance for disarmament. The US president limited himself to rather general declarations, along the following lines: "the accession to the NPT, in and of itself, already includes non-aggression assurances by nuclear powers."[10] Clinton also stated that Ukraine's participation in the Partnership for Peace program, to which he just invited Kravchuk, would further contribute to Ukraine's security.[11]

The US president was more explicit in his answers about financial aid of Ukraine: "Mr. Clinton firmly refuted the rumors of the billions of dollars America was supposedly willing to pay for the destruction of Ukrainian missiles. He once again emphasized that the money for this purpose, as well as for the conversion of Ukraine's military industry, would be coming from the Nunn-Lugar program. That is, we are talking about a few hundred million dollars."[12] Melloan described both the drama and contradictions within the aforementioned *Wall Street Journal* article:

> Ukraine is being pressed hard to become a compliant CIS player. Ukrainian President Leonid I. Kravchuk, steeped in old Communist command and control lore, has hopelessly muddled what little attempt he has made at economic reform. Taking advantage of his ineptitude, the Russians have squeezed him hard on the price of oil. Although Ukraine inherited large numbers of planes and tanks from the old Red Army, it is short on fuel. Its primary source of leverage with the Russians is to hold its nuclear warheads hostage.[13]

Many Ukrainian journalists also shared this opinion on the role of nuclear weapons.

The actual text of the Trilateral Declaration was published just a few days before it was signed.[14] A general alarm, not only heralded a victory for Zhirinovsky in the Russian parliamentary elections, but also underlined Yeltsin's idiosyncratic approaches to democracy. On 11 January 1994, *Holos Ukraïny* wrote: "It will be a while before we forget Boris Nikolaievich [Yeltsin's] joke about there being nuclear strikes between Russia and Ukraine. Do we really have a guarantee that Russia will not use its nuclear strength in the future to force the former republics, for instance, to sign a CIS charter to resuscitate the Soviet Union?"[15]

However, any hints at the prospect of Kremlin being overrun by radicals capable of military aggression against their "brother" republics, were seen as inconceivable in Russia.[16] Meanwhile, on the eve of these events, Poland, Hungary, and the Czech Republic all plead for NATO membership, stressing the Russian threat during the aforementioned NATO summit in Brussels.[17]

The Trilateral Statement: How Kravchuk Violated the Law

After the Kyiv meeting with Clinton, Kravchuk flew to Moscow on 14 January 1994, to sign the so-called Trilateral Statement, which was proposed jointly by Washington and Moscow.[18] The document contradicted every decision Parliament made on nuclear disarmament: from the first bills adopted in 1991 to the most current, adopted in late 1993. Listed below are the key contradictions between the provisions of the Ukrainian law and the Trilateral Statement:

1. *Transfer of weapons to another country instead of their elimination.* The decision of the Ukrainian Parliament to attain a nonnuclear status was based in the following two prerequisites: "the destruction of nuclear weapons," and the existence of "reliable international control ensuring that nuclear warheads' components are not used again to create new weapons or exported to other countries."[19] None of the Parliament decisions gave authority to the executive branch, or the president, to transfer nuclear weapons to another country. The Trilateral Statement, instead of eliminating nuclear warheads, called for transferring them to Russia. It had no mention of control over any further use of nuclear material as a condition of transfer, undermining the very idea of reducing the world's overall nuclear stockpiles, as no one knew for sure that Russia would not use these warheads again for its military.

2. *Destruction of an entire infrastructure, instead of peaceful use and nonnuclear rearmament.* Parliament has never approved the destruction of an entire Strategic Nuclear Forces infrastructure, but only the elimination of nuclear warheads. The Ratification Resolution's Provision 11 mandated the use of missile launch facilities for peaceful purposes, instead of destroying them, and recommended

that the president and the government negotiate possibilities of their use under international control.[20] By signing the Trilateral Statement, Kravchuk agreed to destroy the Strategic Nuclear Forces' entire infrastructure, and not the 36 percent of the warheads as stipulated in START I. This was the most economically disadvantageous, as well as environmentally hazardous, way of demilitarizing these units.

3. *Disarmament without security guarantees.* Provision 5 of Parliament's Ratification Resolution had established that "Ukraine [would] eliminate nuclear weapons gradually, and on the condition of obtaining dependable guarantees of national security," authorizing the government to exchange ratifications documents only upon meeting this condition.[21] The Trilateral Statement, on the other hand, merely offered Ukraine assurances that the US and Russia would "confirm their obligations before Ukraine in accordance with CSCE [now OSCE] Final Act principles." In other words, these were not legally enforceable security guarantees, as was prescribed by Parliament, but only security assurances on paper.[22]

4. *Disarmament at our own expense.* The Ratification Resolution also mandated that Ukraine would commence disarmament only once it secured appropriate financial support for the dismantling of weapons as well as the social welfare of the discharged military personnel, the cost of which was estimated at more than US $3 billion. The Trilateral Statement only guaranteed US aid in the sum of US $175 million under the Nunn-Lugar Cooperative Threat Reduction (CTR) Program, of which US $135 million was intended for the dismantling of missile silos and missile disassembly.[23]

5. *Reduction of weapons in their entirety, rather than their specific share.* The Trilateral Statement called for transporting all nuclear weapons from Ukraine's territory to Russia. START I, which Parliament ratified, obligated Ukraine to eliminate only 42 percent of its nuclear warheads in the course of seven years. The remaining nuclear warheads were only to be destroyed according to procedure "that can be determined by Ukraine."[24] However, the executive power has never developed such procedure, nor any agreements for its implementation.

6. *Disarmament timeframes and Ukraine's implementation conditions.* The Trilateral Statement called for elimination of all weapons "within a seven-year period." The Ratification Resolution specified a seven-year term only for the reduction of 36 percent of carriers and 42 percent of warheads, designating the following terms under which Ukraine would fulfill these START obligations: "provided [Ukraine's] legal, technical, financial, organizational and other capacities, as well as assurance of appropriate nuclear and environmental safety."[25] Considering Ukraine's unfolding economic crisis, START implementation was termed possible only given sufficient international financial and technical assistance specifically allocated for the elimination of strategic arms.

7. *Danger of losing the right of ownership to nuclear weapons.* The Trilateral Statement called for Ukraine's accession to the NPT as a nonnuclear-weapon state, in accordance with Article V of the Lisbon Protocol. The Ratification Resolution specifically stated that Ukraine

78. Trilateral Statement, 14 January 1994
Bill Clinton, Boris Yeltsin, and Leonid Kravchuk after signing the Trilateral Statement in Moscow, January 1994

did not consider itself obligated under this article. To reiterate, the issue was fundamental for Ukraine from the legal standpoint: by joining the NPT as a nonnuclear-weapon state, Ukraine would be losing all rights of ownership to nuclear weapons (as discussed in detail in chapter 1).

8. *A stipulation to surrender nuclear weapons was made without any agreements on returning nuclear material to Ukraine.* The Ratification Resolution named a condition under which Ukraine would disarm, namely, "the return of nuclear weapons' components to Ukraine for peaceful use, or compensation of their value." The conditions with regard to compensation also included nuclear material from tactical nuclear weapons transferred to Russia in 1992.[26]

The Trilateral Statement entirely omitted any mention of tactical nuclear weapons and plutonium. This was despite the language included in the Massandra Accords signed on 3 September 1993, stating, "Ukraine shall receive the value of military-grade plutonium as it is processed, less costs and expenses incurred by the Russian Federation for its utilization."[27] To reiterate, the estimated value of Ukraine's tactical weapons was nearly US $60 billion, and of plutonium in strategic weapons, almost US $40 billion. Instead, the Trilateral Statement mentioned only "a fair and timely compensation to Ukraine by the Russian Federation and the United States for the value of highly enriched uranium upon its transfer from Ukraine to Russia for disassembly."[28] Thus, the only compensation mentioned in the Statement was for HEU from strategic weapons warheads, and its value was estimated at roughly US $6.8 billion.

However, even the latter figure, which should have been the basis of agreement negotiations, was not specified in the Trilateral Statement. The only specifics provided in the document included the 100 tons of HEU which were to be placed in fuel rod assemblies and which Russia was to supply in the course of 10 months, the same period during which Ukraine was to transfer 200 of its nuclear warheads to Russia. In lieu of compensation, Russia would provide Ukrainian NPPs with nuclear fuel. The US, in its part would cover the advance payment, and would then deduct from money owed to Russia under the Uranium Deal.

I shared my analysis of these contradictions with the Ukrainian press, conveying what the Ukrainian state would be losing should it go through with implementation of the Trilateral Statement. This alteration of the parliamentary decisions effected by Kravchuk's signing of the Statement, signified an express violation of the Ukrainian legislation, which in a democratic setting would likely result in impeachment.

Moscow now knew it was about to achieve the goal that it had been pursuing since 1991. The Russian media hastily launched a promotion campaign for the Trilateral Statement, advancing Russia's standpoint.

Information Campaign

Realizing that Kravchuk had clearly violated the law and the Parliament was likely to prevent the document's implementation, *Moskovskie novosti* bestowed an aura of irreversibility in its 16 January 1994 article: "Most analysts on both sides of the Atlantic tend to believe that the Moscow summit finally put an end to the 'Ukrainian nuclear drama.'"[29] They presented a US-Russian version of disarmament history, which traditionally represented Ukraine as a state that habitually reneged on its obligations.

The Russian press pompously reported on the so-called added security guarantees that were extended to Ukraine: "In a joint statement, US and Russian Presidents committed to providing Ukraine with added security guarantees, if Ukraine joins the START I Treaty and accedes to the NPT, including an inviolability of existing borders and the impossibility of their altering except 'in a peaceful manner and by accord,' rejection of 'economic pressures,' and affirmation that they will not be the first to use nuclear weapons."[30]

After the document's signing, the US led its own information campaign, communicating that all Ukraine's obligations were now fulfilled by this Statement, purporting: yes, this was difficult to achieve, but the US accomplished its mission and, thanks to its efforts, Ukraine would now be disarming.[31] This is how the *Washington Post* covered the developments in its editorial piece on 11 January 1994, three days before the Statement was signed: "The new accord emerges as an executive agreement, but these terms are framed to meet the reservations that the parliament had earlier imposed on the government's commitment to abandon a nuclear option. [...] Here it must be said that the delayed or phased manner in which the terms of the new agreement are to be disclosed—in order, it is said, to accommodate President Kravchuk's political requirements."[32]

Russia immediately began using the Kravchuk signature incident to advance its own interests. Taking advantage of the fact that there was no reference to compensation for tactical weapons in the document, it announced in the press: "Ukrainian commentary suggests that Russia must compensate Ukraine for the cost of uranium from tactical nuclear weapons previously transferred from its territory. The Russian side asserts that that the transportation and disassembly of these weapons cost Russia more than the possible value of the enriched uranium contained in them."[33] These commentaries manifestly avoided any mention of plutonium, the monetary share of which was estimated at up to US $55 billion, while, by comparison, the estimated cost of uranium was only US $4 billion.

However, as Konovalov wrote in *Moskovskie novosti*, not everyone in the Ukrainian society shared Kravchuk's point of view, and the opposition movement's leadership called President's actions "a betrayal of national interests."[34] Konovalov continued: "Furthermore, recent sociological polls have been revealing a constant and increasing growth in the part of the electorate that does not support Ukraine's nuclear disarmament. According to estimates, this point of view was shared by almost half of the respondents. Thus, Kravchuk will have to fight hard for these stipulations, and the outcome of this fight is difficult to predict at this time."[35]

Since the document signed by Kravchuk conflicted with nearly a dozen provisions of Parliament's 18 November 1993 Ratification Resolution, there was still a chance that the Parliament would revoke the Statement. Besides, Pliushch was prepared for an open conflict with the president.

Parliament versus Kravchuk

On 20 January 1994, Pliushch ordered several parliamentary committees, including the Defense and National Security Committee, the Law and Constitutional Compliance Committee, and the Committee on International Affairs, together with our Working Group, to prepare an assessment of the Trilateral Statement's compliance with the Ratification Resolution of 18 November 1993. He demanded prompt results, to be submitted to Parliament by 25 January. Based on this document, Parliament was to vote on whether or not Kravchuk's actions conformed to Parliament's prior decisions. To fulfill Pliushch's order, Deputy Speaker Durdynets' called a joint meeting of all Committees and official representatives of the Cabinet of Ministers, the MFA, the MD, and Minmashprom on 24 January.

On the day of 24 January, Kravchuk made a preemptive move. He submitted his draft resolution "On Implementation of Recommendations in Provision 11 of Parliament's Ratification Resolution by the President and the Government of Ukraine" to Pliushch, attaching the text of the Trilateral Statement as an addendum, writing that these documents "reflect a compromise between the three states in solving the problem of nuclear weapons inherited by Ukraine from the former USSR." In his document, Kravchuk acknowledged that the US financial assistance would only constitute $175 million, "with a possibility of the number being increased," and that "all warheads" should be transferred to

Russia in return for fuel rod assemblies for Ukraine's NPPs. At the same time, he asserted that both the Resolution and the attached Trilateral Statement were meeting the requirements of the 18 November Resolution. He suggested that Ukraine should still join the NPT, but with a reservation, stating that "the Treaty does not fully cover the unique situation that arose as a result of USSR's collapse."[36]

The idea from Kravchuk's camp was to substitute the parliamentary deliberations on the Trilateral Statement, by just simply voting on two specific draft resolutions: the presidential and the parliamentary drafts. On 24 January, a meeting previously announced by Durdynets' took place, during which Deputy Prime Minister Shmarov, Defense Minister Radets'kyi, and Foreign Minister Zlenko provided their reports. Neither a quorum, nor a unified conclusion resulted from the meeting. Yet, on 2 February, the eve of the decisive parliamentary session, the MPs received each of the commissions' and the Working Group's conclusions on the issue.

Our Working Group and the Commission on Foreign Affairs held the most radical conclusions. The Working Group asserted that the Trilateral Statement was not in line with the parliamentary resolution, and suggested an amendment to conform with Parliament's decisions, as the latter held the primary power. The Working Group also concluded:

> 1. A number of provisions in the Trilateral Statement by the presidents of Ukraine, the USA and Russia, contradict provisions within the parliamentary resolution:
> a) According to the Statement, "President L. Kravchuk reaffirmed his commitment to Ukraine's accession to the Nonproliferation Treaty as soon as possible, as a state that does not own nuclear weapons." However, Articles 2 and 3 of the Resolution states that Ukraine is the owner of said nuclear weapons, and only the implementation of START "will enable the Parliament of Ukraine to resolve the issue of Ukraine's accession to the Nonproliferation Treaty."
> b) The Trilateral Statement's Addendum states that "Ukraine will ensure the destruction of all nuclear weapons, including strategic offensive weapons located on its territory." On the other hand, Article 6 of the Resolution stipulates destruction of only 36 percent of carriers and 42 percent of warheads. Besides, an institution of the Trilateral Declaration then obligates Ukraine to not only eliminate all of its nuclear weapons within a seven-year period, but also all strategic offensive arms, including their infrastructure. As stated in the US Embassy Memorandum, "any storage of ICBMs SS-24 after the seven-year arms reduction period under START would be in violation of Ukraine's commitments, regardless of whether these missiles are equipped with nuclear or conventional warheads," and the US "will not agree to ease any of these obligations." However, the above obligations are at odds with Articles 5, 6, and 7 of the parliamentary resolution.
> 2. The provisions of the Trilateral Statement do not specify:
> a) terms of compensation for tactical nuclear weapons (Article 10 of the Resolution);
> b) terms of financing the inspection activities (Articles 11–14);
> c) possibility of using launch facilities for peaceful purposes (Articles 11–15);
> d) the use of weapons-grade plutonium (Articles 11–16);

e) compensation guarantees for the value of nuclear weapons' components (Articles 11–17);

f) expenditures for social protections for missile forces' officers and other
military personnel and their transitioning to other occupations.

Taking the above into account, the Working Group suggests bringing the
Trilateral Statement in line with the decisions of Parliament, especially in the
part concerning the authority of the president of Ukraine.[37]

The Commission on Foreign Affairs under Pavlychko's helm likewise underscored the language in the Statement concerning "the destruction of all
strategic offensive arms," including those unrelated to nuclear weapons. In the
Commission's conclusion, "this could be interpreted as an obligation to also
eliminate all of corresponding industrial infrastructure in Ukraine, including
the Pivdenmash plant."[38]

Parliament's legal department emphasized the same aspects as our Working Group above, save for the last point. The department's head, Fedir Burchak,
in co-authorship with Anatolii Matsiuk, the head of Parliament's group of science advisers, pointed to the amount of the reduced weapons prescribed in the
Statement: all, instead of 36 and 42 percent, as well as to the lack of clarity on
"the compensation terms for the tactical nuclear weapons already transferred
from Ukraine in 1992." Additionally, they stressed that the security guarantees
were directly related to the NPT accession as a nonnuclear-weapon state, and
not the other way around. This meant Ukraine would accede to the NPT after
it received the guarantees. Additionally, the Legal Department observed the
Resolution stated that Ukraine did not consider itself bound by Article V of
the Lisbon Protocol. At the same time, they also emphasized that the Trilateral Statement was a form of political concession that could not be deemed
an international agreement requiring parliamentary ratification. Therefore,
if Parliament withdrew its reservations to Article V of the Lisbon Protocol, all
other inconsistencies could be resolved with an execution of corresponding
interstate and intergovernmental agreements.[39]

The latter gave an idea to the supporters of the Trilateral Statement. Immediately following the above conclusions, two committees, the Defense and
National Security Commission, chaired by Valentyn Lemish, and the Law and
Constitutional Compliance Commission, chaired by Oleksandr Kotsiuba, both
proposed that Parliament remove its reservations to Article V of the Lisbon
Protocol. Lemish's proposal at least mentioned the discrepancies with the Resolution, but failed to mention tactical weapons. Kotsiuba's proposal went as
far as welcoming the Trilateral Statement and then proposed to enter into
international arms reduction treaties and accede to the NPT.[40]

On 2 February 1994, MPs received MFA's explanatory brief, naming
Ukraine's accession to the NPT "an urgent task which Ukraine's independence
itself demands," so that "the country does not find itself in international isolation." The MFA literally threatened the deputies with the possibility of Russia
stopping the supply of nuclear fuel for Ukraine's NPPs, but at the same time
also cautioned against Ukraine creating its own nuclear energy cycle. The brief
noted that "each nonnuclear state that is party to START accepts the condition
not to receive any transfers from the suppliers of nuclear weapons or nuclear
explosive devises, as well as not to possess control of or to produce same."
The MFA essentially attested in this statement that by transferring HEU or
plutonium over to Russia for storage, Ukraine would be transferring material

worth millions of dollars, without any possibility of returning it. In other words, Ukraine was agreeing to pay storage fees, while Russia would use the material for its own needs.[41]

Parliament's Consultative Advisory Council subsequently prepared its own brief authored by Ian Brzezinski, which provided an in-depth analysis of the key provisions of the NPT, and included obligations and issues concerning the accession of nuclear and nonnuclear-weapon states. The Council argued that Ukraine's possession of nuclear weapons would not violate the country's accession to the NPT if it did not attempt to obtain positive control of the weapons (meaning its own ability to launch missiles). However, it could limit itself to the so-called negative control (meaning preventing missile launch from its territory). The NPT only obliged Ukraine to not attempt to develop nuclear weapons or gain operational control over them.[42]

Additionally, on the eve of the parliamentary vote on 2 February, a group of MPs opposed to Kravchuk's position, brought forward a radical draft Resolution, "On the Trilateral Statement by the Presidents of Ukraine, the US and Russia." For the first time in the history of an independent Ukraine, legislators demanded approaching "the General Prosecutors Office with a proposal to assess the actions of the president within the framework of Article 56 of Ukraine's Criminal Code (i.e., treason)." They further proposed to deem the Trilateral Statement unenforceable, and for the Cabinet of Ministers to be guided by the 18 November 1993 Resolution when implementing nuclear weapons reduction measures. The deputies asserted that the president stepped outside the boundaries of the above-referenced Resolution, and violated nine of its provisions, including provisions one through eight, and twelve, by which he had exceeded his authority and "caused damage to Ukraine's national interests, diminishing the country's defense capacity and international standing."[43]

The parliamentary deliberation was scheduled for 3 February 1994. While the seats in Parliament had not yet changed, the situation in Parliament overall was now fundamentally different. Early parliamentary elections were to be held March and the election campaign was effectively underway.[44] A few of the parliamentarians went on to seek President Kravchuk's graces with the hope of gaining his administrative support in their majoritarian electoral districts. Among them were the Communists, who just two months prior, voted in line with the hawks. Therefore, the 260-vote majority that adopted the favorable decisions for Ukraine's nuclear disarmament back in June and November 1993, no longer existed. We were aware of this new distribution of powers and understood well enough that the vote was not going to be favorable for the effective disarmament strategy. Nevertheless, we were prepared to fight for it.

Once Again, What is the NPT?

Since the subsequent events unfolded in a heated debate over a dozen-word sentence referring to Ukraine's accession to the NPT, and more specifically around a single word, whether Ukraine would join the NPT as a "nuclear" or a "nonnuclear" state, let us make a quick detour to review NPT's essentials in more detail.

Any mention of the NPT during the early 1990s in Ukraine was shrouded mystery. Few people, even among Ukrainian politicians, knew much of its essence or status. This provided fertile ground for those who tried to speculate on the acronym, so as to force Ukraine into giving up its nuclear weapons quickly and irrevocably for the sake of proverbial "world peace." Meanwhile, the idea of nonproliferation was already under significant global scrutiny.

The NPT was opened for signature on 1 July 1968. Its main goal was to prevent the proliferation of nuclear weapons in the world. As already mentioned, the NPT is the only international document that legally defines what constitutes a nuclear-weapons state: "For the purposes of this Treaty, a nuclear-weapon state is one which has manufactured and exploded a nuclear weapon or other nuclear explosive device prior to 1 January 1967."[45] When the Treaty entered into force on 5 March 1970, only 62 countries had signed. By the early 1990s, this number had grown to 167, but still with nearly 30 countries remaining outside, including Israel, India, Pakistan, Argentina, and Brazil.[46]

The NPT named two categories of participants: states that had nuclear weapons and states that did not. Initially, only the US, the USSR, and the UK had joined the NPT as nuclear-weapon states. It took France all of 24 years to decide to accede to the Treaty. Thus, France along with China, was not party to the NPT until 1992. Although the document conferred nuclear status to only five countries, by 1993, nearly 20 additional states were approaching ownership of their own nuclear weapons. The alleged "clandestine" nuclear-weapon states at the time included Israel, Pakistan, and India.

In April 1993, General Major Oleksandr Kovtunenko, military analyst and Professor of the Kharkiv Military University, prepared a brief at the request of the Working Group, titled "Forecasts for Nuclear Weapons Proliferation, and Prospects of Transitioning to Nonnuclear Weapons." He wrote: "Experts unanimously agree that Israel has developed its own nuclear weapons in the late 1960s. By early 1990s, its nuclear potential, at a minimum, could range between 60 and 100 nuclear warheads [...] and by 2010, it can comprise 210–250 units of weapons."[47] Thus, in more than 20 years of existence, the NPT did not prevent the growth in the number of nuclear-weapon states.

Also, back in 1992, the predictions were that by the dawn of twenty-first century more than 15 countries would join the nuclear club, and by that century's second decade more than 40 countries could be on the threshold of developing their own nuclear weapons. Approximately 20 developing countries would have chemical weapons, over 15 countries would have ballistic missiles, and nearly 10 would have biological weapons. These were the stepping-stones on the path to creating nuclear arms. By 2010, Libya, Syria, Algeria and Iran were on their way to become nuclear-weapon states and were "determined to eliminate Israel's monopoly on nuclear weapons" in the region. Not to mention Brazil, which by that time had the technical capacity to accumulate enough material to produce 15 nuclear explosive devices.[48]

Professor Kovtunenko stressed in his report: "As we can see, no state possessing nuclear weapons is presently in the process of their immediate elimination. [...] Statements made by leaders of various countries that have nuclear weapons or intend to have them in the coming years, are largely declarative, and, for the most part, politically, rather than technically, oriented." As for alternatives to nuclear weapons and their full renunciation in the future, this question was on the agenda, Kovtunenko noted, but not because the world was becoming increasingly committed to peace. The analyst earnestly declared that Ukraine was the only state "actively attempting to accelerate the destruction of its nuclear weapons," while the latter, in fact, remained but "the only deterrent force able to ensure Ukraine's becoming a truly independent state."[49]

The fate of the NPT was once decided every five years during the so-called Review and Extension Conferences. The positive result from this event would involve one of the following three options extending the NPT: to extend indefinitely, to last for a specific period of time, or to last for several specified periods of time. At the time of the events described in this book, the most recent Conference was held in 1990, and the next one was scheduled for April–May 1995 in New York City. Of the 167 states parties to the NPT, only 60 countries voted in favor of an indefinite extension of the NPT in 1990, with a minimum of 84 required for the decision to take force. This validated the notion that the effectiveness of the NPT was already doubtful among most member states. Major General Kovtunenko drew our attention to the fact that all previous Conferences, held in 1975, 1980, 1985, and 1990, have demonstrated that most nonnuclear states were convinced that the US and USSR (and later Russia), were not fulfilling the NPT's provisions.[50] Thus, by early 1990s, the state of the NPT, which was in force until 1995, was rather uncertain.[51]

Among NPT's key problematic aspects was that it did not grasp the international community in its entirety, in addition to having significant limitations in its ability to control activities related to nuclear weapons. The latter issue allowed for several countries to approach procurement of their own nuclear arms, given these countries' scientific and manufacturing potential. What is more, the Treaty was unable to halt the arms race between the countries that already had nuclear weapons. "In the fifty years since the first nuclear explosion, the number of nuclear warheads of all types exceeded 60,000 units. According to expert estimates, their combined payload comprised up to 1,150 tons of highly enriched uranium, 200 tons of weapons-grade plutonium, and nearly 200 kilograms of tritium," reported *Dzerkalo tyzhnia* on 23 June 1995.[52]

To create such volumes of ammunition of various types, yield, and means of delivery, they had to be tested by detonation, which of course, the NPT could not prevent. *Dzerkalo tyzhnia* went on to write: "The IAEA has reported that, over the past fifty years, 2,965 nuclear detonation tests were recorded globally. The greatest number, 1,093, was carried out by the United States (of these, 11 were high altitude detonations, 81 air, 72 above-ground, 36 above-water, 5 underwater, and 888 underground). The Soviet Union is next in the number of detonations, with a total of 715 (of which 8 were high altitude, 176 air, 25 above-ground, 3 above-water, 3 underwater, and 500 underground). France carried out 190 detonations; England, 43; China, 42; and India, one."[53]

Another way of getting nuclear weapons was through the acquisition of weapons by-products and even of warheads, which the NPT also was not able to halt. *Dzerkalo tyzhnia* continued: "These opportunities were permitted by the conditions in the former Soviet territories, mainly, in Russia. By now, Russia's missile bases and special storage units have amassed nearly 20,000 tactical and tactical-operational warheads, which it has shipped in from the countries of its near abroad. As we speak, Russia is transferring warhead systems from the disassembled strategic missiles in Kazakhstan and Ukraine. Their dismantling, with the extraction of distinct components, is underway."[54] The newspaper added:

> Numerous explosions from the artillery arsenal and ammunition depots, as well as mass theft of weaponry, all demonstrate the insufficient and unprofessional security measures and disregard for safety procedure at these sites. [...] The same [...] can be said about the weapons-grade uranium and plutonium production facilities, including their warehouses and laboratories. All this means, is that we cannot entirely exclude the acquisition or theft of these materials, or even of the small-scale nuclear weapons.[55]

Dzerkalo tyzhnia then made a "sufficiently substantiated conclusion" regarding the situation: "the collapse of the 'evil empire' and the post-Soviet circumstances that have developed on the territory of Russia and the nearby countries, objectively contributes to the proliferation of nuclear weapons."[56]

Indeed, the circumstances have changed since the institution of the NPT, demanding changes in the approaches to the very systems of nonproliferation. I conveyed the inevitability of such changes in my interview in *Visti z Ukraïny* in early 1994:

> Legal safeguards must be put in place to guarantee security to those states that have endeavored to eliminate their nuclear weapons, as these are precedents of global significance. Today we keep hearing one thing only: Ukraine is undermining the regime of nuclear weapons nonproliferation. However, Ukraine is undermining nothing. This regime has been undermined long ago, because as of now, nearly 20 countries already have nuclear weapons, despite putting their signatures under the NPT. We need to understand why they do not comply with the NPT. Why do states seek to possess nuclear weapons? They seek nuclear weapons not because they intend to attack. Everyone sees these weapons as an effective deterrent of aggression. Until we see effective regional and global security mechanisms, this process is unstoppable.[57]

Thus, by the early 1990s, the NPT has proven to be neither a mechanism for reducing nuclear weapons and the countries possessing them, nor a mechanism for stopping the arms race. Russia and the US, despite START I and START II, and all other agreements governing reduction of nuclear weapons, such as the Intermediate-Range Nuclear Forces (INF) Treaty, currently possess over 90 percent of all nuclear weapons in the world. Additionally, not a single country possessing nuclear warheads has yet revealed any plans to completely abandon their nuclear programs in the medium term. On the contrary, we can observe a tendency of these countries seeking to modernize their nuclear arsenals. The only reason countries seek nuclear weapons is because it is the best method of deterrence with collective systems of security proving to be highly flawed.

Ukraine has simultaneously surrendered all nuclear weapons and prospects of joining systems of collective security. The MFA exhausted all concerted efforts towards having a signed piece of paper, which ostensibly was to become a guarantee of Ukraine's security. Thus, the country was moving rapidly towards the Budapest Memorandum, and the first palpable step in its direction was Parliament's session on 3 February 1994.

The Gates to the NPT Open: 3 February Defeat

On 3 February 1994, Parliament did not deliberate the legality of Kravchuk's signature on the Trilateral Statement. Instead, another debate transpired regarding whether or not to lift Parliament's reservation to Article V of the Lisbon Protocol. Principally, Article V required Ukraine to accede to the NPT as a nonnuclear state, whereas the reservation in the Parliament's Ratification Resolution of 18 November 1993, declared it not applicable to Ukraine.

The ratification of Ukraine's nonnuclear status within the NPT was the last step in handing over the weapons without compensation or guarantees. After lifting this reservation, all others adopted by the Parliament on 18 November effectively became void, including Ukraine's right of ownership to nuclear weapons, the share of arms reductions, and the demands for compensation and financial assistance. By eliminating this cornerstone of the effective disarmament strategy, our opponents could now continue steering the disarmament process, practically unhindered, towards the *handing weapons to Russia* approach.

The fate of Ukraine's nuclear weapons were once again in the hands of the Parliament. Parliament could not only decide whether or not Ukraine would accede to the NPT, but also on the hypostasis of its accession, as a nuclear or nonnuclear-weapon state.

The Parliament had three options with regard to this decision: first, to postpone the deliberations on the issue of NPT accession until after the 1995 Review Conference in New York City, since the Conference would ostensibly be deciding the future fate of the Treaty at that time (in effect, this is what happened on June 3 and November 18, 1993, and there were legal grounds for doing so again at this session); second, to accede to the NPT as a nuclear-weapons state temporarily, until the completion of liquidation process; and third, to accede to the NPT as a nonnuclear-weapon state, without any security guarantees or legally-enforceable rights to compensation or sufficient funds to cover the costs of disarmament.

The latter was the only option that was indisputably perilous for Ukraine, as without the appropriate legally binding documents, the nuclear disarmament process was befalling the already stunted Ukrainian state budget. Compounding Chernobyl and the social upheavals in the economically devastated Ukraine, these weighty additional expenditures could become the tomb stone on the nascent state's very idea of independence. However, it is this third option that was actively advocated before the Parliament by President Kravchuk, Foreign Minister Zlenko, and Defense Minister Vitalii Radets'kyi.

Arguments in Favor of Handing Nuclear Warheads to Russia

Opponents of the effective disarmament strategy used arguments that conflicted with Parliament's previous decisions on nuclear disarmament. These arguments echoed the ones used by the Russian Foreign Ministry in the information and diplomatic war against Ukraine. Parliamentary official record reveals all three of the above speakers, i.e., Kravchuk, Zlenko, and Radets'kyi, used an identical reasoning to transport all of Ukraine's nuclear weapons to Russia.

First, instead of considering optimal options for disarmament, all three testimonies centered around the argument of why maintaining a nuclear status was economically detrimental and impossible for Ukraine. This thesis without fail was among the first mentioned, with nearly half of the subsequent speech dedicated to it; despite the fact that the issue of Ukraine's preserving its nuclear status indefinitely was never considered by the parliament, only different options for achieving disarmament:

79. Vitalii Radets'kyi, 2005

Vitalii Radets'kyi (b. 1944), minister of defense of Ukraine from 8 October 1993 to 25 August 1994. Radets'kyi graduated from the Kyiv Higher Combined Arms Command School in 1968, the Frunze Military Academy in Moscow in 1975, and the Military Academy of the General Staff of the Armed Forces of the USSR in 1989. From 1968 to 1992, he served in the Baltic, Prykarpattia, Odesa, and Far East military districts, and in the Soviet Forces in Germany. In 1992–1993, he was the commander of the Odesa Military District. From this position he was appointed defense minister. In November 1993, he was promoted to general in the Ukrainian Army. During 1995–2004, he was the Chief Inspector for the Defense Ministry and a member of the Military Sciences Council under the Defense Ministry. He favored Ukraine's immediate disarmament and transfer of all warheads to Russia. He believed that strategic nuclear weapons were not a deterrent against attacks from an aggressor state or a guarantee of Ukraine's security, and he never considered Russia a potential attacker.

> KRAVCHUK L. M. It seems that the question of Ukraine's inevitable accession to the NPT, which has been the focus of everyone's attention, not only here in Ukraine, but also throughout the international community, should be obvious and clear to everyone who lives in the real world. [...] Some people, however, distort the facts, whether consciously or unconsciously, trying to uphold that Ukraine could supposedly become a nuclear state and that this would strengthen its security and international standing, that nuclear weapons can safeguard our independence, territorial integrity, and future prosperity of the Ukrainian people.
>
> [...] I repeat, information on the true state of affairs demonstrates that this argument is simply untenable, that is if one thinks, listens and takes into account the realities that truly exist.

> ZLENKO A. M. [...] It would be a big mistake to also ignore Russia's reaction to a possibility of Ukraine's becoming a nuclear state. You know that such eventuality is not only unacceptable to the political and military leadership of that country, but also plays a role in creating populistic slogans for those who, having just prevailed in the Russian elections, are now advocating for turning Ukraine into new Russian provinces.

> RADETS'KYI V. H. [...] Ukraine is a state with unique missile and space technologies. And we have virtually everything for

developing this industry. All we need is to scrap the nuclear symbolism for the sake of our economy, I am firmly convinced of it.[58]

Second, our opponents would consistently replace the notion of the nuclear warheads' destruction, and their subsequent use for peaceful purposes, with the idea of transferring them to Russia. They used the following rationale: First and foremost, to become a nuclear state meant risking international isolation (which is exactly what happened, they suggested, as a result from the reservations in the 18 November 1993 Ratification Resolution). Additionally, Ukraine could not be a nuclear state, because the manufacturing and testing of nuclear weapons would entail extraordinary costs, and the country was already in crisis. Besides, the existing weapons could not be considered an adequate defense measure for Ukraine as a deterrent, because Ukraine had no operational control of these weapons. Finally, the lifespan of the warheads was nearly expired, practically that very year, and since we could not service them domestically and any inaction could mean another Chernobyl, we had to transfer the weapons to whomever agreed to accept them, i.e., Russia. Listed below are their arguments, as reflected in the following Parliament session statements:

> KRAVCHUK L. M. [...] You know the world's reaction to the Resolution [of 18 November 1993], which we adopted. Following this parliamentary decision, Ukraine was faced with the possibility of complete isolation. [...] In the today's critical economic situation, this carried with it a threat to the very existence of the Ukrainian state.
>
> [...] Beginning with the current year, the guaranteed lifespan, which not only means combat readiness but simply safe storage of a significant number of nuclear warheads is coming to an end. [...] These nuclear warheads must be transported out for dismantling and recycling, and as you know, we do not have the appropriate plants for that. Russia, being the manufacturer of these weapons, does have these plants. If this is not done in time, the threat of a new disaster may become very real. I think we carry a tremendous responsibility for this.

> ZLENKO A. M. [...] I would like to open with the fact that neither Russia nor the United States of America recognize Parliament's ratification of START I, as something that has actually taken place. [...] The very fact of our ratification of START was questioned, and now rumors are spreading that Ukraine is reconsidering its intentions of becoming a nonnuclear state, the very intentions it proclaimed in its founding foreign policy documents. This certainly caused an enormous damage to Ukraine's international reputation.
>
> [...] I would like to come back to the crucial issue pertaining to our security and also to the very future of our independent state, namely, the value of nuclear weapons as a factor of deterrence. Foreign experts, as well as an overwhelming majority of those who are knowledgeable in these matters here, verify that you can only deter aggression with weapons you can actually use, and which would cause unimaginable damage to the aggressor.

> RADETS'KYI V. H. [...] I believe the most important thing here, is the fact that nuclear warheads with which the intercontinental ballistic missiles are equipped, are highly complex technical devices that necessitate

time-consuming and extraordinarily costly process of design, testing and manufacturing. [...] Our country does not have a scientific and manufacturing base for the production of nuclear weapons; we do not have nuclear weapons testing sites, nor proper professionals for utilization of nuclear warheads. However, of utmost importance, dear deputies, is the economic state of our country.[59]

Third, in disregarding all alternatives Western countries were offering Ukraine for peaceful use of nuclear material, the executive branch systematically conserved the status quo of complete energy dependence on Russia. Now, by their statements, they introduced the threat of Russia's terminating the supply of nuclear fuel for Ukrainian NPPs, as the circumstances provided no alternative, but to transfer the warheads to Russia:

KRAVCHUK L. M. [...] We need to assess whether we can preserve our independence, which would be attained at such a price that if we found ourselves in isolation with no friends in the world, we would face a complete termination of [fuel] energy supplies, in particular, nuclear fuel, which on top of everything would freeze all other economic cooperation. How long would we last in such a situation?

ZLENKO A. M. [...] I think each of us [here] understands that nothing is more urgent and vital for us than saving our economy. However, despite our internal turmoil, we are being pushed towards an economic abyss [...]. Today, it is a question of not only upholding Ukraine's reputation in international relations, but also of the survival of our state as such. [...]

RADETS′KYI V. H. [...] We believe that in order to guarantee the implementation of the Trilateral Statement's provisions and an addendum during future negotiations with the Russian Federation; we need to develop and sign a packet of documents that would determine timelines, procedure and conditions of transferring nuclear warheads to Russia.

[...] I believe that the concessions made by the presidents of Ukraine, the United States of America and Russia on 14 January of this year, in Moscow, allow for a generally positive resolution of all military and technical issues concerning nuclear weapons in Ukraine, and in a way that satisfies the conditions of nuclear disarmament determined by Parliament's Declaration of 18 November 1993. I am asking you, honorable members of Parliament, to support our president's propositions.[60]

To top this off, their arguments contained sheer disinformation, which the other deputies pointed out during the debate. The missiles whose guaranteed lifespan was about to expire, were the liquid-propellant SS-19s. However, our officials were proposing to transfer to Russia the solid-fueled SS-24 missiles as a priority, despite their guaranteed lifespan extending beyond the year 2000.

While outlining the prospects of replacing the nuclear weapons with precision-guided munitions, our officials also argued for the destruction of the launch silos and remaining infrastructure. However, the two concepts were entirely incompatible, as precision-guided weapons were the focus of the military doctrine and required the same infrastructure as nuclear weapons.

Thus, the destruction of launch silos virtually eliminated the prospect of having precision-guided weapons.

When drawing on the terrifying prospects of one of our warheads detonating, which could in Kravchuk's words theoretically exceed the aftereffects of the Chernobyl disaster, the president then for some reason, relied on the information supplied by the Russians, which was forwarded to him by Ukraine's Defense Ministry. However, neither he, nor Minister Radets'kyi, appeared to have seen the information supplied to them back in April 1992 by Koniukhov, Pivdenmash's chief designer, an authority on the matter. The Ukrainian expert then conveyed that the prospects of another Chernobyl were more palpable for Russia, should Ukrainian companies refuse service for the Ukraine-produced missiles in Russia. An unsanctioned launch of the SS-18 missile, also known as Satan, may occur if Pivdenmash refused to service it and could pose a threat of apocalyptic proportions. As discussed earlier in the book, the worst-case scenario for Ukraine (in the eventuality that Russia refused to service its missiles in a timely manner) would be a nonnuclear detonation of a bomb with radioactive fallout over a relatively small area, and it would hardly equate to the Chernobyl disaster.[61]

Furthermore, Kravchuk, Zlenko, and Radets'kyi blamed the Parliament for the unfavorable situation, which arose from the Parliament's adoption of the Ratification Resolution on 18 November 1993, and because it was not willing to ratify everything at the same time on conditions that were potentially ruinous for Ukraine. They blamed the legislature, rather than the executive branch, which at the time proved itself to be entirely unfit as an institution upholding national interests, or to successfully convey Ukraine's position to the international community, never realizing that it was a frail opponent in the informational warfare game. Thus, Kravchuk and the MFA, rather than trying to reach a compromise on the conditions defined by Ukraine's parliament, sought to pass the blame without even trying to conceal the sentiment in their statements.

Finally, Minister Zlenko spoke of the catastrophic situation surrounding fuel supply for Ukrainian NPPs of which Russia held a monopoly, while also taking into consideration Russia's refusal to continue supplying the fuel until Ukraine acceded to the NPT. While Zlenko conveyed his deep concern, as Ukraine's economy would not last long in international isolation, he failed to mention the proposals that we received from General Atomics to collaborate in developing Ukraine's own nuclear fuel production, or how his ministry obstructed these proposals.

The Parliament: No Unified Position

The Parliament's position was now presented by Deputy Speaker Vasyl' Durdynets'. He reported that due to a lack of quorum at the joint meeting of the Defense and National Security Committee, the Law and Constitutional Compliance Committee, the Committee on International Affairs, and our Working Group; no unified conclusion was reached on the legality of the Trilateral Statement.

Thus, each Commission and the Working Group provided their own conclusions separately. Parliament's Legal Department and Consultative Advisory Council also prepared their conclusions on the issue. However, this time, Durdynets' unequivocally supported Kravchuk's position, offering a few gentle remarks nevertheless on the apparent contradictions between the Trilateral Statement and several points of the Ratification Resolution:

> DURDYNETS' V. V. [...] Overall, assessing the concessions achieved by the president of Ukraine, we have to understand that this was an international compromise that conformed to Ukraine's interests, because it offers opportunities to receive the necessary economic and financial assistance, reliable international security guarantees, as well as the appropriate fair and timely compensation for the value of nuclear material, and the necessary financial and technical assistance for the implementation of commitments undertaken by Ukraine.
>
> Additionally, it should be noted that the discrepancy between the substance of reservations [in Ratification Resolution] and the concessions made by the presidents of Ukraine, US and Russia [in the Trilateral Statement] can be removed, if Parliament removes these reservation today, taking into account these latest concessions, and approves Ukraine's accession to the Treaty on the Nonproliferation of Nuclear Weapons of 1 July 1968.
>
> This is the approach which the president of Ukraine proposes in his draft Resolution submitted to the Parliament of Ukraine. [...]
>
> At the same time, the second part of Parliament's resolution, which particularly concerns those reservations proposed for the process of accession to the NPT, I propose to add to the list of those reservations the following language: "Ukraine will gradually eliminate nuclear weapons, provided there are reliable guarantees of national security, as stipulated in the Trilateral Statement of the presidents of Ukraine, the US and Russia."
>
> [...] I am asking you now to support the draft resolution you have received.[62]

Differences in MPs' positions were also revealed during the ensuing deliberations. The proponents of effective disarmament were in a clear minority. However, on that day, they used every opportunity they had to make public their argument—that is, why it was detrimental for Ukraine to proceed according to the president's proposition.

Deliberations: *Effective Disarmament* vs. President

One could very well call the deliberations that took place that day in the Parliamentary Session Hall, a battle: one between the president and the ministers, on the one side, and the supporters of the effective disarmament strategy, on the other. The legislators, using their floor time during statements and questions, provided facts in support of the general argument that the Trilateral Statement was an underdeveloped, semi-finished document, and that it did not specify a host of critical for Ukraine issues. They argued most importantly that the Statement obligated Ukraine to so much, without asking for anything in return from the other two signatories. For example, the deputies appealed:

> HOLOVATYI S. P. [...] Hence, if it is a compromise, there have to be certain concessions. I understand what Ukraine gave up—its reservations added to the Ratification Resolution. Why have we heard nothing today of the concessions made by the United States and Russia? What legally binding obligations did the US and Russia take upon themselves? [...]

VALENIA I. I. [...] The text of the Trilateral Statement gives an impression that it was written by the United States and Russia, because it is written from their standpoint. They are declarative and vague and have no legal underpinning to them.[63]

Considering the parties who drafted the document and those who subsequently approved and signed it, this was professional negligence. First, it was reckless to go through with the document in the manner of how everything stood, because there were no agreements in place that would guarantee compensation for the cost of material and damages, insisted MP Pavlo Movchan:

MOVCHAN P. M. [...] Today, the Commission gave us grounds for reconsidering what Ukraine actually needs. Ukraine needs compensation terms for tactical nuclear weapons. We all know how they had you with your back against the wall, so to speak, in Yalta, claiming Ukraine was insolvent [...]. Therefore, I would like to ask, where are the [RUB] 250 billion that Russia owes us as a successor of the USSR? Where is compensation for the tactical weapons that were already taken away from us and who will guarantee this compensation? [...] We need to be a nonnuclear state, but do not need to be stripped naked. I demand that we reconsider all this and use an entirely different approach.[64]

Additionally, the deputies argued, the plutonium from Ukraine's warheads could be reprocessed, as the Americans offered concrete propositions to help organize the process, and Vice Prime Minister Shmarov was aware of it. However, plutonium was omitted entirely from the Trilateral Statement's list of material for which Ukraine was to receive compensation. The deliberations continued:

SHOVKOSHYTNYI V. F. [...] I want to talk briefly on the issue of control over and use of the weapons-grade plutonium we have in our nuclear warheads, which, under the Treaty, will be exported to Russia. I have in my hands the text of the Isaiah Project, which I already mentioned in this hall. [Three American] companies are prepared to allocate $2 billion for the processing of weapons-grade plutonium from the Russia's, Ukraine's, Kazakhstan's, and Belarus' missiles in Russian nuclear reactors, while simultaneously reprocessing their weapons-grade plutonium at two US nuclear power plants. Knowing the enrichment percentages, knowing how much plutonium there is, the international community can completely control every last gram of this dangerous nuclear material. What will Ukraine get in this Project? It will receive compensation in the form of uranium [...].

MOVCHAN P. M. [...] Terms of financing inspections and how it will be done; possibility of using launch facilities for peaceful purposes; the use of weapons-grade plutonium [...]. By the way, this is a very important energy source for us. However, there is not a single provision addressing any of these.

KOSTENKO Y. I. [...] Russia does not recognize our right to ownership of highly enriched uranium and plutonium. It has repeatedly declared so during negotiations, arguing that it would be in violation of the Treaty on the Nonproliferation of Nuclear Weapons. If Ukraine joins the NPT as a nonnuclear state, we automatically lose our legal right, and I repeat, our legal right to ownership of these components.[65]

Furthermore, the discussion continued, Russia and the US have not legally bound themselves into any obligations with regard to security guarantees and financial assistance.[66] Kuchma believed that arms reduction without these guarantees would lead to severe economic consequences for Ukraine:

> KUCHMA L. D. I spoke, and continue to speak, in support of Parliament's decisions: Ukraine, prospectively, is a nonnuclear state. However, we need to study all these issues comprehensively.
>
> [...] I believe (and I spoke of this in May) that we made a mistake by signing this ill-fated protocol in Lisbon. Ukraine, as a state, never took part in the START negotiations. Why in the world did we walk into that? We walked into it and it was a trap. I agree with MP Kostenko that, with this, they are essentially tying our hands, leaving us no options [...]. But, most importantly, as a result, we will instantly turn into a complete ruin in 18 months [...]
>
> I understand: we, our President, made the right step by giving away the tactical weapons. But what did we get in return? Did the international community immediately come in helping us, providing loans, investments? Not in the least![67]

In the course of the arguments, the legislators supporting the effective disarmament strategy focused their responses on the three key points that their opponents in the executive branch had made. First, the MFA and the president insisted that by refusing to ratify the NPT, the Parliament was trying to legally institute Ukraine's status as a nuclear state. However, this came outright as manipulation, because every document adopted by Parliament stated that Ukraine was to be a nonnuclear-weapon state, and not one proposition brought forth by the proponents of effective disarmament asked for a revision.

Secondly, in citing the START provisions, our Executive Branch insisted on destroying and exporting from Ukraine's territory everything that had to do with nuclear weapons. However, START I, only required the elimination of the delivery systems, and not of the warheads themselves. Thus, Ukraine, which did not manufacture the warheads, could and should have avoided the economically detrimental and inexplicable costliness of destroying its launch facilities, as this was the intent of the reservations added to Parliament's START I Ratification Resolution on 18 November 1993.

At last, the MFA and the president insisted on Ukraine joining the NPT as a nonnuclear state, while simultaneously ratifying START. The proponents of the effective disarmament strategy stressed, as noted earlier, Ukraine would lose legal property rights to HEU and plutonium, and the compensation for the value of these materials.

Parliament session transcripts verify that there were no reasonable counterarguments, or evidence-based responses to their questions about the already signed agreements on compensation for nuclear material, and other related questions. All they heard were the elusive comebacks along the following lines:

MOROZ O. O. Dear comrades! Today we got the treatment we deserve. Advocating in support of the president's signing some documents would be improper. Everything he did was right. We simply have no other choice.[68]

KRAVCHUK L. M. [...] First of all, on the question of legally binding documents, addressing compensation for nuclear weapons to Ukraine. As for strategic nuclear weapons, a document was already signed with Russia. It is legally binding. It was signed by Russian Prime Minister Chernomyrdin and Ukrainian Prime Minister Kuchma.

 With regard to compensation for tactical nuclear weapons, such compensation is positively forthcoming. As early as February, experts from Ukraine, Russia, and the United States begin work on determining the value of these weapons. Thus, the question of compensation for strategic nuclear weapons is already legally resolved, and as for the tactical nuclear weapons, it will be resolved after the valuation of these nuclear weapons. I tell you how it really is. [...] I want to assure you, esteemed Members of Parliament, that Ukraine will receive compensation for both tactical and strategic nuclear weapons.

 [...] In all honesty, my actions were guided by our circumstances. I did what I had to do to keep Ukraine from harm, to preserve our very statehood. [...] I believe that this parliament, too, should say its last word on what it initiated earlier.[69]

Thus, on 3 February 1994, Parliament resolved as President Kravchuk had wished, to lift its reservations to Article V of the Lisbon Protocol, approving Ukraine's accession to the NPT as a nonnuclear state. The reservation was the only safeguard against disastrous consequences within the disarmament process for Ukraine's national security and economy. Without the safeguard in place, Ukraine was accepting a pack of obligations while receiving no guarantees in return, either in the form of security or of compensation for nuclear material transported to Russia, or for disarmament costs. Once the vote was confirmed, the government was instructed to exchange START I ratification charters with the other signatories.

This decision nullified part of Parliament's Resolution of 18 November 1993, which was the resolution that specifically defined the conditions under which Ukraine could ratify START I. The path to the ruinous option of disarmament was now wide open.

CHAPTER 6. **Capitulation**

Ratifying the NPT: The Final Battles

The next act in this drama began in the new convocation of Parliament, with a new Speaker, President, and Cabinet; a snap parliamentary election took place in March 1994, as did an early presidential election in June. Parliament had changed: many of the MPs who understood this complex issue and whose efforts had helped prevent a disastrous turn in the nuclear disarmament process favoring Russia, did not make it to the new parliament.[1]

However, the biggest blow was dealt by the new Speaker, Oleksandr Moroz. He refused outright to form a special deputy group to prepare for the NPT ratification. Instead, he delegated it to two parliamentary commissions: Defense and Security, and Foreign Affairs. One was chaired by the Socialist Volodymyr Mukhin, and the other by the Communist Borys Oliinyk. This resulted in a situation where the new parliament no longer had a professional platform that could resist the executive branch, which was under unrelenting influence and pressure from Russia.

On 5 October 1994, Leonid Kuchma, who was now president, submitted a draft ratification bill to Parliament. This bill proposed joining the NPT, but with a caveat. It would state that the provisions of the Treaty did not cover the unique situation that had developed with the collapse of the Soviet Union. It would further declare that Ukraine was the owner of the nuclear weapons and intended to use the materials contained in them only for peaceful purposes, and that the presence of such weapons on Ukraine's territory prior to their destruction was not in violation of the Treaty.

The deputies received an explanatory brief signed by Kuchma. It maintained that by adopting the 3 February 1994 Resolution, entitled "On Implementation of Recommendations from the President of Ukraine and the Government of Ukraine that are Included in Point II of Parliament Resolution 'On Ratification of START,'" in Kuchma's words, "Ukraine had committed itself to join the Nuclear Weapons Nonproliferation Treaty of 1 July 1968 as a participating state that did not own nuclear weapons."[2] On 14 November, the brief was then signed by eight different heads of ministries and agencies.

On 2 November, a letter laying out the same information, which was also signed by Kuchma, was delivered to Speaker Moroz. Using the arguments that were typical of those applied by officials from both Ukrainian and Russian MFAs, it insisted on the need to accede to the NPT as soon as possible. The reason stated was, supposedly, to not slow down the global disarmament process by halting the enactment of START I and START II.

80. Kuchma and Kravchuk, 1994

Leonid Kravchuk (right) congratulates Leonid Kuchma on his victory in the presidential election, 19 July 1994.

The letter, signed by the president, also confirmed that the issue of compensation had been resolved, although only tactical weapons were mentioned, and without any numbers. Meanwhile, financial assistance for disarmament was only mentioned within the framework of the Nunn-Lugar Act, at the clearly inadequate sum of US $175 million. Kuchma himself had criticized this amount when he was prime minister. However, now his letter stated: "The issue of compensation for tactical nuclear weapons that were shipped out earlier has been essentially resolved. There have been significant shifts in organizing effective assistance from the US under the Nunn-Lugar Act."[3]

At the same time, the letter effectively acknowledged that Ukraine had still not been provided with any security guarantees: "Today, active measures are being taken to get Russia, the US, Great Britain, and France to agree to the text of a multilateral Memorandum providing Ukraine with national security guarantees. The results of the latest talks on this issue by the Russian and American side give reason to believe that such a document could quite possibly be signed."[4]

All the documents distributed to MPs by the office of the president and executive branch, contained arguments claiming that Ukraine had no options other than to ratify the NPT as a nonnuclear state. They also claimed that the agreement between Russia and the US reached after the signing of the Trilateral Statement was enough to satisfy the precautions in the Ratification Resolution of 18 November 1993. One interesting observation, was although our opponents continually referred to this resolution as the source of all unpleasant eventualities, including international isolation for Ukraine, none of them dared to publicly propose any actions which violated the requirements of the resolution. These arguments were echoed in reports and speeches by supporters of immediate disarmament.

It must be said that at this point, after the signing of the Trilateral Statement, the strategic nuclear weapons were being shipped out of Ukraine enmasse, even though START I had not yet come into effect. The latter required an

81. *Oleksandr Moroz (left, at the microphone), 1995*

Oleksandr Moroz (b. 1944), Ukrainian politician. Moroz was the leader of the Communist majority, known as the Group of 239, in the first democratic convocation of the parliament (1990–1994), and later Speaker of the parliament (1994–1998 and 2006–2007). Moroz graduated in 1965 from the Ukrainian Agricultural Academy as a mechanical engineer. In 1983, he graduated from the Higher Party School at the Central Committee of the CPU. From 1976 to 1990, he worked within the CPU apparatus. He was later the founder and leader of the Socialist Party of Ukraine (SPU) after the Soviet-era Communist Party of Ukraine was banned in 1991. He ran for president of Ukraine as member of the SPU in 1994, 1999, and 2004. During the Orange Revolution, he supported Viktor Yushchenko in the first run-off. In 2006, he was elected to the legislature based on democratic slogans and promises to join the democratic majority, but his party instead joined with the so-called Anti-Crisis Coalition of Viktor Yanukovych and the Party of the Regions. Because of this, he was appointed as Speaker of the parliament for the second time.

exchange of ratification documents by the participating parties to gain force. Thus, the process was becoming irreversible as time carried on. Kuchma had come to power in the midst of this process but had moved on to a different objective: to establish a relationship with the US. By 1994, the US was once again working in tandem with Russia, and deliberately moving the situation towards a complete and immediate disarmament of Ukraine.

By November 1994, the situation in Ukraine was already one where President Kuchma, Premier Vitalii Masol, and Speaker Moroz all held a common position in favor of the immediate ratification of the NPT. There was no longer any organized counterweight to this position in the legislature, which in the past had been centered around the Working Group. Still, some of the MPs, even if every man was for himself, were preparing for the final battle, gathering facts for their arguments, and registering alternative draft resolutions.

Prior to the vote, which was scheduled for 16 November, four draft resolutions were registered in Parliament, in addition to the presidential one. A group of deputies from the democratic opposition camp, including Oleh Vitovych, Iurii Tyma, Stepan Khmara, Iaroslav Iliasevych, and Bohdan Iaroshyns'kyi, proposed a Resolution to remove the debate of the NPT ratification from the agenda "until the next global changes in the nuclear standoff." They argued, because Ukraine could not afford to independently finance nuclear disarmament, the conditions under which the NPT had been written had changed, and that, with another 10–15 countries having, in fact, developed nuclear weapons, the NPT had ceased to be an effective instrument for nonproliferation policy.[5]

MP Andrii Mostys'kyi introduced an amendment to one of the draft resolutions by adding: "The process of reducing and eliminating nuclear arms needs to be comprehensive. Moreover, the reduction and elimination of nuclear arms needs to take place in other countries simultaneously with Ukraine. We consider the demand of some states for unilateral disarmament in Ukraine unacceptable."[6] The Iednist [Unity] deputy group proposed an amendment to this declaration by which Ukraine "should be recognized as a temporary nuclear-weapon state."[7]

Both of Parliament's Commissions for Foreign Affairs and for Defense and Security presented a draft that was nearly identical to the presidential one. They proposed joining the Treaty with a declaration that stated:

> 1. The provisions of the Treaty do not encompass the unique situation with regard to Ukraine in its entirety, which resulted from the collapse of a nuclear state, the USSR. The presence of nuclear weapons on Ukraine's territory prior to their destruction and the related effort maintaining, servicing, and destroying them by the Armed Forces of Ukraine are not in violation of Articles I and II of the Treaty.
> 2. Ukraine owns nuclear weapons that it inherited from the former Soviet Union. After the last of these weapons is disassembled and destroyed under its control, following procedures that include the option of recycling the nuclear materials that are in these weapons, in line with their primary use, Ukraine intends to use these materials exclusively for peaceful purposes.[8]

Despite the change in personnel that took place after February, the alignment of rhetoric by the president and the foreign minister remained unchanged. On 16 November, Kuchma repeated the same arguments that Kravchuk had announced before him from the same podium, on 3 February 1994:

> Ukraine has no choice today whether to be nuclear or nonnuclear. The choice has been made by the actions of Parliament of the previous and current convocations, by Ukraine's international commitments; and ultimately, by the situation itself.
>
> And that situation is thus. Our decisions today will affect the global process of nuclear disarmament. If we decide to cut a deal with the world community, then we are choosing the most unpromising time and the most inappropriate item over which to bargain. We will surprise, to put it mildly, the civilized world, as the decision to accede to the Treaty is the concluding phase, which is logically driven by all the previous political acts passed in Parliament from 1990 to 1994. These are: the Declaration of State Sovereignty of Ukraine, Parliament's Declaration of Ukraine's Nonnuclear Status, and Parliament's Resolution to ratify START I and the Lisbon Protocol, which provide for Ukraine to join the NPT.[9]

Meanwhile, the new foreign minister, Hennadii Udovenko, repeated statements that echoed his predecessor, Anatolii Zlenko, from 3 June 1993 and 3 February 1994, and without changing any sentences:

> The question of Ukraine's position regarding the Treaty on the Nonproliferation of Nuclear Weapons (NPT) has been debated in Parliament many a time. Various opinions on this issue have been presented, often diametrically opposed ones, and also more than once. Therefore, I find myself repeating some arguments as well.

1. The 1968 Nuclear Nonproliferation Treaty is one of the most significant factors in maintaining strategic stability in the world. [...] We should not forget that the only way to really influence the fate of this Treaty is to become a full-fledged participant.

2. [G]iven that control over these weapons has never belonged to Ukraine, the country cannot be considered a nuclear state in the strict sense of this term.

3. On the other hand, by not acceding to the NPT, Ukraine will effectively remain outside international economic, scientific and technological cooperation, even in the peaceful use of atomic energy [...]

4. Ukraine's plans to join the NPT as a nuclear state can only exist on a strictly theoretical level. In addition, if Ukraine had intended to go down this path from the very start, it would have considerably complicated international recognition of our country's independence [...]

5. Today, it is possible to state that, after Parliament ratified START I and approved the Trilateral Statement of the presidents of Ukraine, the US and Russia, the signing of related bilateral agreements with Russia has definitely resolved the practical aspects of Ukraine's intention to become a nonnuclear state.

6. The question of moving nuclear weapons to Russia to be destroyed under Ukraine's control has been decided and is being implemented, as has providing compensation in the form of nuclear fuel rods for Ukraine's nuclear power plants [...]

7. Thanks to the focused, determined efforts of the president and the executive and legislative branches of government, it has been possible to achieve real shifts in deciding the extremely important, long-standing problem of increasing the volume and the effectiveness of international aid to Ukraine.

8. Yesterday, ambassadors from 14 key Western countries and the European Union announced at the meeting with the president of Ukraine that their governments intended to allocate additional funding worth more than US $230 million in the current fiscal year.

9. Over the last two and a half years, we have worked hard on another issue that is fundamental to us: getting security guarantees for Ukraine from the nuclear states in connection with accession to the NPT.

10. I am convinced that for us to get stuck on illusory nuclear weapons at this decisive point in our history is a losing proposition [...][10]

The content of their speeches came down to an already familiar logic. First, once again the discussion was raised with imaginary supporters of producing nuclear weapons in Ukraine although, in fact, no one had proposed anything resembling this position in the past. Udovenko stated: "Even if Ukraine were to gain access to the 'nuclear button,' the problem of reproducing nuclear warheads would arise, and without solving that, Ukraine would only remain a nuclear state for a relatively short time."[11]

Second, they stated that, since Ukraine could not actually use its existing nuclear weapons, it was not an effective deterrent. In his speech, Kuchma insisted: "I would like to remind those who have been roused by artificial patriotism that they seem to have forgotten that Ukraine not only lacks the manufacturing capacities necessary and cannot use its nuclear weapons for defensive purposes, but also cannot use the nuclear warheads that it has inherited to generate power."[12]

Another argument used was that Ukraine could not be a nuclear state because it lacked the resources and the status in the eyes of the world community. Kuchma argued:

> Just setting up and upgrading infrastructure for the safe storage of the nuclear weapons on our territory today [...] could cost US $10–30 billion [...]. As to our own manufacturing facilities, setting up a closed production cycle for nuclear warheads would cost at the least, and this was mentioned already today, US $160–200 billion over a decade [...]. Do we really have a choice? Who of those who favor nuclear games can stand up right now and tell us to whom we should sell or hand over all of Ukraine's assets in order to bring the country the pleasure of having its own nuclear arsenal?[13]

They continued; Ukraine had to come through on its promise to become nonnuclear. This was clear in Udovenko's speech: "The real situation is such that our tactical and strategic national interests force us to come to a single possible decision: for Ukraine to accede to the NPT. This will indubitably increase Ukraine's reputation as a state that fulfills its international commitments, and as a trustworthy and reliable partner that is prepared to abandon 'Byzantine diplomacy.'" For Ukraine not to fulfill its commitments would raise the threat of economic sanctions, something Ukraine's economy would not be able to withstand. Udovenko went on: "On the other hand, by not acceding to the NPT, Ukraine would effectively find itself outside the bounds of international economic, scientific and technological cooperation, even in the case of the peaceful use of the atom. The consequences of this kind of isolation could include, for instance, the end to Russia's deliveries of nuclear fuel rods and the equipment necessary for Ukraine's nuclear power plants to keep operating."[14]

In addition, according to Kuchma and Udovenko, there was no time to bargain because the lifespan of the nuclear weapons was coming to an end. To this argument, they added the commitments in the Trilateral Statement: "In ratifying the Lisbon Protocol, Ukraine legally committed itself to do this, which is confirmed in the Trilateral Statement of 14 January 1994." The conclusion was that all warheads needed to be immediately moved to Russia.[15]

The supporters of the effective disarmament strategy insisted on four basic points.

First, a Memorandum could not be a guarantee of national security.[16] As Oleksandr Kozhushko, chair of the parliamentary Fuel and Energy Complex, Transport and Communications Commission, underscored: "What we need to see signed is not a memorandum but an international legal document 'guaranteeing Ukraine's security.' We can come to terms with that, as they say, but a memorandum is just 'we sat and had a chat and that is that.'" In addition, as MP Vitalii Shevchenko pointed out, the Duma had not withdrawn Russia's claims to Sevastopol: "I am interested in whether Russia will give us any guarantees. You know that the Black Sea Fleet is based in Crimea, the Duma has not withdrawn its claim that Sevastopol is a Russian city, and so on. I want to know what guarantees we will have from a country that has real leverage over Ukraine's economic and political fate."[17]

Second, Ukraine was already reaping the fruits of its unilateral steps; now, steps from the international community were needed, but in response we were only getting pressure. One of the members of Parliament's Standing Committee on Defense and National Security, Iaroslav Iliasevych, made this appeal:

Why is the issue of signing the NPT being presented only from one side? Why are the only arguments presented here, all in favor of Ukraine disarming? Not a single argument has been presented in favor of Ukraine not signing the Treaty because it will lose so much [...]. We just heard over the radio that Russia is already blackmailing, threatening to cut deliveries of fuel rods if we do not sign this Treaty. What awaits us if we do sign it? We already know what happens when we sign agreements of this nature with our neighbor [...]. Why has the Foreign Affairs Commission not looked into this issue properly? Why are we treated like mushrooms, being tossed willy-nilly into a basket or a hot frying pan? Why are we being pressured? Why has the commission not reported on this and why are you also not reporting on this now?[18]

Third, these MPs pointed out, the tactical weapons had already been shipped out, but no compensation had been given. Where were the guarantees that Ukraine would be compensated for the weapons once the rest were shipped out? Ivan Kirimov, chairman of the parliamentary Subcommittee on the Chernobyl Disaster, noted: "As we know, there is supposed to be some compensation after the nuclear weapons in Ukraine are destroyed. So far, there has been no such compensation, as we all know, and even if something was planned, it was only after many requests."[19]

Last, but not least, was the fourth point and the leitmotif of the discussion: the world respects the strong, and Ukraine needed to project strength, both internally and externally. As MP Mykhailo Ratushnyi summed it up:

82. Vitalii Masol, 1985

Vitalii Masol (1928–2018), chairman of the Council of Ministers of the Ukrainian SSR (1987–1990), MP of the Ukrainian parliament (1994–1998), and prime minister of Ukraine (1994–1995). Beginning in 1972, he was the first deputy director of Derzhplan of the Ukrainian SSR, the state planning committee. In 1979, he was promoted to director of Derzhplan. On 23 October 1990, in response to one of the demands from the students on hunger strike during the Revolution on Granite, the parliament dismissed Masol. At the request of President Leonid Kravchuk, he was appointed prime minister on 16 June 1994 and served until March 1995.

Here, it was often noted that the strong are respected around the world. But the primary reason for military or economic power is political will and the political strength of the leadership of a given country to carry out its own policies, not policies that are dictated by external forces, regardless of where they come from, be it the East or the West. [...] If we are able to demonstrate this firstly to our own people, not to the world, economic power and military strength will follow.[20]

The ensuing vote demonstrated just how radically the balance in Parliament had changed. Where on 18 November 1993, only 162 deputies had voted to join the NPT, this time, 301 did. The newly formed legislature had ultimately decided the fate of Ukraine's nuclear weapons.

Thus, on 16 November 1994, Parliament under Speaker Moroz ratified the NPT. However, the battle continued "to the last bullet." The MPs managed to add to the text of the ratification, a reservation: that Ukraine was nevertheless "the owner of nuclear weapons, which it had inherited from the former Soviet Union." The last point stated: "This Law shall come into effect after Ukraine receives security guarantees from the nuclear states in the form of an international legal document to this effect signed by them."[21]

Debate on the issue began on 16 November at 12:30 pm and ended, according to the official transcript, at 15:43 pm. The MPs took three hours and 13 minutes to make a decision that was to determine Ukraine's geopolitical fate. *Holos Ukrainy* published this landmark decision next to the Law "On Amending Article 32 of the Law of Ukraine 'On Road Traffic.'"[22]

The Story of the Budapest Memorandum

Ukraine's political class had two points-of-view on how "security guarantees" for a nonnuclear state were supposed to look. One was that Ukraine would accede to European and Euroatlantic security organizations, and this was supposed to become one of the conditions for nuclear disarmament. The other was that some kind of collective document would be signed in which the nuclear states testified that they "guaranteed" Ukraine's "security." Among the MPs, the latter was referred to as "paper guarantees."

The first position was held by Parliament, the second by the Foreign Ministry. Over 1993–1994, these positions were openly antagonistic. Yet, at the early stages of the disarmament process, there did not appear to be any divergences, as the MFA based its position on wording adopted by the legislature.

The need to see the renunciation of nuclear arms as part and a parcel of providing Ukraine with security guarantees was first established by Parliament in its Resolution of 9 April 1992, "On Additional Measures to Ensure That Ukraine Gains Nonnuclear Status." At the time, the parliamentary commissions made a commitment to "consider the entire complex of issues concerning nuclear disarmament from the viewpoint of security guarantees and Ukraine's foreign policy interests, with the involvement of specialists from the ministries, agencies and Academy of Sciences of Ukraine."[23]

At that point, the MFA's position, as evidenced in official documents, was formally in line with these demands. Thus, Point 2 of the "Written Statements by the Ukrainian Side Related to the Signing of the Protocol and Treaty Between the USSR and the US on the Reduction and Limitation of Strategic Offensive Arms," signed by Minister Zlenko on 23 May 1992, in Lisbon, mentions "Ukraine's right to guarantee its national security" and notes that "the Ukrainian side expects to get such guarantees prior to ratifying START."[24]

In short, the way the issue of "security guarantees" was presented, gave the impression that the positions of Parliament and the MFA were in line with Ukraine's national interests. As things developed, however, problems arose in the interpretation of this phrase.

World practice has several systems that protect the security of countries effectively. For the most part, this is military strength or a high level of integration in world financial and economic systems. The classic example of the latter is Switzerland. After World War II, the most effective collective security system proved to be NATO, which is a reliable guarantor of the borders and sovereignty of its member countries. And it was precisely this form of guarantee for Ukraine that was meant during the previously mentioned discussion between Ukrainian Ambassador to the US Bilorus and US Under Secretary of State for International Security Affairs Wisner. At the time, the American official stated that, for Ukraine, the only guarantee of security was "to join the Transatlantic structures and cooperate with them."[25] (See chapter 3.)

However, instead of developing economic and military cooperation with the US and Europe, the MFA launched the search for some form of document that they could call security guarantees. This process intensified towards the end of 1992 and early 1993, because President Kravchuk was counting on bringing up the ratification of START and the NPT in Parliament that February. Given the requirements in Parliament's previous decisions, together with the draft ratification resolution, deputies were supposed to receive something that could at least nominally be considered security guarantees.

It would have been logical for the MFA to form at least the outline of such guarantees as Ukraine found acceptable, taking as an example the case of a country like Austria. However, as events unfolded, it was clear that Ukraine's diplomats were unable to draft anything of their own for negotiating with the potential signatories and insisted that this task be delegated to other countries.

The vagueness of the demands of the Ukrainian side was effectively confirmed by Kravchuk during his meeting with US senators Nunn and Lugar. Clearly based on a formulation by the MFA and not him personally, he presented this view of the guarantees: "Let us have the Western states prepare the necessary statement to the effect that Ukraine and other states that are voluntarily withdrawing from the possession of nuclear weapons will have guarantees that other countries will not use nuclear force against them. Moreover, this declaration should stay in effect until such time as all nuclear weapons in the world have been destroyed."[26]

Because the MFA had no clear concept of what the essence of such guarantees might be, the US press was unable to clearly formulate for what purpose Ukraine's diplomats, led by Deputy Minister Tarasiuk, had come to Washington at the beginning of January 1993. The press reported, "Tarasiuk was looking for a formal 'political declaration at the highest level.'"[27] The American press also revealed that the MFA needed this document in order to present it to the legislature, specifically in the run-up to the debate on ratifying the nuclear treaties. The press reported that as a result of the visit, the Ukrainian official expressed satisfaction with the letter signed by Bush, which he had received, and was confident that this letter would help overcome Ukraine's lawmakers' fears of possible aggression from neighboring Russia.[28]

In January 1993, the MFA received a draft of the supposed guarantees proposed by Russia. It was not submitted to Parliament, because it did not even satisfy the MFA.[29]

Speaking at the Davos forum in early February 1993, Kravchuk first confirmed the position of Parliament that, "in rejecting nuclear arms [...] Ukraine had the indubitable right to demand from the nuclear states clear guarantees of its national security." Commenting on the draft of the guarantees received from Russia, he stated: "These commitments merely repeat the general principles of international law, specifically established in the CSCE's Concluding Act (1975) and in the Paris Charter for a New Europe."[30] However, in the very next sentence he effectively stated that it would be entirely enough to simply repeat them once more: "It is important for us that these be confirmed in the context of Ukraine becoming a nonnuclear-weapon state in the future." With this, Ukraine's president testified before the prestigious Davos gathering what Ukraine demanded as "security guarantees," in other words, repeating generally accepted norms of international law. Going forward, that would, in fact, be what other countries used as guidance.[31]

An illustrative statement in Kravchuk's speech, which echoed Tarasiuk's opinion was this: "We should persuade deputies with facts and not words." The President seemed to be placing himself, and some unnamed group of others, in opposition to Ukraine's lawmakers, for whose sake he was trying to get the country national security guarantees, because only these would persuade Parliament to ratify START and the NPT.[32]

On 25 February 1993, Moscow sent the next draft "Declaration on Security Guarantees," in which it once again simply compiled articles from international conventions to which Ukraine was already a party, as was Russia and the other nuclear states. These included broad guarantees of territorial integrity, sovereignty and the intent to resolve all disputes through negotiations. Thus, the document stated, Russia, "confirms its commitments to Ukraine in the context of the Treaty on the Founding of the Commonwealth of Independent States, recognizes and respects Ukraine's territorial integrity and the inviolability of existing borders within the framework of the Commonwealth." In addition, it "recognizes that changes to borders may only take place by peaceful and consensual means, and acknowledges its duty to refrain from the threat or use of force against the territorial integrity or political independence of any state, and that none of its weapons would be used other than for the purposes of self-defense or in any other form in line with the UN Charter."[33]

On 6 May 1993, the French Foreign Ministry also sent its draft "Provision of Security Guarantees to Ukraine," and on 20 May the Chinese MFA followed suit. These were also compilations of existing international agreements and conventions whose repetition added nothing new. Thus, when Ukraine acceded to the NPT, France was prepared to guarantee that it would not use nuclear weapons against Ukraine, which was what Kravchuk had asked for at Davos. Additionally, France guaranteed that it would try to get the UN Security Council to act immediately to help Ukraine if it became victim to an act of aggression, respect Ukraine's independence and sovereignty, and "decisively refrain from using or threatening to use force against Ukraine's territorial integrity."[34] Parliament did not bother to review such guarantees, the MFA offered nothing in response, and the process ground to a halt.

On 18 November 1993, Parliament passed a resolution on ratifying START I that made the provision of national security guarantees a mandatory condition for Ukraine to accede to the Treaty. This meant that until Ukraine received such guarantees, it would not be taking any steps to exchange ratification documents and renounce nuclear weapons. This, of course, implied that Ukrainian officials had no right to do so on behalf of Ukraine.

Still, Kravchuk reported on 3 February 1994 the following: "We shipped out a number of warheads at the end of December for dismantling because their guaranteed lifespan expired."[35] The very fact that the removal of nuclear weapons from Ukrainian territory continued even without any security guarantees and that this was well known to the country's leadership was confirmed on 16 November 1994, from the parliamentary podium by Foreign Minister Udovenko: "The question of removing nuclear warheads to Russia is being handled," and by President Kuchma: "They are being shipped out to Russia."[36]

After the snap parliamentary election of March 1994, the new legislature no longer had the equivalent of our Working Group. Now, the MFA worked on the text of the "security guarantees" outside Parliament, and without its input. Things were now moving quickly.

Almost immediately after the summer recess, on 30 September 1994, the new president of Ukraine sent a letter to the presidents of Russia, the US and France, and to the British prime minister "proposing to put together a multilateral Memorandum" among all five countries. The Ukrainian MFA wrote about this in its 15 November brief: "On Ukraine's Accession to the NPT: Progress with Implementing the Decisions of Parliament dated 18 November 1993 and 3 February 1994."[37] At this point it became quite clear, with this very document Ukraine's MFA confirmed that it was the agency on the Ukrainian side that ultimately agreed to the text later signed in Budapest; the brief stated: "Ukraine's executive is taking the necessary steps to develop a final version of such guarantees that will be acceptable to Ukraine [...]. For this purpose, the Ministry of Foreign Affairs of Ukraine held talks with representatives of the designated countries, the result of which was an agreed text for the Memorandum between Russia, the US, the UK and Ukraine."[38] It must be said that the principle on which this "agreed text" was built and even the selection of articles from international documents were almost identical to the "Declaration on Security Guarantees" proposed by Russia a year and a half earlier.

The MFA reported that on 7–9 November 1994, it received official confirmation from Russia, the US and the UK about their "readiness to sign the text of a multilateral Memorandum on security guarantees for Ukraine in connection with its accession to the Nuclear Nonproliferation Treaty." The ministry was convinced that "the extraordinary political significance of the provision of security guarantees to Ukraine was the very fact that they were in the form of a multilateral international legal document."[39]

However, the former US ambassador to Ukraine Steven Pifer recalled in 2014: "When the negotiations were ongoing over the adoption of this document in 1994, the terminology used was 'security assurances,' not guarantees [...]. And this led to misunderstandings. The difference between these two concepts for the American side was not important because America understood the use of this term, that is, the American side would not ever use force in this case."[40]

As the 2014 and 2022 events in Ukraine were to show, Russia was also quite aware of what it was signing at the time.[41] Based on these events, it seemed that only Ukraine's own diplomats, proud of the fact that the Budapest Memorandum was signed, never quite understood what they were handing to the president of Ukraine for signature in 1994; or even worse, they understood it very well.[42] Thus, the Memorandum was signed during the CSCE summit in Budapest on 5 December 1994.

The Budapest Summit. The Final Capitulation

Nevertheless, even after the 16 November vote, Moscow would not subside with its pressure. Russian diplomat Iurii Dubinin wrote about this in his 2004 article:

> On 17 November, Russia's Foreign Ministry already issued a statement: Moscow was pleased that the Parliament of Ukraine had adopted the law on accession to the NPT. At the same time, the statement noted: "There remains one unresolved issue regarding what kind of state Ukraine intends to be when acceding to the NPT: a nuclear state, or one that does not own such weapons [...]. At this time, NPT depositary states were developing a document on security guarantees for Ukraine, which they intended to present to the country as a nonnuclear-weapon state. It is obvious that some clarity on this matter would be very helpful."[43]

Moscow was worried because Parliament's Resolution still left Ukraine a reserve parachute, so to speak. Based on that decision, Ukraine's MFA was supposed to present a note with the content established by Parliament, that the provisions of the Treaty did not reflect the situation of Ukraine and that Ukraine would be acceding as a country that had inherited nuclear weapons.

In the end, however, the worst possible option was chosen for Ukraine in Budapest. Dubinin wrote: "On 5 December, after tense all-night negotiations with participation of the US, Ukraine presented a note from the Foreign Ministry regarding acceding to the NPT as a nonnuclear-weapon state."[44] Dubinin then quoted *Izvestiia*: "When Ukrainian President Leonid Kuchma handed the document to Boris Yeltsin, Bill Clinton, and John Major regarding his country's accession to the Treaty on the Nonproliferation of Nuclear Weapons, the room in which the Budapest CSCE summit was taking place, let out a collective sigh of relief."[45]

Consequently, on 5 December 1994, Ukraine, the US, Russia, Belarus, and Kazakhstan exchanged ratification certificates for START I. That same day, Ukraine handed over documents for accession to the NPT. Meanwhile, the US, Russia, and Great Britain signed the "Memorandum on Security Assurances in Connection with Ukraine's Accession to the Treaty on the Nonproliferation of Nuclear Weapons," otherwise known as the Budapest Memorandum. Later, France and China added their diplomatic assurances of security.

The deed was done. The scheme that unfolded with the May 1992 Lisbon Protocol, was finished with the completion of all the objectives Russia had placed for itself. Yet even this event was not the final defeat that Ukraine would face on the nuclear issue. As already mentioned, in ratifying START I and the Lisbon Protocol, Parliament established that Ukraine would destroy only 36 percent of its missiles and 42 percent of its nuclear warheads. The Trilateral Statement, which did not undergo ratification in the Ukrainian legislature, also

83. The loading of a nuclear warhead into a container for transport (1), 1992

established a seven-year term, although not to eliminate these percentages, but to hand over all of Ukraine's nuclear weapons to Russia. In both cases, the countdown was to start only after ratification certificates were exchanged, which meant 5 December 1994. Thus, even in this situation, which resulted from the signing of the Budapest Memorandum, Ukraine was to finish shipping out its warheads by the end of 2001. The guaranteed lifespans of the strategic nuclear weapons, which ended in 2003–2005, also permitted this time frame.

This was the time frame for disarmament, in which Ukraine, still possessing nuclear warheads on its territory, could have verified for itself just how these security guarantees might work in action and the international assistance to undertake reforms, which the democratic world kept promising so

84. *Signing the Budapest Memorandum, 5 December 1994*
Left to right, Boris Yeltsin, Bill Clinton, Leonid Kuchma, John Major

insistently; and to adjust its actions accordingly. Nevertheless, the process of clearing out Ukraine's nuclear weapons kept accelerating, something that was not required by any agreements. The timeframes kept being set and altered exclusively based on external circumstances, ignoring national legislation and international agreements. The seven-year deadline was changed to four years and then to one and a half years. On 8 December 1995, *Dzerkalo tyzhnia* wrote:

> According to [General Volodymyr] Mykhtiuk, Commander of Ukraine's 43rd Rocket Army, the nuclear warheads are being shipped out to Russia according to the previously established schedule and the last of them will cross our country's border by the end of 1998. However, it must be pointed out that the original schedule was for the shipment to finish by April–May 1996, but it turned out that Russia was unable to ensure the quality and safety of storage and dismantling of the weapons shipped to it. In connection with this, the completion of nuclear disarmament was extended to 1998. Nevertheless, the majority of our nuclear warheads are already in Russia now.[46]

Then, even this date, was soon adjusted.[47] Quoting the Defense Ministry figures, *Dzerkalo tyzhnia* wrote that as of the end of October 1995, which was less than a year after the Budapest Memorandum was signed, all 46 SS-24 missiles had been decommissioned, along with 80 of the country's 120 SS-19s. Altogether, 90 percent of the nuclear arsenal that had been in Ukraine at the beginning of 1994 had already been deactivated.[48]

By 1995, as the nuclear weapons kept being shipped out, interest in Ukraine faded sharply, and along with it, any hope for the much-promised assistance. As *Dzerkalo tyzhnia* reported on 5 January 1996:

> Having chosen the path to a nonnuclear future, Ukraine nevertheless expected the world community to support its contribution to reducing the global

nuclear threat. These expectations were reflected in a 15 November 1994 Joint Statement from Belgium, Canada, Denmark, Finland, France, Germany, Italy, Japan, the Netherlands, Norway, Spain, Sweden, the UK, and the US. The governments of these countries expressed their intentions to provide assistance jointly with the European Union for a total of US $234 million to compensate to Ukraine the costs of destroying its nuclear weapons. However, other than the US, only Canada, Germany, Japan, the Netherlands, and Norway actually made good on their promise. [...] On 4–5 January, during an official visit by the US and Russian Federation defense ministers, Ukraine received another round of moral support that, as folk sayings put it, you cannot spread on your slice of bread.[49]

Accelerated denuclearization did not "open all doors," as Minister Udovenko had promised; on the contrary, it led to their closure. The world had lost interest in Ukraine and the country's armed forces were not even invited to participate in international peacekeeping missions. *Dzerkalo tyzhnia* wrote about this humiliating situation in that same issue: "Nevertheless, despite the unexpectedness and seriousness of the situation, Ukraine's leadership has not lost hope that its special battalion will remain in former Yugoslavia. As competent sources report, this topic was seriously discussed during a meeting between President Kuchma and US Defense Secretary William Perry."[50] It then confirmed: "The more nuclear warheads cross the Ukrainian border, the less our country shows up in the headlines of the Western press. Participating in the process of monitoring the restoration of peace in former Yugoslavia could more-or-less become a significant factor for Ukraine's presence in the global decision-making process. We think our country deserves this much. In any case, hope dies last."[51]

After the Budapest summit, Russia never did withdraw it territorial claims on Ukraine's Sevastopol. Moreover, Moscow now turned to energy blackmail by threatening to stop deliveries of nuclear fuel rods to Ukraine's NPPs and the removal of spent fuel. It used the price of gas to exert political pressure on the division of the Black Sea Fleet, and it succeeded in stopping Ukraine's movement towards Eurointegration, subsequently launching a series of trade wars.

Despite all of this, Ukraine continued to rapidly ship its nuclear weapons to its northern neighbor, way ahead of its own time commitments under START. On 1 June 1996, the last echelon of about 200 nuclear warheads contained in strategic weapons left Ukrainian territory forever. As the press then put it, this kind of urgency was tied not to the end of their guaranteed lifespans, but with the upcoming presidential elections in the US. For the Clinton Administration, it was exceedingly important, as *Dzerkalo tyzhnia* wrote on 7 June 1996, "to be able to report that the nuclear threat to the United States had been reduced."[52]

Basically, within 18 months of the exchange of ratification documents in Budapest, and not seven years, Ukraine had handed all of its nuclear weapons over to Russia. Why Ukraine's leadership did not react to the obvious political and economic pressure coming from Russia, is a question yet to be answered. Altogether, between March 1994 and June 1996, nearly 2,000 nuclear warheads in the strategic weapons category were shipped to Russia for the supposed dismantling. Adding in the tactical weapons already in Russia, a total of approximately 5,000 nuclear warheads had been moved to Russian territory in the 1990s.[53]

85. The loading of a nuclear warhead into a container for transport (2), 1992

Events of the following years were to show both the substance of the written guarantees in the Budapest Memorandum and the value of the verbal promises to draw Ukraine more closely to the civilized world after renouncing nuclear status. In 2003, the next territorial threat arose: Russia decided to claim the Ukrainian island of Tuzla in the Kerch Strait. The world community did not react at all.

In July 2008, Henry Kissinger, a national security advisor to several American presidents, called on the US to stop NATO expansion to the east, in order to avoid a worsening of relations with Russia, whom he called a great sovereign state with corresponding national interests.[54] His aim was to remove the granting of the Membership Action Plan (MAP) to Kyiv from the NATO agenda.

As with the Tuzla gambit in 2003, the Budapest Memorandum proved its complete uselessness when Russia took over Crimea in March 2014, then launched a military intervention in the Donbas, and the full-scale invasion of Ukraine on 24 February 2022.[55] The document was little more than a piece of paper, which did not bind any of the signatories to action with respect to defending the territorial integrity of a country that had willingly given up the world's third largest nuclear arsenal.

Summing It Up: Everybody Loses

The main argument most often used to pressure Ukraine during disarmament process was that it was not giving humanity a chance to reduce the nuclear standoff and thus increase the level of global security. In the end, however, Ukraine eliminated the third largest arsenal in the world in record time, at its own cost, and yet the level of nuclear warheads around the planet is more than enough to destroy all of humanity several times over. What was achieved by the three main participants in the process, and the world, in forcing Ukraine to eliminate its nuclear weapons?

The World

According to the Stockholm International Peace Research Institute (SIPRI), in 2014, the US and Russia controlled over 93 percent of all nuclear weapons in the world, despite START I and other nuclear arsenal reduction treaties.[56] Even as the specific number of nuclear warheads has gone down, the process of improving them has picked up pace. Thus, the US plans to allocate US $350 billion over the 2020s to modernize and support its nuclear arsenal.[57] Russia is also actively investing in its nuclear program. According to Russia's former deputy prime minister Dmitry Rogozin, quoted in the 22 September 2014 issue of *Ukraïns'ka pravda*, Russia was planning to "amaze" the world with a 100-percent upgrade of its strategic nuclear forces.[58]

In its report, SIPRI also pointed out that not a single country that currently possesses nuclear warheads plans to completely renounce nuclear weapons in the medium term. On the contrary, it is clear that most of them want to upgrade their arsenals. The most ambitious plans in this field are in the US, which has been spending over US $60 billion annually just to maintain its strategic nuclear forces. By comparison, Russia spent only US $14.8 billion on its nuclear weapons in 2011.

By maintaining its one-sided nuclear leadership during the first years of the Cold War, the United States used this enormous strategic advantage to resolve many of its geopolitical issues from a position of power. After the USSR collapsed, the US gained a significantly increased military and political advantage, and strengthened the balance in its own favor, in part by pressing Ukraine to give up its nuclear arsenal. Many experts assert, according to the 2 March 2006 issue of *Foreign Affairs*, that Washington continues to maintain its nuclear advantage over other countries.[59]

86. *"The Value of the Budapest Memorandum," cartoon by Oleh Smal'*

This is evident, for example, from the programs and pathways to improve the US nuclear arsenal, under which the "US nuclear force [...] seems designed to carry out a preemptive disarming strike against Russia or China."[60] The article goes on to say:

> The Bush Administration's 2002 National Security Strategy explicitly states that the United States aims to establish military superiority: "Our forces will be strong enough to dissuade potential adversaries from pursuing a military build-up in hopes of surpassing, or equaling, the power of the United States." To this end, the United States is openly seeking primacy in every dimension of modern military technology, both in its conventional arsenal and in its nuclear forces.[61]

In other words, more than fifty years of mutual deterrence against aggression among nuclear countries, will no longer operate as a deterrent to a large-scale military conflict in the world. Strange as it may seem, Ukraine's nuclear disarmament, despite eliminating an enormous nuclear arsenal, led to barely any improvement in global security. On the contrary, it effectively eliminated the nuclear deterrent of the previous fifty years, leading to a new race for nuclear advantage, the sole leader of which is the US. Will the world become safer as a result? Humanity should receive an answer soon, most likely within the next decade.

The United States

Consequently, after the collapse of the Soviet Union, Washington ostensibly achieved its key objectives. First, it disarmed Ukraine and opened the way to a nuclear advantage; second, it spent the minimum amount of money by shifting all the main expenditures onto Ukraine itself; and third, it preserved the parity of Russian-American relations in Europe after the Warsaw Pact fell apart, at

87. *The destruction of the twenty-first ICBM silo (following the dismantling of the command post) under the terms of START I, Khmelnytskyi Oblast, 1996*

Ukraine's cost in order not to "upset" Russia, by refusing Ukraine the prospect of both Eurointegration and accession to NATO.

However, as Russia's takeover of Crimea and military aggression against eastern Ukraine in 2014 and the full-scale invasion in 2022 demonstrated, this proved a Pyrrhic victory for the US. Weakened through the combined efforts of the Americans and the Russians, Ukraine became the first victim to the Kremlin's neo-totalitarian regime, which resumed its military and political confrontation with the West, even by threatening to use nuclear weapons. One problem had simply been replaced by another, much more difficult one.

The US had failed to consider that, after the fall of the USSR, Russia might nevertheless preserve a totalitarian regime that, by the start of the 21st century had taken on a form far more dangerous to the world. The notion that "Russia is no longer what it was," which American strategists liked to quote, proved deeply erroneous.[62] Instead of placing their stake on Ukraine as the main factor in democratizing Russia, they left the Kremlin alone, allowing it to pull Ukraine back into its sphere of interests. The situation in 2022, which put the entire world on the brink of World War III, was a direct consequence of the US geopolitical miscalculation at the beginning of the 1990s. How this will end, is truly anyone's guess.[63]

Russia

Russia's goal was to keep Ukraine in its sphere of influence after the collapse of the Soviet Union. This goal was met through two key objectives. The first of these entailed maximally weakening Ukraine, militarily, economically and by fostering social unrest and dissatisfaction with independence; and the second, by blocking the country from getting closer to the EU and NATO. Moscow succeeded in reaching these objectives. Ukraine was left without nuclear weapons, having not only shipped all its costly nuclear material to Russia, but also paying

88. *The last RS-22 ICBM launcher is destroyed on the territory of the 43rd Rocket Army of Ukraine, Pervomaisk, Mykolaïv Oblast, 2001*

from its own state budget over the period of two decades to eliminate these nuclear capacities.

The suspended path to NATO enabled the eventual deployment of the Russian Black Sea Fleet in Crimea and the gradual disintegration of the Ukrainian Army.[64] Meanwhile, the border between the two countries remained porous for all kinds of Russian influence in Ukraine, including the provision of military equipment to Ukrainian separatists.[65]

The suspension of the path to Eurointegration made it possible to stop the refurbishing of those parts of Ukraine's economic sector that remained oriented towards the Russian market and dependent on Russian orders and Russian natural gas. This led to "milk," "meat," "chocolate," "cheese" and other trade wars through which Russia steered Ukraine's political course until the end of 2013.[66]

Still, even for Russia, victory was not complete. Although Moscow had appropriated all former Soviet strategic nuclear weapons from Ukraine, Kazakhstan and Belarus, its own nuclear capacities began to deteriorate rapidly. According to figures in a *Foreign Affairs* article entitled "The Rise of US Nuclear Primacy" published in spring 2006, at that point, Russia had "39 percent fewer long-range bombers, 58 percent fewer ICBMs, and 80 percent fewer SSBNs than the Soviet Union fielded during its last days."[67] Despite the Russian Federation's declarations that it intended to complete a 100 percent upgrade strategic nuclear forces by 2020, in fact, even the numbers in the *Foreign Affairs* article did not reflect the real degree of collapse of Russia's nuclear power. Strategic bombers were only deployed at two bases, making them highly vulnerable to attack. They were also rarely used in military exercises, while the nuclear warheads for their missiles were actually kept at other sites. More than 80 percent of Russia's strategic silo-based missiles have come to the end of their guaranteed lifespans and plans to replace them are continually disrupted.[68]

However, the most vulnerable was the third component of Russia's nuclear power, specifically, the ballistic-missile-launching submarines. Since 2000, they have carried out only two patrols per year, compared to 60 in 1990. By contrast, US nuclear submarines carry out 40 patrols per year. One indicator of the state of Russia's submarine-based strategic missiles was the complete failure of the ballistic missiles' test launches from a nuclear submarine in 2004, where President Putin himself was present. Every one of the missiles either failed to launch or went off-course. The next series of test launches in 2005 was also largely a failure.[69]

The more Russia's nuclear arsenal is reduced, the more capable the US becomes of making the first nuclear strike on Russian territory. In this way, having forced Ukraine into complete denuclearization together with the US, and thus having reduced the most effective component in Ukraine's national security at the time, Russia has now turned into the number one nuclear target in the world. In other words, by having cleaned out the nuclear arsenals of Ukraine, Belarus and Kazakhstan, and becoming the only nuclear state in the post-Soviet territory, Russia gave the complete nuclear advantage to the United States.

This is the conclusion drawn in the article entitled "The End of MAD? The Nuclear Dimension of US Primacy," which was published in *International Security* in spring 2006. In the article military analysts used computer modeling to establish that the US already had enough potential capacity to destroy all of Russia's strategic bomber bases, all its nuclear submarines, and all Russian strategic missile complexes without the threat of a retaliatory nuclear strike.[70] As the folk saying goes, "Don't go digging holes for the other guy."

Ukraine

In contrast to the US and RF, Ukraine had not established goals shared by all branches of power when it came to the question of its own nuclear disarmament. The objectives of the old and new generation of politicians were diametrically opposed. The former acted according to the Russian plan, while the latter resisted them, seeing Ukraine's and Russia's interests as antagonistic. Under these conditions, Ukraine did not have a chance to achieve advantageous results as a sovereign state.

Thus, our assessment of the results of Ukraine's nuclear disarmament policy is based on the contrast of what the country should have received to what it actually got:

Politically: Ukraine could have used its nuclear weapons as a significant source of seed capital that would have gradually transformed into large-scale cooperation with the US and the EU, including accession to both the EU and NATO. Ukraine's movement towards nonnuclear status, based on Russia's scenario, drove the country to the status of a post-colonial state in Russia's sphere of influence. Sadly, this remained the case until the start of the Revolution of Dignity in November 2013.

Economically: The cost of nuclear disarmament laid at Ukraine's feet added up to astronomical figures. At a time when its state budget was a paltry US $7–8 billion, Ukraine turned over assets, including nuclear material, missiles and silo equipment, worth more than US $100 billion to Russia. In addition, it destroyed 100 percent of its nuclear weapons at its own cost, which amounted to an additional nearly US $6 billion.

Militarily: Due to the losses Ukraine suffered as a result of destroying its own strategic weapons, and then by effectively slashing its military activity, the country turned from being one of the top military powers in Europe, to suffering a crushing defeat during Russia's military aggression in summer 2014.

Major General Robert P. Bongiovi (left), deputy director of the US Defense Threat Reduction Agency, and Lieutenant General Leonid Fursa, first deputy commander of the Ukrainian Air Force, cut the ribbon at the opening ceremony.

89. Opening ceremony for a nuclear weapon processing facility at the Ozerne Airfield, Zhytomyr Oblast, Ukraine, 2002

In short, everyone lost, particularly, the US, because it cannot independently ensure stability in the world today. Despite any Western diplomatic pressure and economic sanctions on Russia, they have not forced Moscow to return Crimea to Ukraine, or to withdraw its forces from the occupied territories of Donbas. The US has not provided Ukraine with any significant military support, fearing that this supposed spat among neighbors could lead to WWIII.

This weakness from the main guarantor of Ukraine's territorial integrity, according to the Budapest Memorandum, has shown that the global security system, which humanity so laboriously built up after World War II, has burst like a soap bubble. Paper agreements and friendly pats on the back do not work in the contemporary world.

Russia lost, too. The absence of a powerful, democratic influence in the region via Ukraine, has led to a regime unlike any the world has known so far. As irony would have it, the West saw Ukraine as a major threat to world peace in the early 1990s. The *Times* wrote in 1993: "Brandishing nuclear arms may win attention, but not plaudits from the West."[71] Today, the UN, NATO, OSCE and other international security institutions proclaim in their resolutions that it is now Russia that poses a major threat to world peace.

Nevertheless, the greatest losses fell upon Ukraine. Not just in the counting of the billions of dollars, the tremendous brain drain, the loss of economic opportunities, an unjustified decline in the standard of living, or even the disintegration of its armed forces; they could never match the loss of thousands of Ukrainian patriots who have fallen in battle against the Russian aggressor or the civilians who have died because of Russian shelling.[72]

AFTERWORD

Paul J. D'Anieri

Russia's brutal attack on Ukraine, beginning 24 February 2022, was almost unimaginable. Even at the height of Ukraine's debate over denuclearization, not even those most committed to retaining the weapons predicted that this would happen. A Ukrainian nuclear arsenal would have been intended to deter Russia, but the Ukrainians had not foreseen an all-out conventional assault along four major axes using upwards of 150,000 military forces. Only with Russia's invasion of Donbas and annexation of Crimea in 2014 did Ukraine deploy significant forces in eastern and southern Ukraine.

It may appear obvious in retrospect that Ukraine should never have given up its nuclear weapons. Russia would almost certainly not have attacked a nuclear-armed Ukraine; if it did, Ukraine would have had the potential to impose a devastating cost on Russia, as long as it were willing to endure nuclear retaliation. Instead, Ukraine finds its Western partners withholding needed weapons due to fears that Russia will escalate to use nuclear weapons. Rather than Ukraine deterring Russia from attacking, Russia is deterring the West from aiding Ukraine against a Russian attack.

While this line of thought is sound, it does not answer the question of what would have happened in the early to mid-1990s had Ukraine sought to retain nuclear weapons. Would Russia have attacked then, to take out the weapons before they became operational? Would the US have approved such an attack, or even encouraged it? If not military attack, would the US and Russia have collaborated on the kind of sanctions later imposed on Iran and North Korea? While Iran had its oil wealth and North Korea had the support of China, it is not clear who or what would have sustained Ukraine through sanctions. Even without sanctions, Ukraine's GDP dropped more in the years following independence than the US GDP during the Great Depression. If there was a solution to Ukraine's security problem, retaining nuclear weapons was not a straightforward one.

Following Russia's invasion, one has the impression that Ukrainian resentment stems less from the West coercing Ukraine to give up its nuclear weapons than from the West's failure to meet the commitments made in the 1994 Budapest Memorandum. While the text was carefully worded to avoid any legal commitment on the part of the US, Ukraine was clearly led to believe that the US and UK would protect it against Russia. While Western weapons have been essential in Ukraine's ability to fight the war, Ukraine has asked for much more than it has received, and far short of what the security assurances of 1994 implied.

When this war ends, as long as Ukraine remains intact in some form, the task of deterring Russia and preparing for a potential invasion will remain. While a thorough discussion of how to do that is beyond the scope of this short essay, one option (in theory more than in practice) would be for Ukraine to develop a nuclear deterrent. Such a plan would likely run into the same problems that it ran into thirty years ago: namely, the likelihood that it would alienate the West when Ukraine is desperate for support, and the possibility that it would provoke a Russian first strike (potentially with nuclear weapons). The failure of the existing security commitments pushes Ukraine toward a massive armament campaign, but the effects of war and the need for reconstruction will leave it dependent on the West for money and weapons.

What are the lessons? Unfortunately, other countries watching can only conclude that if they can find a way to getting nuclear weapons without being attacked, they should do it. More generally, Ukraine's experience is evidence for the realist reminder that international politics is a "self-help" world: those who rely on the good will or promises of others to ensure their security put their survival at grave risk.

August 2022
Riverside, California

Appendices

Appendix A

Declaration
On Ukraine's Nonnuclear Status
(VVR [Vidomosti Verkhovnoï Rady Ukraïny], no. 51 [1991]: 742)

In confirmation of Ukraine's intention, expressed in the Declaration of State Sovereignty of Ukraine dated 16 July 1990 (http://zakon3.rada.gov.ua/laws/show/55–12), to uphold three nonnuclear principles: to not accept, produce or acquire nuclear weapons,

recognizing the need for strict adherence to the 1968 Treaty on the Non-proliferation of Nuclear Weapons,

aspiring to contribute to the strengthening the international regime for the nonproliferation of nuclear weapons,

the Parliament of Ukraine **hereby declares**:

1. The presence of nuclear weapons of the former USSR on Ukrainian territory is temporary.
2. These nuclear weapons are currently under the control of the appropriate agencies of the former USSR.

Ukraine reserves the right to control the non-use of the nuclear weapons deployed on its territory.
3. Ukraine will carry out a policy aimed at the complete elimination of nuclear weapons and their components based on the territory of the Ukrainian state. Ukraine intends to do so in the shortest time, based on legal, technical, financial, organizational, and other capacities, with the proper provision of environmental safety.

Ukraine will roll out a broad-based program to convert its defense industry by reengineering part of its military industrial potential to the needs of economic and social development.
4. As one of the successor states of the former USSR, Ukraine will uphold the provisions of the 1991 Treaty between the US and the USSR on the reduction of strategic offensive weapons in those aspects that are related to the nuclear weapons deployed on its territory.

Ukraine is prepared to begin negotiations with the Republic of Belarus, the Kazakh SSR, and the Russian SFSR, with the participation of the related entities of the former USSR, about destroying strategic nuclear weapons that are governed by this Treaty.

5. Ukraine will take measures to destroy all other nuclear weapons that are deployed on its territory and, with this purpose in mind, if necessary, is prepared to participate in negotiations with all interested parties, including through existing multilateral disarmament mechanisms.

6. Ukraine will take the necessary measures to ensure the physical security of the nuclear weapons that are deployed on its territory until such time as they are completely destroyed.

7. Ukraine intends to accede to the Nuclear Nonproliferation Treaty (http://zakon3.rada.gov.ua/laws/show/995_098) as a nonnuclear-weapon state and to sign a related agreement with the IAEA on safeguards.

Parliament of Ukraine
Kyiv, 24 October 1991
No. 1697-XII

Appendix B

Agreement
on Strategic Forces Among Participating States
of the Commonwealth of Independent States

Date of Ukraine's signature: 30 December 1991

Date of coming into effect in Ukraine: 30 December 1991

(For additional changes see Agreement dated 22.01.1993, http://zakon. rada.gov.ua/laws/show/997_348)

Recognizing the need for a consensual and organized resolution to the management of Strategic Forces and for consolidated control over nuclear weapons, the republics of Armenia, Azerbaijan, Belarus, Kazakhstan, Kyrgyzstan, Moldova, Tajikistan, Turkmenistan, Uzbekistan, Ukraine, and the Russian Federation, further "Participating Commonwealth States," have agreed as follows:

Article 1
The term "Strategic Forces" shall mean: the associations, unions, units, institutions, military training institutions of the Strategic Rocket Army, the Army Air Force, the Naval Fleet, the Air Defense, the Office of the Chief of Aerospace Vehicles, the Airborne Troops, Strategic and Operational Intelligence, Nuclear-Technological Units, as well as the forces, equipment and other military objects designated for the management and provision of the Strategic Forces of the former USSR (a list is provided for each participating commonwealth state in a separate protocol).

Article 2
The Participating Commonwealth States obligate themselves to uphold the international treaties of the USSR and to carry out a coordinated policy regarding international security, disarmament and control over armament, to participate in the preparation and implementation of arms and Armed Forces reduction programs. The Participating Commonwealth States shall immediately undertake negotiations amongst them and with other states that were formerly part of the USSR and did not join the Commonwealth, for the purpose of ensuring guarantees and developing mechanisms for implementing the designated treaties.

Article 3
The Participating Commonwealth States recognize the need for joint command over the Strategic Forces and the maintenance of consolidated control over nuclear weapons and other weapons of mass destruction of the Armed Forces of the former USSR.

Article 4

Until such time as the nuclear weapons are destroyed, any decision regarding the need to deploy them shall be made by the president of the Russian Federation with the agreement of the heads of the Republic of Belarus, the Republic of Kazakhstan, and Ukraine, after consultations with the heads of other Participating Commonwealth States.

During the time prior to their complete destruction, the nuclear weapons deployed on the territory of Ukraine shall be under the control of the joint command of the Strategic Forces with the purpose of preventing their use and dismantling them by the end of 1994, including tactical nuclear weapons, by 1 July 1992.

The process of destroying nuclear weapons deployed on the territories of the Republic of Belarus and Ukraine shall be undertaken with the participation of the Republic of Belarus, the Russian Federation and Ukraine under the joint control of the Commonwealth States.

Article 5

The status of the Strategic Forces and the procedure for serving in them shall be established in a special agreement.

Article 6

This Agreement shall come into effect from the time it is signed and shall cease to have effect on the decision of the Participating Commonwealth States or the Council of Commonwealth Heads of State.

The effect of this Agreement shall cease to have effect in those participating states from whose territories all Strategic Forces have been withdrawn or nuclear weapons removed.

Concluded in the City of Minsk, 30 December 1991, in a single copy in the state languages of the Participating Commonwealth States.

All texts shall have equal force. The original copy shall be kept in the archives of the government of the Republic of Belarus, which shall send a certified copy to each state participating in this Agreement.

For the Republic of Armenia *(signature)*
For the Republic of Azerbaijan *(signature)*
(Except for funding)
For the Republic of Belarus *(signature)*
For the Republic of Kazakhstan *(signature)*
For the Republic of Kyrgyzstan *(signature)*
For the Republic of Moldova *(signature)*
(Except for funding)
For the Republic of Tajikistan *(signature)*
For the Republic of Turkmenistan *(signature)*
For the Republic of Uzbekistan *(signature)*
For the Russian Federation *(signature)*
For Ukraine *(signature)*

Appendix C

**Agreement
on Joint Measures Regarding Nuclear Weapons**
(Alma-Ata, 21 December 1991)

The Republic of Belarus, the Republic of Kazakhstan, the Russian Federation (RSFSR) and Ukraine, hereafter "the States Parties,"
    affirming their commitment to the nonproliferation of nuclear weapons,
    seeking to eliminate all nuclear weapons,
    aspiring to contribute to the strengthening the international stability, have agreed as follows:

Article 1
The nuclear weapons that are part of the Joint Strategic Armed Forces guarantee the collective security of all members of the Commonwealth of Independent States.

Article 2
The States Parties to this Agreement confirm their obligation not to use nuclear weapons first.

Article 3
The States Parties to this Agreement shall jointly develop policies regarding nuclear issues.

Article 4
Until the complete elimination of nuclear weapons on the territories of the Republic of Belarus and Ukraine, any decision on the need to use them shall be agreed with the heads of the States Parties by the president of the RSFSR based on procedures established jointly with the States Parties.

Article 5
1. The Republic of Belarus and Ukraine pledge to accede to the 1968 Treaty on the Nonproliferation of Nuclear Weapons as nonnuclear-weapon states and to sign a related agreement with the IAEA regarding safeguards.
2. The States Parties to this Agreement pledge not to transfer any nuclear weapons or other nuclear explosive devices and technologies to any third parties, as well as control over such nuclear and explosive devices, whether directly or indirectly; and equally not to assist, or encourage any other state that does not have nuclear weapons to manufacture or acquire in any way nuclear weapons or other nuclear explosive devices or take control of such weapons or explosive devices.

3. The provisions of Point 2 of this article shall not prevent the transfer of nuclear weapons from the territories of the Republic of Belarus, the Republic of Kazakhstan and Ukraine to the territory of the RSFSR for the purpose of destroying them.

Article 6
The States Parties to this Agreement shall cooperate in the destruction of nuclear weapons in line with the international Treaty. The Republic of Belarus, the Republic of Kazakhstan and Ukraine shall, by 1 July 1992, ensure that their tactical nuclear weapons are removed to central plant facilities for dismantling under joint control.

Article 7
The governments of the Republic of Belarus, the Republic of Kazakhstan, the Russian Federation (RSFSR), and Ukraine pledge to present the Strategic Arms Reduction Treaty (START) for ratification to the legislatures of their countries.

Article 8
This Agreement is subject to ratification. It shall enter into force 30 days after all ratification instruments have been deposited with the government of the RSFSR.

Concluded in Alma-Ata in one original copy in the Belarusian, Kazakh, Russian, and Ukrainian languages. All texts shall have equal force.

(Signatures)

Appendix D

Resolution of the Parliament of Ukraine
On Additional Measures to Ensure That Ukraine Gains
Nonnuclear Status

(VVR, no. 29 [1992]: 405)

The Parliament of Ukraine, adhering to the Declaration on State Sovereignty of Ukraine dated 16 July 1990 (https://zakon.rada.gov.ua/laws/show/55–12), the Statement of the Parliament of Ukraine on Ukraine's nonnuclear-weapon status dated 24 October 1991 (https://zakon.rada.gov.ua/laws/show/1697–12), which announced Ukraine's intention to uphold nonnuclear principles in the future and the right of Ukraine to control the non-use of nuclear weapons deployed on its territory,

confirming Ukraine's intentions to accede to the 1968 Treaty on the Nonproliferation of Nuclear Weapons (https://zakon.rada.gov.ua/laws/show/995_098),

recognizing that the government of the Russian Federation and the Command of the Strategic Forces have not established systems for implementing workable technical oversight, including also by Ukraine, of the non-use of nuclear weapons deployed on Ukraine's territory, as required by the Agreement on Joint Measures Regarding Nuclear Weapons dated 21 December 1991,

considering Ukraine's great responsibility for ensuring that the nuclear warheads that are transferred from its territory to the territory of the Russian Federation are destroyed under reliable international control which is supposed to ensure that the nuclear components of these warheads are not used to make new weapons and that they are not exported to third countries,

emphasizing that the destruction of nuclear weapons deployed on the territory of Ukraine should take place on condition of national security guarantees for Ukraine,

deeming it necessary for Ukraine as an independent nation to carry out a comprehensive study of the political, economic, financial, environmental, and other consequences of destroying nuclear weapons,

resolves:

1. To confirm Ukraine's adopted course towards peaceful cooperation with the entire world community, its non-bloc status, neutrality, and adherence to the three nonnuclear principles in the future.
2. To consider that tactical nuclear weapons should not be removed from Ukraine's territory until such time as a mechanism has been devised and instituted for international oversight of their destruction, with the participation of Ukraine.

3. To have the Cabinet of Ministers of Ukraine immediately take the necessary steps to ensure Ukraine's effective technical control over the non-use of nuclear weapons deployed on its territory.

4. To recommend to the president of Ukraine that he enter into negotiations with the leaders of world nuclear states regarding a comprehensive resolution of the problems connected to the destruction of nuclear weapons, keeping in mind the need to ensure the speediest possible bringing into effect of the 1991 Strategic Arms Reduction Treaty.

5. To have the Parliament of Ukraine Commissions on defense and national security, on foreign affairs, on planning, budgeting, finances and prices, on the development of basic industries in the domestic economy, and on the environment and rational resource utilization, review from the standpoint of security guarantees and Ukraine's foreign policy interests the entire range of issues on nuclear disarmament, including economic, financial, environmental, organizational and other aspects of destroying the nuclear weapons deployed on Ukrainian territory, including the reutilization of their components for peaceful purposes, by engaging specialists from the ministries, agencies and Academy of Sciences of Ukraine, and also, if necessary, independent experts, this April.

6. To have the government of Ukraine submit the Agreement on Joint Measures Regarding Nuclear Weapons dated 21 December 1991 (998_086), the Agreement on Strategic Forces Among CIS Member States dated 30 December 1991 (997_082), and the Agreement on the Status of Strategic Forces Among CIS Member States dated 14 February 1992, (997_104) to the Parliament for ratification

7. To have the Ministry of Defense of Ukraine take steps to staff the Strategic Forces deployed on Ukraine's territory by the service personnel of the Armed Forces of Ukraine.

8. That control over the implementation of this Resolution be delegated to the Presidium of the Parliament of Ukraine.

Parliament of Ukraine Speaker I. Pliushch
Kyiv, 9 April 1992
No. 2264-XII

Appendix E

Protocol
to the Treaty between the United States of America and the Union
of Soviet Socialist Republics on the Reduction and Limitation and
Limitation of Strategic Offensive Arms

The Republic of Belarus, the Republic of Kazakhstan, the Russian Federation, Ukraine, and the United States of America, hereinafter referred to as the Parties,

Reaffirming their support for the Treaty Between the United States of America and the Union of Soviet Socialist Republics on the Reduction and Limitation of Strategic Offensive Arms of 31 July 1991, hereinafter referred to as the Treaty,

Recognizing the altered political situation resulting from the replacement of the former Union of Soviet Socialist Republics with a number of independent states,

Recalling the commitment of the member states of the Commonwealth of Independent States that the nuclear weapons of the former Union of Soviet Socialist Republics will be maintained under the safe, secure, and reliable control of a single unified authority,

Desiring to facilitate implementation of the Treaty in this altered situation, Have agreed as follows:

Article I
The Republic of Belarus, the Republic of Kazakhstan, the Russian Federation, and Ukraine, as successor states of the former Union of Soviet Socialist Republics in connection with the Treaty, shall assume the obligations of the former Union of Soviet Socialist Republics under the Treaty.

Article II
The Republic of Belarus, the Republic of Kazakhstan, the Russian Federation, and Ukraine shall make such arrangements among themselves as are required to implement the Treaty's limits and restrictions; to allow functioning of the verification provisions of the Treaty equally and consistently throughout the territory of the Republic of Belarus, the Republic of Kazakhstan, the Russian Federation, and Ukraine; and to allocate costs.

Article III
1. For purposes of Treaty implementation, the phrase, "Union of Soviet Socialist Republics" shall be interpreted to mean the Republic of Belarus, the Republic of Kazakhstan, the Russian Federation, and Ukraine.
2. For purposes of Treaty implementation, the phrase, "national territory," when used in the Treaty to refer to the Union of Soviet Socialist Republics,

shall be interpreted to mean the combined national territories of the Republic of Belarus, the Republic of Kazakhstan, the Russian Federation, and Ukraine.

3. For inspections and continuous monitoring activities on the territory of the Republic of Belarus, the Republic of Kazakhstan, the Russian Federation, or Ukraine, that state shall provide communications from the inspection site or continuous monitoring site to the Embassy of the United States in the respective capital.

4. For purposes of Treaty implementation, the embassy of the Inspecting Party referred to in Section XVI of the Protocol on Inspections and Continuous Monitoring Activities Relating to the Treaty between the United States of America and the Union of Soviet Socialist Republics on the Reduction and Limitation of Strategic Offensive Arms shall be construed to be the embassy of the respective state in Washington or the embassy of the United States of America in the respective capital.

5. The working languages for Treaty activities shall be English and Russian.

Article IV

Representatives of the Republic of Belarus, the Republic of Kazakhstan, the Russian Federation, and Ukraine will participate in the Joint Compliance and Inspection Commission on a basis to be worked out consistent with Article I of this Protocol.

Article V

The Republic of Belarus, the Republic of Kazakhstan, and Ukraine shall adhere to the Treaty on the Nonproliferation of Nuclear Weapons of 1 July 1968 as nonnuclear weapon states Parties in the shortest possible time, and shall begin immediately to take all necessary action to this end in accordance with their constitutional practices.

Article VI

1. Each Party shall ratify the Treaty together with this Protocol in accordance with its own constitutional procedures. The Republic of Belarus, the Republic of Kazakhstan, the Russian Federation, and Ukraine shall exchange instruments of ratification with the United States of America. The Treaty shall enter into force on the date of the final exchange of instruments of ratification.

2. This Protocol shall be an integral part of the Treaty and shall remain in force throughout the duration of the Treaty.

Done at Lisbon on 23 May 1992, in five copies, each in the Belarusian, English, Kazakh, Russian, and Ukrainian languages, all texts being equally authentic.

For the Republic of Belarus: P. Kravchanka
For the Republic of Kazakhstan: T. Zhukeev
For the Russian Federation: A. Kozyrev
For Ukraine: A. Zlenko
For the United States of America: James A. Baker, III

Appendix F

Agreement
between the Government of the Russian Federation and the
Government of Ukraine on the Disposal of Nuclear Warheads
(Yalta, 3 September 1993)

The government of the Russian Federation and the government of Ukraine, hereinafter referred to as "the Parties,"

in reaffirmation of the commitment of the Parties to strengthen the international regime for the nonproliferation of nuclear weapons based on the Treaty on the Nonproliferation of Nuclear Weapons of 1 July 1968, (http://zakon2.rada.gov.ua/laws/show/995_098),

in consideration of the Treaty between the Union of Soviet Socialist Republics and the United States of America on the Reduction and Limitation of Strategic Offensive Arms of 31 July 1991 (http://zakon2.rada.gov.ua/laws/show/840_050), and documents related to it,

in consideration of the Agreement between the Russian Federation and Ukraine on the Procedure for Transferring Nuclear Warheads from the Territory of Ukraine to a central plant facilities of the Russian Federation with the purpose of dismantling and destroying them, of 1 April 1992 (http://zakon2.rada.gov.ua/laws/show/643_016), have agreed as follows:

Article 1

For the purposes of this Agreement, the terms used below shall have the following meaning:

"nuclear warhead"—the armed component of weapons deployed on Ukrainian territory that consists of nuclear material contained in a single construction;

"nuclear fuel for nuclear power plants (NPPs)"—material containing fissile nuclides with uranium-235 concentration of no more than 4.4 percent that, if placed inside a nuclear reactor, will make possible a nuclear chain reaction;

"highly enriched uranium (HEU)—uranium containing the uranium-235 isotope enriched to or greater than 20 percent;

"low enriched uranium (LEU)—uranium containing the uranium-235 isotope enriched to over 4.4 percent and below 20 percent;

"fuel assembly"—a set of nuclear fuel rods that are a single unit and cannot be detached from one another when loaded into the reactor core, irradiated, and removed from the core;

"fuel rods"—elements of the fuel assembly that consist of nuclear fuel and cladding and that provide the reliable transfer of heat from the fuel to the coolant;

"disposal of nuclear warheads"—the process of dismantling, transporting, storage and processing to utilize the components of nuclear warheads for peaceful purposes.

Article 2

1. The Russian Federation shall provide for the disposal of all nuclear warheads deployed in Ukraine.
2. The Russian Federation shall handle the processing of the highly enriched uranium that results from the disposal of the nuclear warheads into low enriched uranium for the manufacture of fuel assemblies that are used for atomic energy stations in Ukraine or other peaceful purposes.

Article 3

1. The delivery of fuel assemblies from the Russian Federation to Ukraine's atomic energy stations shall take place with the understanding that Ukraine shall place its nuclear activities under the safeguards of the International Atomic Energy Agency (IAEA).
2. To achieve the purposes set out in Article 2 of this Agreement, the Parties shall enter into a contract. This contract shall be concluded within six months after this Agreement comes into effect. If necessary, the Parties may conclude additional contracts in accordance with this Agreement.
3. In the case of any discrepancies between this Agreement and any contracts concluded in accordance with this Agreement, the provisions of this Agreement shall prevail.

Article 4

Each of the Parties shall designate an executive body to implement this Agreement. For the Russian Federation, the executive body responsible for the dismantling and transportation of nuclear warheads shall be the Ministry of Defense of the Russian Federation, and for the processing, storage and accounting shall be the Ministry of Atomic Energy of the Russian Federation; for Ukraine, they shall be the Ministry of Defense and the State Committee for the Utilization of Atomic Energy of Ukraine. Each of the Parties shall have the right to change the executive body by notifying the other Party in writing within 30 days.

Article 5

To implement this Agreement, the Parties shall ensure the unimpeded entry and exit of personnel of the Parties and shall offer reciprocal tax incentives as agreed between the Parties.

Article 6

1. This Agreement shall enter force on the date that the Parties exchange notifications that they have completed the internal procedures necessary for the Agreement to enter into force. This Agreement shall be in force based on the Parties international commitments and shall remain in force for 30 years. The term of this Agreement may be altered by written agreement of its Parties.
2. This Agreement may be supplemented or revised by written agreement of its Parties.
3. The effect of this Agreement may be stopped one year after receiving written notification to this effect from one of the Parties.

4. Concluded in Yalta on 3 September 1993, in two copies each, in the Russian and Ukrainian languages, and both texts shall have equal force.

   (Signatures)

Appendix G

Basic Principles
for the Disposal of Strategic Nuclear Forces Warheads Deployed in Ukraine
(Yalta, 3 September 1993)

The government of the Russian Federation and the government of Ukraine, hereinafter referred to as "the Parties," have agreed to the Basic Principles for the utilization of strategic nuclear forces warheads deployed in Ukraine:

1. Dismantling and transportation
    1.a. The Parties shall uphold the conditions for the Parties to remove all types of nuclear ICBM warheads and transfer such warheads, together with the main components of nuclear ALCMs to the territory of the Russian Federation with the purpose of dismantling and utilizing them.

    1.b. The Parties' executive bodies shall cooperate in the removal, transportation and storage of such warheads, in line with the provisions of the 1992 Agreement between Ukraine and the Russian Federation on the procedure for moving nuclear warheads from the territory of Ukraine to central plant bases in the Russian Federation for the purpose of dismantling and destroying them (http://zakon2.rada.gov.ua/laws/show/643_016).

    1.c. The Parties shall ensure the safety of the process of exploiting and removing nuclear warheads from missiles and transporting them, in compliance with the requirements and rules of current regulatory and technical documentation for nuclear weapons. At the same time, the transportation of such nuclear warheads shall be handled by the forces and resources of the Russian side. The Ukrainian side shall be responsible for ensuring safe conditions for the movement of nuclear warheads on its own territory.

    1.d. The implementation of Point 1 shall be carried out according to a schedule that shall be developed by the Parties and approved by the Heads of government of the Parties. This shall consider the need to first dismantle nuclear warheads whose service life has expired or storing warheads and their individual components. Compensation for each nuclear warhead exported from Ukraine to the Russian Federation shall be provided within one year of the date that the warhead crossed the border between the Parties.

2. The procedure for compensating the value of nuclear material:
    2.a. all highly enriched uranium (approximately 50 tons of HEU) recovered in accordance with current agreements from

nuclear warheads, shall be processed in the Russian Federation into low enriched uranium (LEU).

2.b. The principles for settling for the uranium shall be based on the following:

b.1. Neither of the Parties shall subsidize the other.

b.2. Commercial deals—no profits or losses, just breaking even.

b.3. Prices and costs reflecting inflation and changes in the world market.

b.4. The Russian Federation shall supply Ukraine with fuel assemblies for its atomic energy stations (AESs). Compensation for the costs to the Russian Federation for the delivery of fuel assemblies shall be covered partly from the sale of uranium removed from the strategic nuclear warheads deployed in Ukraine.

2.c. Weapons-grade plutonium recovered from strategic warheads deployed in Ukraine shall be stored in the Russian Federation until the Parties make a decision regarding its utilization. Ukraine shall receive the value of the weapons-grade plutonium as soon as it is sold, minus the costs and the expenses of the Russian Federation for disposing of it.

2.d. Nonnuclear components of strategic nuclear weapons that have been removed from nuclear warheads deployed in Ukraine shall be disposed of as agreed by the Parties.

2.e. The Parties shall provide transparent measures in terms of accounting for nuclear materials and their sale prices in accordance with Points a, b 4, and c.

2.f. The supply of fuel assemblies to Ukraine's AESs shall take place on the basis of annual contracts between the executive bodies of the Parties.

2.g. Should the uranium released from warheads located on Ukrainian territory be sold on the world market, tripartite contracts can be executed to guarantee that the Parties receive their appropriate share of the value of the products.

These Basic Principles have equal force in accordance with the Agreement between the Government of the Russian Federation and the Government of Ukraine on the disposal of nuclear warheads dated September 1993 (http://zakon2.rada.gov.ua/laws/show/643_132).

(Signatures)

Appendix H

Agreement
between Ukraine and the Russian Federation on the Provision of Warranty and Manufacturer Oversight for the Operation of the Strategic Forces' Strategic Missile Systems Deployed on their Territories (Ukr/Rus)

Date of signing: 3 September 1993

Date of entering into force: 3 September 1993

Ukraine and the Russian Federation, hereafter "the Parties,"
guided by the need to ensure nuclear and environmental safety, to exclude unsanctioned use of nuclear weapons and to maintain the Strategic Forces' strategic missile systems deployed on their territories in the proper technical condition,
have agreed as follows:

Article 1

Work on warranty and manufacturer oversight for the operation of strategic missile systems of the Strategic Forces deployed on the Parties' territories shall be carried out by specialists at companies that develop and manufacture such systems, as well as the systems and units that constitute them, regardless of which Party's territory they are deployed on.

Article 2

The Parties' firms undertaking the warranty and manufacturer oversight in accordance with this Agreement are governed by the laws of the Parties as well as the following documents:
  – Basic conditions for the delivery of products for military organizations;
  – Provisions for the warranty oversight for the technical condition of OS-type increased-security launch silos;
  – Provisions on warranty oversight over the technical condition of 15P645-type systems (for 15P158 systems);
  – Provisions on warranty oversight over the technical condition of 15P961-type systems;
  – Provisions on warranty oversight over the technical condition of R-type systems;
  – Provisions on the development of bulletins, providing documentation and material, and planning and carrying out improvements in the Armed Forces;
  – GOST B15.704–83. Manufacturer oversight of operations.

- Basic provisions;
- GOST B15.703–78. Procedure for the presentation and satisfaction of claims;
- Provisions on the types, procedures and organization of the work of manufacturer oversight over type-R systems at Naval facilities;
- Provisions on the delivery of products for industrial and technical purposes;
- GOST B22.027–82. Strategic missile and space systems,
- Claim procedures at industrial enterprises;
- Provisions on the participation of industry representatives in work on Rocket Army sites.

Article 3

The functions that were previously carried out by the USSR Ministries of General Machine-building and of the Defense Industry in terms of warranty and manufacturer oversight shall be delegated:

for Ukraine—the Ministry of Machine-Building for the Military Industrial Complex and Conversion;

for the Russian Federation—the Military Industrial Commission.

Article 4

The volume of work and costs for warranty and manufacturer oversight for the operation of strategic missile systems are established by the plan for warranty and manufacturer oversight drafted by the Russian Federation Military Industrial Commission, based on the initial data from the Strategic Missile Forces, the Navy and key production enterprises, and are agreed with the Ministry of Defense and the Ministry of Machine-Building for the Military Industrial Complex and Conversion of Ukraine for strategic missile systems deployed on its territory.

Financing for the work of warranty and manufacturer oversight over strategic missile systems is handled by the Strategic Missile Forces Command, which shall receive money for this purpose from the two Parties, in volumes established in the approved work plan for warranty and manufacturer oversight.

The Parties shall settle accounts for completed works in accordance with procedures that they establish. The allocated funds shall be indexed on a quarterly basis, taking into account real inflation.

Strategic Missile Forces and Navy Command shall conclude a contract with key development and production enterprises to carry out the work of warranty and manufacturer oversight over the operations of strategic missile systems, taking into account the funding for supporting maintenance bases, and the engineering and technical personnel at enterprises in the industries that will carry out these works.

Article 5

Access to sites where warranty and manufacturer oversight are taking place shall be arranged with the approval of the Ministry of Defense of the Party on whose territory these sites are located. The Parties agree to ensure the unimpeded removal/delivery of equipment and materials needed to carry out this work on the territory of the Parties.

Article 6

This Agreement shall take effect from the moment it is signed and shall remain in force until the disposal of strategic missile systems is complete.

Concluded in the city of Yalta on Septembern3, 1993 in two copies, each in the Ukrainian and Russian languages, with both versions having equal force.

For Ukraine *(signature)*
For the Russian Federation *(signature)*

Appendix I

Resolution of the Parliament of Ukraine
on Ratifying the Treaty between the Union of Soviet Socialist
Republics and the United States of America on the Reduction
and Limitation of Strategic Offensive Arms, Signed in Moscow on
31 July 1991, and the Protocol Thereto, Signed in Lisbon on Behalf of
Ukraine on 23 May 1992

(VVR, no. 49 [1993]: 464)

The Parliament of Ukraine **resolves**:

To ratify on behalf of Ukraine, one of the successors of the former USSR, the Treaty between the **Union of Soviet Socialist Republics and the United States of America on the reduction and limitation of strategic offensive arms** (http://zakon3.rada.gov.ua/laws/show/840_050) (further "the Treaty"), signed in Moscow on 31 July 1991, which includes the following component documents to the Treaty:

Memorandum on establishing initial data related to the Treaty;

Protocol on the procedures that regulate the re-equipping or destruction of articles that are governed by the Treaty;

Protocol on inspections and uninterrupted monitoring in connection with the Treaty;

Protocol on notifications in connection with the Treaty;

Protocol on the weight launched by ICBMs and SLBMs in connection with the Treaty;

Protocol on telemetric data in connection with the Treaty;

Protocol on the Joint Execution and Inspection Commission in connection with the Treaty;

Addendum of Agreed Statements;

Addendum of Terms and Definitions;

Protocol to the Treaty, signed in Lisbon on behalf of Ukraine 23 May 1992 (http://zakon3.rada.gov.ua/laws/show/998_070) (except for Art. V), with the following reservations regarding the Treaty and its component documents:

1. In accordance with the Vienna Convention on Succession of States in respect to State Property, Archives and Debts (http://zakon3.rada.gov.ua/laws/show/995_072), the Law of Ukraine "On the Enterprises, Entities and Organizations Subordinated to the Union and Located on the Territory of Ukraine" (http://zakon3.rada.gov.ua/laws/show/1540–12) dated 10 September 1991, and the "Key Directions of Foreign Policy of Ukraine" (http://zakon3.rada.gov.ua/laws/show/3360–12), all assets located on the territory of Ukraine belonging to strategic and tactical nuclear forces, including their nuclear warheads, are state property of Ukraine.

2. Ukraine does not consider Art. V of the Lisbon Protocol binding on the country.

3. Having become the owner of nuclear weapons inherited from the former USSR, Ukraine takes administrative control over the strategic nuclear forces on its territory.

4. Having experienced the disastrous consequences of the Chernobyl nuclear accident, the people of Ukraine are aware of their great responsibility before the nations of the world to make sure that a nuclear war does not start on Ukrainian soil. For this reason, Ukraine is taking the necessary steps to prevent the use of the nuclear weapons deployed on its territory.

5. As a state possessing nuclear weapons, Ukraine shall move towards non-nuclear status and rid itself of the nuclear weapons deployed on its territory in stages, on condition that it receives reliable national security guarantees in which nuclear states shall bind themselves to never use nuclear weapons against Ukraine, use regular armed forces against Ukraine, or resort to threats of force, to respect Ukraine's territorial integrity and the inviolability of its borders, and to refrain from using economic pressure for the purpose of resolving any disputed matters.

6. Reduction followed by the destruction of strategic offensive arms deployed on the territory of Ukraine shall take place in accordance with the Treaty on the Reduction and Limitation of Strategic Offensive Arms (http://zakon3.rada.gov.ua/laws/show/840_050) dated 31 July 1991, and Article 11 of the Lisbon Protocol, on the understanding that 36 percent of carriers and 42 percent of warheads shall be destroyed. This does not preclude the possibility that additional carriers and warheads might be destroyed according to procedures determined by Ukraine.

7. Ukraine shall carry out its obligations under the Treaty in the timeframes established therein, based on legal, technical, financial, organizational, and other capacities, and the necessary assurance of nuclear and environmental safety. Given the current crisis state of Ukraine's economy, these obligations can only be met provided that sufficient international financial and technical assistance is provided.

8. The coming into force of the Treaty and its implementation shall not establish any conditions for the participating states to attempt to gain unilateral advantages that could harm the national interests of Ukraine, on high-technology markets, in scientific and technological exchanges, or in cooperation in sectors using nuclear energy for peaceful purposes, and applying missile technology.

9. If the dismantling and destruction of nuclear warheads deployed on the territory of Ukraine take place beyond its borders, Ukraine shall exercise direct oversight over these processes in order to ensure that the nuclear components of such warheads are not utilized again to produce nuclear weapons.

10. The conditions and priorities for transferring nuclear warheads for further dismantling and destruction shall be established in a special agreement or agreements that provide for the components of these nuclear weapons to be returned to Ukraine for further application to peaceful purposes or for their value to be compensated. The terms for compensation shall extend to tactical nuclear weapons that were shipped to Russia from Ukrainian territory in 1992.

11. Given that Ukraine did not directly participate in the negotiations to draw up the Treaty, it is recommended that the president and

government of Ukraine enter into negotiations with the appropriate states and international organizations:

> *a)* regarding international guarantees of Ukraine's national security;
>
> *b)* regarding the conditions for economic, financial, and scientific and technological assistance to implement the Treaty;
>
> *c)* regarding safeguard and design servicing of the nuclear warheads and missile launching systems;
>
> *d)* regarding revising the terms of financing inspection activity provided for in the Treaty;
>
> *e)* regarding the possibility of using silos for peaceful purposes under proper control;
>
> *f)* regarding the conditions for utilizing weapons-grade fissionable material extracted during the destruction of nuclear weapons;
>
> *g)* regarding guarantees of compensation for the material value of the components of nuclear weapons.

12. It is recommended that the president of Ukraine approve the schedule for destroying the strategic offensive arms designated in this Resolution and to ensure control over implementation.

13. When drafting the Budget for 1994, the Cabinet of Ministers of Ukraine should anticipate a separate article for spending on fulfilling Ukraine's commitments under this Treaty.

* * *

Ukraine shall exchange ratification certificates only after the conditions in Points 5, 6, 7, 9, 10, and 11 have been met.

The Parliament of Ukraine anticipates that nuclear states that are not parties to the Treaty will join Ukraine's efforts and those of other successor states of the former USSR, as well as the US, and begin to reduce their nuclear arsenals.

The coming into effect of the Treaty and its implementation open the pathway for the Parliament of Ukraine to resolve the question of joining the Treaty on the Nonproliferation of Nuclear Weapons dated 1 July 1968.

Speaker of the Parliament of Ukraine I. Pliushch
Kyiv, 18 November 1993
No. 3624-XII

Appendix J

Resolution of the Parliament of Ukraine
On the Implementation by the President of Ukraine and Government
of Ukraine of Recommendations in Point 11 of the Resolution of the
Parliament of Ukraine "On Ratifying the Treaty between the Union
of Soviet Socialist Republics and the United States of America on
the Reduction and Limitation of Strategic Offensive Arms, Signed in
Moscow on 31 July 1991, and the Protocol Thereto, Signed in Lisbon
on Behalf of Ukraine on 23 May 1992"

(VVR, no. 22 [1994]: 147)

The Parliament of Ukraine,
– keeping in mind the specific measures taken by the president and
government of Ukraine between November 1993 and January 1994
with regard to implementing the provisions of the Resolution of the
Parliament of Ukraine dated 18 November 1993;
– considering the results of the meeting between the presidents of
Ukraine, the United States of America and the Russian Federation in
Moscow on 14 January 1994, the Trilateral Statement signed by them
(http://zakon.rada.gov.ua/laws/show/998_300) and the Addendum
to it;
– taking into account the fact that Ukraine received confirmation
from the presidents of the US and Russia about their readiness to
provide Ukraine with national security guarantees after START 1
(http://zakon.rada.gov.ua/laws/show/840_050) comes into effect
and Ukraine joins the Treaty on the Nonproliferation of Nuclear
Weapons (NPT) as a state that does not possess nuclear weapons, as
well as commitments on the part of the United States of America, the
Russian Federation and Great Britain regarding Ukraine, to respect
its independence, sovereignty and existing borders, to refrain from
threats or use of force against its territorial integrity or political independence, to refrain from economic pressure and taking into account
the commitment not to use any weapons against Ukraine;
– considering the confirmation from the presidents of Ukraine, the
US and Russia that relations between them shall be built on the basis
of respect for the independence, sovereignty and territorial integrity of each state, and the confirmation of their readiness to provide
assistance in establishing an effective market economy in Ukraine;
– recognizing the fact that the United States of America have assured
Ukraine that technical and financial assistance will be provided for
the reliable and safe dismantling of nuclear weapons and the storage

of fissionable material, as well as a rapid implementation of existing agreements regarding such assistance fostered;

– taking into account that, in accordance with the Protocol "On the procedure for control over the destruction of nuclear warheads that are being removed from the territory of Ukraine to industrial enterprises in the Russian Federation," officials from the Ministry of Defense of Ukraine shall exercise control over the dismantling and destruction of strategic nuclear warheads on the territory of Russia, which shall exclude the reutilization of the components of these warheads for their primary purpose;

– taking into account, also, Russia's commitments to provide technical servicing and safe use of nuclear warheads;

– considering that Ukraine is being given fair compensation for the value of the highly enriched uranium and other components of all the nuclear weapons of which Ukraine is the owner;

– taking into account the agreement on the provision of fair and timely compensation to Ukraine for the value of its highly enriched uranium by the Russian Federation and the United States of America, in line with the nuclear warheads being moved from Ukraine to Russia for dismantling, and the fact that the measures for removal and compensation shall be applied at the same time;

– considering that the United States of America, the Russian Federation and Ukraine shall strictly adhere to the agreements contained in the Trilateral Statement of the presidents and the Addendum to it, and other agreements that exist among them and those that remain to be concluded regarding nuclear weapons deployed on the territory of Ukraine;

– considering that the above makes it possible to implement the conditions and reservations that were included in the Resolution dated 18 November 1993 (http://zakon.rada.gov.ua/laws/show/3624–12), resolves:

1. To remove the reservations concerning Art. V of the Protocol to START signed in Lisbon on 23 May 1992, based on the specific measures taken by the president and government of Ukraine with regard to implementing the provisions of the Resolution of Parliament dated 18 November 1993, and compromises on the part of the US and Russia.

2. To instruct the government of Ukraine to exchange certificates on the ratification of START I (http://zakon.rada.gov.ua/laws/show/840_050) and set in motion the conclusion of specific international agreements that arise out of the reservations of the Resolution of the Parliament of Ukraine on ratifying START I.

Speaker of the Parliament of Ukraine I. Pliushch
Kyiv, 3 February 1994
No. 3919-XII

Appendix K

Law of Ukraine
On Ukraine's Accession to the Treaty on the Nonproliferation of
Nuclear Weapons of 1 July 1968

(VVR, no. 47 [1994]: 421)

Based on the provisions of the Declaration of State Sovereignty of Ukraine dated 16 July 1990 (http://zakon.rada.gov.ua/laws/show/55–12), the Declaration of the Parliament of Ukraine "On Ukraine's nonnuclear-weapon status" (http://zakon.rada.gov.ua/laws/show/1697–12) dated 24 October 1991, the Resolution of the Parliament of Ukraine "On additional measures to ensure that Ukraine gains nonnuclear-weapon status" dated 9 April 1992, the Resolution of the Parliament of Ukraine "On ratifying the Treaty between the Union of Soviet Socialist Republics and the United States of America on the reduction and limitation of strategic offensive arms (http://zakon.rada.gov.ua/laws/show/840_050), signed in Moscow on 31 July 1991, and the Protocol thereto, signed in Lisbon on behalf of Ukraine on 23 May 1992 (http://zakon.rada.gov.ua/laws/show/3624–12) dated 18 November 1993, and the Resolution of the Parliament of Ukraine "On the Implementation by the President of Ukraine and Government of Ukraine of Recommendations in Point 11 of the Resolution of the Parliament of Ukraine 'On Ratifying the Treaty between the Union of Soviet Socialist Republics and the United States of America on the Reduction and Limitation of Strategic Offensive Arms, Signed in Moscow on 31 July 1991, and the Protocol Thereto, Signed in Lisbon on Behalf of Ukraine on 23 May 1992'" (http://zakon.rada.gov.ua/laws/show/3919–12) dated 3 February 1994, the Parliament of Ukraine **resolves:**

To accede to the Treaty on the Nonproliferation of Nuclear Weapons dated 1 July 1968 (http://zakon.rada.gov.ua/laws/show/995_098) with the following reservations:

1. The provisions of the Treaty do not fully cover the unique situation that has arisen as a result of the collapse of a nuclear state, the Soviet Union.

2. Ukraine is the ownder of nuclear weapons that it inherited from the former USSR. After dismantling and destroying these weapons under Ukraine's control and following procedures that exclude the option of reusing the nuclear materials that are components of these weapons for their primary purpose, Ukraine intends to use such materials exclusively for peaceful purposes.

3. The presence of nuclear weapons prior to their disposal and related work maintaining, servicing and disposing of them on the territory

of Ukraine are not in violation of the provisions of Articles I and II of the Treaty.

4. The threat or use of force against the territorial integrity and the inviolability of the borders or political independence of Ukraine on the part of any nuclear state, similarly to the use of economic pressure with the intent to subjugate Ukraine's exercise of the rights inherent in its sovereignty to such state's own interests shall be treated by Ukraine as exceptional circumstances that jeopardize its own highest interests.

5. Documents on Ukraine's accession to the Treaty shall be transferred to the depositaries of the Treaty after this Law enters into force.

6. This Law shall enter into force after Ukraine is provided with security guarantees, in the form of an appropriate, signed international legal document, by the nuclear states.

Speaker of Parliament O. Moroz
Kyiv, 16 November 1994
No. 248/94-BP

Appendix L

Memorandum on Security Assurances in Connection with Ukraine's Accession to the Treaty on the Nonproliferation of Nuclear Weapons

Ukraine, the Russian Federation, the United Kingdom of Great Britain and Northern Ireland and the United States of America,

Welcoming the accession of Ukraine to the Treaty on the Nonproliferation of Nuclear Weapons as a nonnuclear-weapon State,

Taking into account the commitment of Ukraine to eliminate all nuclear weapons from its territory within a specified period of time,

Noting the changes in the world-wide security situation, including the end of the cold war, which have brought about conditions for deep reductions in nuclear forces,

Confirm the following:

1. The Russian Federation, the United Kingdom of Great Britain and Northern Ireland and the United States of America reaffirm their commitment to Ukraine, in accordance with the principles of the Final Act of the Conference on Security and Cooperation in Europe, to respect the independence and sovereignty and the existing borders of Ukraine;

2. The Russian Federation, the United Kingdom of Great Britain and Northern Ireland and the United States of America reaffirm their obligation to refrain from the threat or use of force against the territorial integrity or political independence of Ukraine, and that none of their weapons will ever be used against Ukraine except in self-defense or otherwise in accordance with the Charter of the United Nations;

3. The Russian Federation, the United Kingdom of Great Britain and Northern Ireland and the United States of America reaffirm their commitment to Ukraine, in accordance with the principles of the Final Act of the Conference on Security and Cooperation in Europe, to refrain from economic coercion designed to subordinate to their own interest the exercise by Ukraine of the rights inherent in its sovereignty and thus to secure advantages of any kind;

4. The Russian Federation, the United Kingdom of Great Britain and Northern Ireland and the United States of America reaffirm their commitment to seek immediate United Nations Security Council action to provide assistance to Ukraine, as a nonnuclear-weapon State party to the Treaty on the Nonproliferation of Nuclear Weapons, if Ukraine should become a victim of an act of aggression or an object of a threat of aggression in which nuclear weapons are used;

5. The Russian Federation, the United Kingdom of Great Britain and Northern Ireland and the United States of America reaffirm, in

the case of Ukraine, their commitment not to use nuclear weapons against any nonnuclear-weapon state party to the Treaty on the Non-proliferation of Nuclear Weapons, except in the case of an attack on themselves, their territories or dependent territories, their armed forces, or their allies, by such a State in association or alliance with a nuclear-weapon State;

6. Ukraine, the Russian Federation, the United Kingdom of Great Britain and Northern Ireland and the United States of America will consult in the event a situation arises that raises a question concerning these commitments.

This Memorandum will become applicable upon signature.

Signed in four copies having equal validity in the Ukrainian, English and Russian languages.

For Ukraine: *(Signed)* Leonid D. Kuchma
For the Russian Federation: *(Signed)* Boris N. Yeltsin
For the United Kingdom of Great Britain and Northern Ireland: *(Signed)* John Major
For the United States of America: *(Signed)* William J. Clinton

Appendix M

Brief
on the Results of Negotiations with the Russian Delegation on the Dismantling, Transportation and Disposal of Nuclear Warheads of the Strategic Forces of Ukraine

1. The Russian Federation shall determine the nomenclature of the materials contained in the warheads;

2. The Russian Federation shall determine the quantity of nuclear material contained in the strategic warheads and present them, together with evidence confirming the accuracy of the data;

3. The Russian Federation shall present an estimation of the costs of transporting, storing and dismantling the warheads and disposing of individual elements that shall be paid by Ukraine.

4. The Russian Federation agrees to review and evaluate the costs for the nuclear materials based on the following options:

4.1. Highly enriched uranium (HEU) shall be stored in the Russian Federation and, at the request of Ukraine, in accordance with an agreed schedule, shall be processed into materials that Ukraine needs to use in its domestic economy;

4.2. HEU shall be depleted to the level of VVER-1000 fuel, out of which fuel rods and fuel assemblies shall be produced that will then be shipped to Ukraine at a price that includes the cost of processing the warheads, depleting the fuel and preparing the fuel assemblies;

4.3. The Russian Federation agrees to consider the share of Ukraine in any profits from possible sale of uranium to third party countries. However, the Russian Federation considers that the sale of HEU is extremely inexpedient and that it should be used for the development of nuclear energy or left to descendants for the future when fuel resources are limited;

4.4. Over the next 10–20 years, the Russian Federation does not see any options for the broad-based application of plutonium to nuclear energy. There is no commercial demand in the world for weapons-grade plutonium. The only option, therefore, is to store it in Russia until a solution can be found. Moreover, the Russian Federation believes that, currently, weapons-grade plutonium cannot be handed over to Ukraine because of the basic principle of the nonproliferation of nuclear weapons.

5. The Russian Federation believes that the problems with developing nuclear energy and a nuclear fuel cycle are in many respects similar for both Russia and Ukraine. Therefore, for Ukraine to make strategic decisions in this area it should participate in the specific work being carried out by the Russian Federation.

Appendix N

Statement by the Government of the Russian Federation, 5 April 1993

The government of Russia proposes that Ukraine immediately remove the nuclear weapons deployed on its territory. The government of Russia notes that the situation around nuclear weapons deployed on the territory of Ukraine has sharply escalated lately. "Ukrainian officials are declaring openly that these nuclear weapons belong to Ukraine. Such statements cannot be seen as anything but claims by Ukraine to own nuclear weapons," notes a statement issued by the government of the Russian Federation in Moscow on Monday.

This kind of tack by the Ukrainian leadership indicates a direct violation of the decision of CIS Heads of State on 6 July 1992, regarding the participation of the Commonwealth members in the Treaty on the Nonproliferation of Nuclear Weapons. According to that decision, of the successor states to the USSR only the Russian Federation is a nuclear weapons country, as indicated in the document.

Kyiv's claims to possess nuclear weapons indicates a violation of the Lisbon Protocol to START I, according to which Ukraine committed itself to accede to the Treaty on the Nonproliferation of Nuclear Weapons in the shortest possible time as a nonnuclear-weapon state. Kyiv's tack also does not match other commitments Ukraine has made in terms of removing nuclear weapons from its territory with the purpose of dismantling and disposing of them by the end of 1994.

On the contrary, the statement of the Russian Federation government points out, the Ukrainian side is taking practical steps to take the nuclear weapons deployed on its territory into its own hands. For instance, back in April 1992, the Rocket Army and Air Force with their combat units and subunits deployed on the territory of Ukraine were added to the Ukrainian Armed Forces. After this, a new staff structure was added to the Armed Forces: the Center for Administrative Control of the Strategic Forces Troops under the Ministry of Defense of Ukraine. It is in charge of all nuclear technology units deployed on Ukrainian territory.

Russia strongly advocates for all the nuclear weapons temporarily deployed on Ukraine's territory to be placed under the jurisdiction of the Russian Federation. As the Russian Federation government statement points out, Kyiv's position is against this, in violation of the commitments it has made, and could lead to extremely dangerous consequences. "The effectiveness of the nuclear weapon nonproliferation project is under threat."

As the Russian government points out, nuclear weapons cannot belong to a nonnuclear-weapon state: the security of nuclear weapons is indivisible. It can only be ensured by a system of consistently connected links, starting with unified command and control. Nuclear weapons also cannot and should not be a subject of political games.

Notes

Introduction

1. Polina Sinovets and Mariana Budjeryn complicate the notion that Ukraine saw the weapons in terms of deterrence, arguing that, at the outset, Ukrainian leaders "had little appreciation for the military-strategic function of a nuclear deterrent." See Polina Sinovets and Mariana Budjeryn, "Interpreting the Bomb: Ownership and Deterrence in Ukraine's Nuclear Discourse," NPIHP Working paper #12 (Woodrow Wilson International Center for Scholars, December 13, 2017), https://www.wilsoncenter.org/publication/interpreting-the-bomb-ownership-and-deterrence-ukraines-nuclear-discourse.

2. John J. Mearsheimer, "The Case for a Ukrainian Nuclear Deterrent," *Foreign Affairs* 72, no. 3 (Summer 1993): 50–66.

3. Ibid., 50–51.

4. John J. Mearsheimer, "Why the Ukraine Crisis Is the West's Fault: The Liberal Delusions That Provoked Putin," *Foreign Affairs* 93, no. 5 (September/October 2014): 77–89.

5. Steven E. Miller, "The Case against a Ukrainian Nuclear Deterrent," *Foreign Affairs* 72, no. 3 (Summer 1993): 68. A similar argument was made by William H. Kincade, "Nuclear Weapons in Ukraine: Hollow Threat, Wasting Asset," *Arms Control Today* 23, no. 6 (August 1993): 13–18.

6. Paul D'Anieri, *Economic Interdependence in Ukrainian-Russian Relations* (Albany: SUNY Press, 1999).

7. Christopher A. Stevens, "Identity Politics and Nuclear Disarmament: The Case of Ukraine," *Nonproliferation Review* 15, no. 1, (March 2008): 43.

8. Steven Pifer, *The Eagle and the Trident: U.S.–Ukraine Relations in Turbulent Times* (Washington, DC: Brookings Institution, 2017). See also Steven Pifer, "The Trilateral Process: The United States, Ukraine, Russia and Nuclear Weapons," Arms Control Series (Brookings, May 2011), http://www.brookings.edu/~/media/research/files/papers/2011/5/trilateral%20process%20pifer/05_trilateral_process_pifer.pdf.

9. Pifer, *Eagle and the Trident*, 72.

10. Ibid., 76.

11. Mariana Budjeryn, *Inheriting the Bomb: The Collapse of the USSR and the Nuclear Disarmament of Ukraine* (Baltimore: Johns Hopkins University Press, 2022). See also Budjeryn's "The Power of the NPT: International Norms and Nuclear Disarmament of Belarus, Kazakhstan and Ukraine, 1990–1994," Ph. D. Dissertation, Central European University, 2016. I am grateful to Dr. Budjeryn for pointing me to some of the sources discussed in this review.

12. Nadiya Kravets, "Domestic Sources of Ukraine's Foreign Policy: Examining Key Cases of Policy towards Russia, 1991–2009," Ph. D. Dissertation, Oxford University, 2012.

13. Robert D. Putnam, "Diplomacy and Domestic Politics: The Logic of Two-Level Games, *International Organization*, Vol. 42, No. 3. (Summer, 1988): 427–460.

14. Sherman W. Garnett, *Keystone in the Arch: Ukraine in the Emerging Security Environment of Central and Eastern Europe* (Washington, DC: Carnegie Endowment for International Peace, 1997).

15. Zbigniew Brzezinski, "The Premature Partnership," *Foreign Affairs* 73, no. 2 (March–April 1994): 80.

16. James D. Fearon, "Counterfactuals and Hypothesis Testing in Political Science," *World Politics* 43, no. 2 (January 1991): 169. See also Jack S. Levy, "Counterfactuals, Causal Inference, and Historical Analysis," *Security Studies* 24:3 (2015): 378–402.

17. Robert Powell, "War as a Commitment Problem," *International Organization* 60, no. 1 (Winter 2006): 169–203.

Author's Note

1. Ukraine's reaction to Russia's seizure of Georgian territories in August 2008 was demonstrative. President Victor Yushchenko made an official statement immediately, in which he condemned Russia's annexation of Georgian territories and called these events the first war within the borders of the Commonwealth of Independent States (CIS). On 2 September, the Parliament did not support a single one of the eight draft resolutions that would call these events what they really were. I registered a draft declaration, entitled "Concerning Russia's Military

Aggression Against Georgia." Only 66 MPs voted in the affirmative (out of 226 necessary for it to pass). A statement with a milder title, "Concerning Russian-Georgian Conflict," was proposed by V'iacheslav Kyrylenko. It gained affirmative 72 votes. One of the regional newspapers published the former Speaker Volodymyr Lytvyn's position, with the headline: "Lytvyn Accuses Yushchenko of a Provocative Stance on the Russian-Georgian Conflict." In it, Lytvyn "considers the president's position to be perilous for the state's internal affairs, and for Ukraine's international relations. [...] Judging by Yushchenko's statement in Tbilisi, he is prepared to drag Ukraine into a conflict with Russia" ("Ukrains'kyi politykum–Lytvyn zvynuvachuie Iushchenka u provokatsiinii pozytsiï u rosiisko-hruzynskomu konflikti," *Odes'ki visti*, 6 September 2008, https://izvestiya.odessa.ua/ru/2008/09/06/ukrainskiy-politikum-litvin-obvinyaet-yushchenko-v-provokacionnoy-pozicii-v-rossiyskoy). Within a week, my interview on the subject, with an alternative view on the situation, was published in the *Halychyna* newspaper ("Movchazna zhoda z vyznanniam Pivdennoi Osetiï ta Abkhaziï zahrozhuie tretioiu svitovoiu viinoiu," *Halychyna*, 13 September 2008, http://www.galychyna.if.ua/publication/policy/jurii-kostenko-movchazna-zgoda-z-viznannjam-pivdennoji-os/).

2. In 2012, a mere 19 percent of Ukrainians supported the idea of Ukraine's joining NATO, while 41 percent supported the idea of "creating a unified state composed of Ukraine, Russia, and Belarus" ("Dynamika ideolohichnykh markeriv," *Rating Group Ukraine*, 3–4 December 2012, https://bit.ly/3YP7nfk). The same report noted decreasing support for NATO since 2011, when it was at 24–26 percent (Ibid). Western regions expressed the greatest level of support for Ukraine's NATO membership, at nearly half the population (Ibid). In October of 2013, on the eve of the Revolution of Dignity, public support for joining NATO was at 20 percent in Ukraine, while in April 2014–38 percent, and in September 2014–43 percent ("Vse bilshe ukraïntsiv khochut v Ievropu, a ne v Mytnyi soiuz" [Increasingly More Ukrainians Want to Join Europe, and not the Customs Union], *Teksty.org.ua*, 17 October 2013, https://bit.ly/3k1rvMy; "Dynamika suspilno-politychnykh pohliadiv v Ukraïni" [The Dynamics of Sociopolitical Viewpoints in Ukraine], *Rating Group Ukraine*, 7–21 September 2015, http://ratinggroup.ua/files/ratinggroup/reg_files/survey_of_ukrainian_public_opinion_september_7–21_2015_ua_0001.pdf). In November 2014, over half (51%) of the population expressed support for joining NATO, while a quarter (25%) reported to be against it ("Otsinka sytuatsiï na Skhodi. Zovnishniopolitychni oriientatsiï naselennia," *Rating Group Ukraine*, November 2014, https://bit.ly/3Ii6GVB). By 2016, 62 percent of the population reported they would take part in a national referendum, with 72 percent of those, voting in favor of NATO membership ("Ukraïna 2016–2017: Oznaky prohresu ta symptom rozcharuvannia [analitychni otsinky]," *Razumkov Centre*, 93, http://razumkov.org.ua/uploads/article/2016–2017_Pidsumky.pdf). Surveys after 2014 did include populations in Crimea, Donetsk, and Luhansk oblasts.

3. The Orhanizatsiia Ukraïns'kykh Natsionalistiv (Organization of Ukrainian Nationalists–OUN) and the Ukraïns'ka Povstans'ka Armiia (Ukrainian Insurgent Army–UPA) were two organizations aiming to re-establish a Ukrainian independent state. The OUN was established in 1929, while the UPA served as its armed wing, a military-political establishment, between 1942 and 1952. After World War II, UPA fighters conducted partisan warfare against the Soviet regime, fighting against the USSR's special forces. The names of these organizations were removed from official Soviet history. Affiliation with OUN and UPA, and even mentioning them, was punishable by exile, imprisonment, or execution. As a result of Soviet propaganda, the majority of the Ukrainian population held negative attitudes towards OUN and UPA even after twenty years of independence.

4. The Declaration of State Sovereignty of Ukraine was the first document of the Ukrainian SSR that outlined the Ukrainian state's development strategy. The document envisioned Ukraine's independence in political, economic, and humitarian spheres, as well as all the attributes of statehood: its own borders, citizenship, national statehood, the development of its own military, currency, and independent foreign policy. Contrary to Article 6 of the Constitution of the USSR (which defined the Communist Party of the Soviet Union [CPSU] as the "nucleus of the political system," underlying USSR's totalitarian, one-party system), the Declaration proclaimed a multiparty system. At the same time, the document contained some fundamental contradictions. For example, despite using the term "Ukraine" in its title, it went on using the term "Ukrainian SSR" in the text of the document. Another was the issue of citizenship: to those who wished, the state guaranteed preservation of the USSR citizenship. The Declaration was adopted by parliamentary vote on 16 July 1990, with the First Deputy Chairman of the Parliament, Ivan Pliushch, presiding. While it was a revolutionary document for its time, the Declaration did not declare Ukraine's independence (*Parliament of Ukrainian SSR*, "Declaration of State Sovereignty of Ukraine," 16 July 1990, no. 55-XII, http://zakon1.rada.gov.ua/laws/show/55–12). Ukraine's independence was adopted a year later, on 24 August 1991.

5. People's Council, an opposition minority of the Parliament's first convocation, was established on 8 June 1990. At first, the People's Council contained 80 MPs, most of whom were representatives of the Narodnyi Rukh (People's Movement of Ukraine); later its membership grew to 130 MPs, including a number of former communists who left the Communist Party of Ukraine (CPU) after the collapse of the USSR. It was the intellectual center for the development of Ukraine's first legistlative documents, including the Declaration of State Soveignty of Ukraine, Declaration of Independence of Ukraine, and a set of laws outlining the newly independent state's political and economics reforms.

6. Between 1992 and 1995, Ukraine experienced a period of hyperinflation. In 1992, there was a 2,100 percent annual inflation, in 1993–10,256 percent, in 1994–501 percent, and in 1995–281,7 percent. In 1992, the highest currency

denomination was 100 *kupon-karbovanets*, while, in 1995, the highest denomination was 1 million *kupon-karbovanets*. In the span of just one month, in January 1992, inflation constituted 385,2 percent ("Indeks infliatsii" (Inflation Index), *Bankovskiye novosti Ukrainy*, http://banknews.com.ua/spravochnaya/indeks-inflyacii).

7. "Treaty on the Non-Proliferation of Nuclear Weapons," entered into force 5 March 1970, *United Nations Office for Disarmament Affairs*, Treaties Database, http://disarmament. un.org/treaties/t/npt.

8. "Modernization of Nuclear Weapons Continues; Number of Peacekeepers Declines: New SIPRI Yearbook out Now," *Stockhold International Peace Research Institute*, 18 June 2018, https://www.sipri.org/media/press-release/2018/modernization-nuclear-weapons-continues-number-peacekeepers-declines-new-sipri-yearbook-out-now.

Chapter 1

1. The People's Movement of Ukraine for Reconstruction (Narodnyi Rukh Ukraïny za perebudovu or Rukh) led the political opposition against Soviet totalitarianism and, as a civil-political association, advocated for Ukraine's independence. Founded by numerous pro-democratic groups, it united people of various political beliefs: from liberal communists to those who professed nationalist beliefs. It was the domain of dissidents, writers, and scholars. Rukh was formally established during the constituent congress of 8–10 September 1989, electing Ivan Drach chairman. During its second congress, in October of 1990, the organization's charter introduced a new provision concerning Rukh's chief goal–the achievement of Ukraine's independence–and the words "for Reconstruction" were then removed from the name. By then, two parties began coalescing within Rukh: the Ukrainian Republican Party and the Democratic Party of Ukraine. Subsequently, a faction of its leaders, including Ivan Drach, Dmytro Pavlychko, and Mykhailo Horyn´, advocated for the preservation of Rukh as an association of democratic organizations and parties, whose primary task would be to create a political base for the development of a new Ukrainian state. The other faction, chaired by V'iacheslav Chornovil, believed that the new state would still be controlled by the old communist nomenclature, thus, that real reforms would not be possible until the nomenclature was removed from power. This faction called for the transformation of Rukh into a political opposition party. In 1993, Rukh became a political party. Chornovil was elected its sole leader. Later, a group of Rukh members left the party, founding a civic organization, the All-National Movement of Ukraine, as well as a number of opposition parties.

2. Parliament of the Ukrainian SSR, Session 63, transcript of the plenary session, 12 July 1990, http://iportal.rada.gov.ua/meeting/stenogr/show/4406.html.

3. Parliament of the Ukrainian SSR, Session 66, transcript of the plenary session, 13 July 1990, http://static.rada.gov.ua/zakon/skl1/BUL11/130790_66.htm.

4. START I (Strategic Arms Reduction Treaty) was the USSR-US Treaty on the reduction of strategic offensive arms. The treaty obligated the USSR to reducing its carriers from 2,500 to 1,600 units, and its warheads from 10,271 to 6,000 units. The parties stipulated that, within the seven-year period from the date of the Treaty's coming into effect, and in the future, the total weight of nuclear warheads would not exceed 3,600 tons.

5. Parliament of Ukraine, "Declaration of the Parliament Regarding Ukraine's Nonnuclear Status," adopted on 24 November, 1991, no. 1697-XII, no. 51:741, http://zakon.rada.gov.ua/laws/show/1697–12.

6. Parliamentary Consultative-Expert Council, Report of July 1996, 2. Author's personal archive.

7. The August Putsch was a Soviet coup d'état of 18–22 August 1991, carried out by the State Committee on the State of Emergency (GKChP). The Committee was headed by USSR's vice president Gennadii Ianaiev, Defense Minister Dmitrii Iazov, KGB Chairman Vladimir Kriuchkov, and Internal Affairs Minister Boris Pugo (who later committed suicide). GKChP attempted to remove the President and General Secretary of the USSR, Mikhail Gorbachev, from power before he were to sign the new Union Treaty decentralizing USSR's central power, on 20 August 1991. A state of emergency was declared throughout the USSR; on 19 August, the Taman Division entered Moscow. On the morning of 19 August, the Secretary of Defense Valentin Varennikov arrived in Kyiv demanding the GKChP decisions be carried out. The Secretariat of the Central Committee of the Communist Party of Ukraine (CPU) distributed a circular to local authorities. The document insisted that the measures taken by the country's leadership were in the interests of the working class and were fully in line with the position of the Communist Party of Ukraine. However, the republics did not support the Putsch. This accelerated the collapse of the Soviet Union and the republics' proclamations of independence.

8. Georgii Arbatov, "The Nature of Key Threats to State Security in the Transition Period," 1991, MS for the *Institute for US and Canadian Studies*, p. 10. Author's personal archive.

9. Ibid., 9.

10. Ibid, 10. The total amount of USSR's debt was nearly 60 billion dollars. See N. Danilov, "My ne nastol'ko bogaty, chtoby torgovat' boegolovkami," *Komsomol'skaia pravda*, February 13, 1992, 3.

11. Arbatov, "Nature of Key Threats," 10–11.

12. By the time of its collapse, the USSR had signed several arms reduction treaties. In addition to the Intermediate-Range Nuclear Forces Treaty, the Anti-Ballistic Missile Treaty, START, and the Treaty on Open Skies, the USSR had signed the Treaty on Conventional Armed Forces in Europe. The essence of the latter was that each

participating state agreed to limits in the following five categories of conventional forces: battle tanks, armored combat vehicles, artillery, combat aircraft, and attack helicopters. The Ukrainian parliament ratified this treaty, as well as the agreement on the principles and procedure of its implementation, in 1992. In accordance with the treaty, Ukraine could have no more than 4,080 tanks, 5,050 armored vehicles, 4,040 artillery systems, 1,080 planes, and 330 attack helicopters on its territory after the arms reduction.

13. Arbatov, "Nature of Key Threats," 9.

14. For Russia, a demise of the military-industrial complex meant economic collapse. Arbatov wrote: "One of the key problems that can lead to a deep societal crisis is the aforementioned increase in the disintegration of MIC structures. Under the circumstances of reduction and even termination of financing to certain programs and manufacturers, some units that are unique in their professional qualities and qualifications are already falling out of existence, with hundreds of thousands of people in danger of becoming unemployed (the MIC system employs 14,400,000)." The MIC consumed a lion's share of the goods produced in the USSR (60% of ferrous metals, and all non-ferrous metals). Therefore, Russia attempted to monopolize Western funding aimed at the elimination of the former USSR's nuclear weapons in order to prevent the collapse of its MIC. Arbatov, "Nature of Key Threats," 5.

15. John J. Mearsheimer, "The Case for a Ukrainian Nuclear Deterrent," *Foreign Affairs* 72, no. 3 (1993): 60.

16. Anne Applebaum, "The Comfort of Missiles," *Spectator*, June 26, 1993, 12–13, http://archive.spectator.co.uk/article/26th-june-1993/12/the-comfort-of-missiles.

17. President George H. W. Bush's opposition to Ukrainian independence irked Ukrainian nationalists and Ukraine's supporters in the US. *New York Times* columnist William Safire dubbed it the "Chicken Kiev" speech.

18. William C. Potter, "Ukraine's Nuclear Trigger," *New York Times*, November 10, 1992, A23.

19. US paid particular attention to Ukraine because Washington was aware of the extent of Ukraine's development in science and production. In a smoke screen created by a massive campaign in the world press aimed at discrediting Ukraine, the country was characterized as a barbarian with a nuclear mace, intellectually inferior to its Russian brother. However, as the *Financial Times* stated in January 1993, "in view of Western military specialists, Ukraine is the only non-Russian republic with the capacity to become a nuclear power" (Chrystia Freeland, "Ukraine Puts Brakes on Start: Kiev Might Obstruct Nuclear Cuts," *Financial Times*, January 8, 1993, 12). At roughly the same time, the *Washington Post* conveyed the Central Intelligence Agency (CIA) leadership's assessment of Ukraine's intellectual potential: "CIA director R. James Woolsey testified before Congress in February that 'a substantial number of former Soviet scientists involved in weapons of mass destruction research and development are of Ukrainian origin'" (Steve Coll and R. Jeffrey Smith, "Ukraine Could Seize Control Over Nuclear Arms," *Washington Post*, June 3, 1993, A01). This

circumstance, ostensibly, had a substantial impact on the CIA's predictions that Ukraine could potentially keep its nuclear weapons: "Moreover, Ukrainians were well represented in the senior officer ranks of the Soviet Strategic Rocket Forces, which managed most of the Soviet nuclear arsenal" (ibid).

20. Mearsheimer, "Case for a Ukrainian Nuclear Deterrent," 60.

21. Jonathan Eyal, "Still Hugged by the Russian Bear: Jonathan Eyal Accuses the West of Betraying the Former Soviet Republics," *Independent*, May 11, 1993, https://www.independent.co.uk/voices/still-hugged-by-the-russian-bear-jonathan-eyal-accuses-the-west-of-betraying-the-former-soviet-2322228.html.

22. Ibid.

23. Virginia I. Foran, "The New Nuclear World: The Nuclear Safety Catch: What Should Follow START? Ukrainian Holdout: The Real Problem with the Treaty," *Washington Post*, January 3, 1993, C03.

24. R. Jeffrey Smith, "U. S. Fears Ukrainian-Russian Clash; Steps Weighed to Head Off Confrontation Over Nuclear Weapons," *Washington Post*, June 6, 1993, A32.

25. Eyal, "Still Hugged by the Russian Bear."

26. "Pozytsiia Ukraïny chitka i iasna: my rozzbroiuiemosia, iakshcho..." *Ukraïns'ka hazeta*, no. 20, December 2–15, 1993, 7.

27. "Pro bez'iadernyi status Ukraïny," adopted on 24 October 1991, published in *VVR* (*Vidomosti Verkhovoï Rady Ukraïny*), no. 51 (1991): 742, http://zakon.rada.gov.ua/laws/show/1697–12.

28. Ibid.

29. The Parliamentary Working Group for Addressing the Issues Related to Refining Ukraine's System of National Security was established on 23 December 1991 by the decree of the chairman of the parliament on my initiative. It consisted of 19 MPs, representing different political views. The Group developed a draft national security policy; however, it was blocked from parliamentary review.

30. William J. Broad, "Nuclear Accords Bring New Fears on Arms Disposal," *New York Times*, July 6, 1992, 1, https://www.nytimes.com/1992/07/06/world/nuclear-accords-bring-new-fears-on-arms-disposal.html.

31. Douglas L. Clarke, "The Impact of START-2 on Russian Strategic Forces," *Radio Free Europe/RL Research Report* 2, no. 8, February 19, 1993, 68.

32. On 8 December 1991, Belarus, the Russian Federation, and Ukraine signed the Agreement Establishing the Commonwealth of Independent States (CIS) in Minsk, Belarus, "as founding states of the Union of Soviet Socialist Republics, having signed the 1922 Treaty on the Creation of the USSR." The document thus confirmed the dissolution of the USSR: "After this Agreement comes into effect, third-party laws can no longer be applied on the territories of the signatory states, including those of the former USSR. [...] The functions of state organs of the former USSR on the territories of the member states of the Commonwealth are hereby terminated." Article 5 of the agreement affirmed that "the High Contracting Parties recognize and respect the territorial integrity of each other and the inviolability of the existing borders within

the Commonwealth." For Ukraine, the Agreement came into force on 10 December 1991, after it was ratified by the Parliament of Ukraine. The Agreement Establishing the CIS regulated four aspects important for Russia: 1) USSR's international obligations; 2) o nuclear disarmament; 3) former republics' neutral status; and 4) the joint command of the Strategic Armed Forces (Parliament of Ukraine, "Agreement Establishing the Commonwealth of Independent States," ratified with reservations, no. 1958-XII, signed by Ukraine on 8 December 8, 1991, effective as of 10 December 1991, http://zakon.rada.gov.ua/laws/show/997_077).

33. "O sovmestnykh merakh v otnoshenii iadernogo oruzhiia Sodruzhestvo Nezavisimykh Gosudarstv Soglashenie ot 21 dekabria 1991 goda." Author's personal archive. Full text in Russian is available at: https://zakon.uchet.kz/rus/docs/H910000006_.

34. Ibid.

35. "Treaty Between Member States of the Commonwealth of Independent States on Strategic Forces," signed on behalf of Ukraine on 30 December 1991, not ratified. Author's personal archive.

36. Ibid.

37. The Working Group for Addressing Issues Related to Ratification of START and to Ukraine's Attaining a Nonnuclear Status was established on 20 July 1992, via parliamentary Speaker Ivan Pliushch's decree. It initially consisted of thirteen people. On 5 January 1993, the Group's status was elevated by the parliamentary Executive Committee's mandate, expanding it to 23 members. Its direct objectives included preparations to ratification of START I, the Lisbon Protocol, and the NPT. The Group's activities resulted in subsequent parliamentary decisions through which Ukraine retained its right of ownership to nuclear weapons, and its right to compensation for nuclear material. It collected unique data on nuclear weapons inherited by Ukraine. Based on these data and the analysis of nuclear agreements' texts, the Group drafted the Resolution of the Parliament on Ratification of START I and the Lisbon Protocol with Reservations, adopted by Parliament on 18 November 1993. Due to the Working Group's definitive position, the ratifications of START and of the NPT were decoupled, preventing Russia's scenario of establishing control over strategic arms located in Ukraine. The Group existed until May 1994.

38. Vladimir Kiselev, "Kogo obogatit obogashchennyi uran?" *Moskovskie novosti* 43, October 25, 1992, 21.

39. Ibid.

40. Ibid.

41. Ibid.

42. Ibid.

43. Borys Tarasiuk, deputy minister, head of Ukraine's National Committee on Disarmament, to Yuri Kostenko, minister of environment and natural resources, Ukrainian Translation of US Embassy Memorandum, UKOR/26–243, 29 March 1993, 1. Author's personal archive.

44. Roman Popadiuk, consul general of the United States of America, to Anatolii Zlenko, minister of foreign affairs, Ukrainian Translation of US Embassy Memorandum, 23 April 1993, 2–3. Author's personal archive.

45. Conclusions from the meeting of experts in Kharkiv, discussing options proposed by Russia for Ukraine's nuclear disarmament, 23–25 February 1993. Author's personal archive.

46. At the beginning of the 1990s, the KGB had significant influence on the USSR's governmental structures, enabling it to preserve its monopoly on national security. In 1990, a group was established within the Committee of the Supreme Soviet of the USSR with the purpose of developing a new concept of national security for the USSR; it convened regularly, with academician Iurii Ryzhov at the helm. During the group's discussion of the draft law on KGB structures, the head of USSR's KGB, Vladimir Kriuchkov, asserted that any new concept that was not exclusively centered on force was essentially a waste of time. The incumbent deputy speaker, Anatolii Luk'ianov, accusing the group of engaging in subversive activity, disbanded the group by official decree; subsequently, members of the group were persecuted. Nevertheless, the KGB continued to work on national security policy and, in 1991, the president of the USSR decreed that it resume its activity.

47. *VVR*, no. 49 (September 20, 1991): 689, https://zakon.rada.gov.ua/laws/show/1581–12.

48. The Defense Council of Ukraine (DCU) was established by a parliamentary Resolution of 11 October 1991 as the highest state body providing collective leadership on issues of defense and security. Among its goals were "development of defense and state security strategies, comprehensive scientific assessment of military threats, defining a stance on contemporary wars, and effective control over implementation of the country's tasks and institutions dedicated to maintaining sufficient levels of Ukraine's defense capabilities" ("Resolution on the Defense Council of Ukraine," Appendix N. 2 to the Resolution of the Parliament of Ukraine of October 11, 1991 no. 1658-XII, I.1, http://zakon.rada.gov.ua/laws/show/1658–12). On 3 July 1992, President Kravchuk reformed the DCU into the National Security Council of Ukraine (NSCU), thus demoting its status to a "consultative and advisory body under the president of Ukraine, whose main task is to prepare proposals and draft bills for the president of Ukraine on implementation of policy on safeguarding the national interests and providing national security of Ukraine" ("Provisional Conditions of the National Security Council of Ukraine," adopted by the decree of the president of Ukraine on 3 July 1992, no. 117/92-rp, http://zakon.rada.gov.ua/laws/show/117/92-%D1%80%D0%BF). The latter Council existed until 30 August 1996, when President Kuchma established the National Security and Defense Council of Ukraine (NSDCU) in its stead after the adoption of Ukraine's Constitution as a permanent constitutional body on coordination and control over the activities of executive authorities in the field of national security and defense.

49. Yuri Kostenko, head of the Working Group, to Leonid Kravchuk, president of Ukraine, 10 January 1992, 3. Author's personal archive.

50. "Decree of the President of Ukraine no. 41," January 15, 1992 [not for publication], lost effect with the issuing of Presidential Decree no. 772/96, August 30, 1996. Author's personal archive.

51. "Pro zatverdzhennia skladu Konsul'tatyvno-ekspertnoï hrupy po pidhotovtsi kontseptsiï natsional'noï bezpeky Ukraïny," document 31/92-rp, adopted February 17, 1992, http://zakon.rada.gov.ua/laws/show/en/31/92-%D1%80%D0%BF.

52. Yuri Kostenko, "Natsional'na bezpeka–zasib zberezhennia ukraïns'koho narodu," *Rozbudova derzhavy*, no. 1, (June 1992): 20–23.

53. In 1992, inflation was intensifying. Ukraine did not have its own currency, and the transitional monetary unit, the coupon-karbovanets, rapidly depreciated: "The disposable coupons had been in use by the population since November 1990. [...] These were simple pieces of paper with scattered rectangles, with 1 krb, 2 krb, and so on, printed on them. They had no watermarks, not to mention other levels of protection. The coupons were only a problem for buyers in the initial weeks. Soon after, people learned to counterfeit them on Xerox machines. There was a general sense that there was no point in issuing real money since the hryvnia was about to be introduced. [...] If on 12 November 1992 the National Bank of Ukraine exchanged 403 coupon-karbovantsi for one dollar, on 1 February 1995 one dollar was equivalent to 116,900 coupon-karbovantsi, that is, the karbovanets had fallen 290-fold" (Alla Kovtun, "Misiia vykonana, 15 rokiv tomu Ukraïna zavershyla vykhid iz rubliovoï zony," *Ekonomichna pravda*, November 12, 2007, http://www.epravda.com.ua/columns/2007/11/12/144068/).

54. At the end of 1991, I had a long conversation with Leonid Kuchma, then director of Pivdenmash. Since I had taken part in the development of the national security policy, he suggested I visit the plant to familiarize myself with problems inherent in the production of strategic nuclear weapons. It was during this informal communication that I gained a fundamental understanding of the complexity of problems we faced with strategic nuclear weapons. I visited Pivdenmash for the second time in 1992. Kuchma was already prime minister, and I was accompanied by Leonid Derkach, who later became the head of the SBU, but at the time was responsible for the security of Pivdenmash. This is when I learned a great deal about the technological control over the implementation of START. I made the key discovery that the Treaty did not require the destruction of the warhead itself, but only of its delivery mechanism, as well as requiring control over manufacturing new delivery systems. Later, information received at Pivdenmash would allow me to avoid taking a superficial approach to the problem of nuclear disarmament.

55. When the national security policy was being developed, the borders with Russia were not delimited; this allowed for the Tuzla Island conflict to happen in 2003, as well as for Russia's military aggression in 2014. Furthermore, Russia's Black Sea Fleet military bases were stationed in Crimea, which made the 2014 Crimea annexation possible. Russia led frequent energy and economic wars with Ukraine, demanding changes in its political course, blackmailing Ukraine by imminent termination of gas or oil supplies as well as arbitrarily raising their prices, all of which devastated Ukraine's national budget. However, the national security policy, which was still in force in 2014, listed none of these threats.

56. Under a 1997 agreement, Russia paid less than $100 million annually for the use of military bases in Crimea up until 2007. The 4,590 units of Ukrainian property that were used, but never paid for, by Russia, included airports, military bases, houses, lighthouses, and radio navigation stations. The 1997 land lease price was tied to the $26 per 1,000 cubic meters price of gas that Ukraine was purchasing from Russia. However, the government of Yulia Tymoshenko, signing a new gas deal in January 2009 and agreeing to pay $450 per 1,000 cubic meters of gas, did not link it with the revision of the land lease in Crimea. Since 2008, Ukraine had possessed the right to establish payment prices according to Ukrainian laws and using international precedents. At international price levels, Russia should have paid Ukraine from between $10 and $20 billion annually for leasing the Black Sea Fleet bases.

57. Yuri Kostenko, MP of the Parliament of Ukraine, interview with Zbigniew Brzezinski, former US National Security Adviseser, transcription in Ukrainian provided by the Ukrainian Embassy in the US, 4 February 1992, 1–3. Author's personal archive.

58. The production association Southern Machine-Building Plant (Pivdenmash, aka Yuzhmash) is the leading Ukrainian manufacturer of rocket and space technology, as well as technologies designed for defense and scientific purposes and national economy. In 1954, Pivdenmash commenced its production of strategic ballistic missiles. In the 1960s, in addition to ballistic missiles, it also began manufacturing a number of unique space carriers and spacecraft. It was the producer of unique strategic carriers SS-18 Satan and SS-24 Scalpel, carrying ten independently targetable nuclear warheads, which had no analogy worldwide. It carried out a serial production of satellites and space systems Zenit, a space rocket system able to launch payloads weighing up to 12 tons to the Earth's orbit. As a member of Sea Launch, a multinational spacecraft launch program, Pivdenmash participated in 28 successful Zenit-3 SI missile launches loaded with foreign commercial satellites.

59. Ivan Vyshnevs'kyi and Arkadii Trofymenko, "Report," Academy of Sciences, National Commission on Scientific Relations with the IAEA, April 6, 1992, 1. Author's personal archive.

60. Ibid.

61. Russian press frequently used the image of Chernobyl in its attempts to persuade the public of how dangerous nuclear weapons were when they were under Ukraine's jurisdiction: "'No army in the world has experienced such disarray since the appearance of nuclear weapons,' said one high-ranking officer who wished to remain anonymous. 'The second Chernobyl is ripe in Ukraine's launch facilities, and people must know about it'" ("Na raketnykh shakhtakh Ukrainy zreet vtoroi Chernobyl,"

Izvestiia, February 16, 1993, 4). Referring to the anonymous officers and experts, Russian propaganda insisted that nuclear weapons were not yet Ukrainian property and that the world would be safer if they never were: "The command and financing of rocket armies should be unified. [They should] come from a single center. The responsibility for the safety of the nuclear missile forces also must be in the same hands" (ibid.).

62. On 5 April 1992, President Kravchuk signed the Decree on Urgent Measures for the Construction of the Armed Forces of Ukraine, which subordinated to the Ukrainian Ministry of Defense "all military formations stationed on the territory of Ukraine that are not specified in Article I of the Resolution of the President of Ukraine on Armed Forces of Ukraine dated 12 December 1991." Unspecified in the latter Resolution were only the Strategic Nuclear Forces. The Defense Minister of Ukraine was charged with "establishing direct (for Strategic Forces, administrative) control of all troops deployed on the territory of Ukraine, ensuring their continuing combat readiness and preservation of military discipline." Rather harsh and uncharacteristic of Kravchuk was the part in which, among other reasons provided for Ukraine's taking control over the strategic nuclear forces located on its territory, was the following: "in connection with the intervention of Russian Federation's leadership and the Chief Command of the United Armed Forces of the Commonwealth into the Ukraine's internal affairs, and the consequent exacerbation of the sociopolitical situation in the armed forces stationed on the territory of Ukraine but not subordinated to the Armed Forces of Ukraine, and the absence of proper military command" ("Pro nevidkladni zakhody po budivnytstvu Zbroinykh Syl Ukraïny," no. 209, April 5, 1992, https://zakon.rada.gov.ua/laws/show/209/92).

63. *VVR*, no. 29 (1992): 405, http://zakon4.rada.gov.ua/laws/show/2264-12. An English translation of the document is available in Appendix D.

64. Ibid.

65. Ibid.

66. "Pozytsiia Ukraïny chitka i iasna," 7.

67. Yuri Kostenko, head of the Working Group, to Leonid Kravchuk, president of Ukraine, 10 January 1992, p. 3. Author's personal archive.

68. The absence of any guiding documents that explain (as well as the lack of any explaining as such) Ukraine's position to foreign representatives in the process of nuclear disarmament was noted by Ukrainian diplomats not only at the onset of MFA's establishment, but also in mid-1993. Counselor Andrii Veselovs'kyi's analytical memo of 18 May attests to this lack of information: "Recommendations. To continue and to strengthen the coverage of Ukraine's position. As often as possible, to send relevant guiding documents to MFA's Office for the Control of Armament and Disarmament, demanding that embassies distribute them. To compile the latest of Ukraine's propositions in English and French, and print and distribute them. To organize visits to nuclear military units for the Western military attaché in Kyiv with explanatory work" (Andrii Veselovs'kyi, Ukraine's Consulate General in Canada, "Nuclear Weapons in Ukraine:

Canadian Position/Political Letter," May 18, 1993, 3–4. Author's personal archive).

69. Ukraine's tactical nuclear weapons comprised approximately 3,500 warheads. The exact figure was unknown to us and was in the sole possession of the general staff of the Armed Forces of the USSR. *New York Times* reported that Ukraine had 2,800 warheads (Potter, "Ukraine's Nuclear Trigger," A23). Anne Applebaum wrote in the *Spectator* on 26 June 1993 that Ukraine "had already sent at least 4,200 tactical weapons to Russia for destruction" ("The Comfort of Missiles," *Spectator*, 12–13). In his article, the former head of the Russian nuclear negotiations delegation to Ukraine, diplomat Iurii Dubinin, cited a figure of nearly 3,000 warheads (Iurii Dubinin, "Iadernyi dreif Ukrainy," *Rossiia v global'noi politike*, April 7, 2004, http://www.globalaffairs.ru/number/n_2860). Ukraine's tactical nuclear weapons were deployed in the Black Sea Fleet, the army, the air force and the air defense forces. According to estimates described in detail in the next chapter, the tactical nuclear weapons contained 30–40 tons of HEU (over 95 percent), and 30–55 tons of plutonium (over 95 percent); the total estimated value of these ranged between $18 and $59 billion.

70. Ukraine's Ministry of Defense, Report Memorandum to the Working Group, "Who Made the Decision and Organized the Urgent Export of Tactical Nuclear Weapons from Ukraine?" 1992, 1. Author's personal archive.

71. Valerii Izmalkov, "Iaderna raketa–ne kam'iana sokyra," *Holos Ukraïny*, December 12, 1992, 7.

72. "'START'–tse nadto ser'iozno, shchob pospishaty," *Holos Ukraïny*, January 11, 1993, 1, 3.

73. The Lisbon Protocol, like START I, came into force only after all of its parties exchanged ratification documents. This event took place on 5 December 1994. In other words, the nuclear arms liquidation process would have to commence in 1994 and continue for at least seven years; that is, until 2001 or 2002. However, in practice, the process unfolded in contradiction to these documents. Washington's new tactics and immense pressure from Russia produced swift results on nuclear weapons stationed in Belarus and Kazakhstan: already at the end of 1993, both of these countries ratified START I as well as acceded to the NPT as nonnuclear-weapon states. Their nuclear weapons essentially ended up under Russia's jurisdiction. Moreover, even after the signing of the Budapest Memorandum and the exchange of ratification documents on 5 December 1994, Ukraine had until 2002 to eliminate its nuclear weapons. Nuclear weapons' shelf life allowed the process to take place until as late as 2005. Yet, all of Ukraine's strategic nuclear weapons were shipped to Russia by 1 June 1996.

74. Vitalii Kriukov, "Ukraine's Nuclear Status: Legal and Political Issues," *Institute of Government and Law, Academy of Sciences of Ukraine*, 1992, 8. Author's personal archive.

75. Ibid, 8.

76. Ministry of Foreign Affairs, "Ukraine's Written Declaration in Connection with the Signing of the Protocol to the Agreement between USSR and USA Regarding the

Reduction and Limitation of Strategic Arms," May 23, 1992, 2. Author's personal archive.

77. Ibid., 3.

78. Andrei Kozyrev, Embassy of the Russian Federation, Lisbon, "Russia's Written Declaration in Connection to Signing of the Protocol to the Treaty on Strategic Arms on 23 May 1992 in Lisbon," Author's personal archive.

79. Freeland, "Ukraine Puts Brakes on Start," 12.

80. In June 1992, Ukraine suggested holding quadrilateral talks on the distribution of quotas under the Lisbon Protocol, as well as to sign a Memorandum regarding the distribution of limits and restrictions. Russia insisted on bilateral talks, even though the document was to be signed by four parties. Its plan entailed breaking each of the countries individually, which became all too evident during the state delegation negotiations in January 1993. The talks on the distribution of quotas never took place.

81. P. Chernenko and O. Odnokolenko, "Eshche do obsuzh-deniia voiennyie voprosy priobretaiut osobuiu ostrotu," *Krasnaia zvezda*, February 13, 1992, 1.

82. V. Smelov, "Gosudarstva sodruzhestva rasshiriaiut sviazi s vneshnim mirom," *Krasnaia zvezda*, February 21, 1992, 3.

83. Nikolai Burbygi, "Boris Yeltsin: U Rossii net kakoi-to osoboi, tainoi politiki v iadernykh voprosakh," *Izvestiia*, February 24, 1992, 1, 3.

84. Gennadii Charodeiev, "'Nichego ne sobiraius´ pros-it´ u amerikantsev,' zaiavil prezident Kazakhstana N. Nazarbaiev pered otletom v Vashington," *Izvestiia*, January 15, 1992, 4.

85. Ibid.

86. The shipment of USSR's nuclear arsenal to Russia began with Belarus and Kazakhstan, neither of which resisted the process: "However, as of late, thanks to Russia's efforts and under the international community's pressure (first and foremost, USA, who are generously financing the process) the nuclear arsenal of the former Soviet republics has been visibly reduced. First of all, we should note the dismantling and transfer to Russia of all ICBM warhead stations from Kazakhstan, which, in this manner, has lost all of its strategic nuclear potential. All property from two of Belarus's Rocket Army units and all weapons from one of Belarus's units were transferred to Russia, and now no more than 42 warheads on rocket carriers remain on the territory of Belarus" (Mark Shtein-berg, "Iadernoie oruzhiie–perspektivy nerasprostraneni-ia," *Dzerkalo tyzhnia*, June 23, 1995, https://zn.ua/POLITICS/yadernoe_oruzhie__perspektivy_neraspros-traneniya.html).

87. Collective Security Treaty (CST), or the Tashkent Pact, was signed by Armenia, Kazakhstan, Kyrgyzstan, Russia, Tajikistan and Uzbekistan on 15 May 1992 in Tashkent. Within one year, Azerbaijan, Georgia and Belarus had also joined the treaty. It came into effect on 20 April 1994. Moldova, Turkmenistan and Ukraine did not participate in the treaty. On 2 April 1999, Presidents of Armenia, Belarus, Kazakhstan, Russia and Tajikistan signed a protocol extending the term of the treaty for another five-year period, while Azerbaijan, Georgia and Uzbekistan refused and withdrew. On 14 May 2002, the

CST was reorganized into the Collective Security Treaty Organization (CSTO). Today it is an intergovernmen-tal military alliance, whose members include Russia, Armenia, Kazakhstan, Kyrgyzstan and Tajikistan. Russia's government is attempting to present CSTO as a real al-ternative to NATO and demands giving greater weight to it in the international relations systems. On 2 December 2004, the UN General Assembly adopted a resolution granting CSTO an observer status.

88. Stephen S. Rosenfeld, "Playing Nuclear Cop," *Washington Post*, March 26, 1993, A25.

Chapter 2

1. *VVR*, no. 39 (1992): 582, http://zakon.rada.gov.ua/laws/show/2562–12.

2. "Speech by the State President, Mr. F. W. De Klerk, to a Joint Session of Parliament, 24 March 1993: Nuclear Non-Proliferation Treaty," *Wilson Center, Digital Archive, International History Declassified*, https://digitalarchive.wilsoncenter.org/document/116789.pdf?v=08917bf1d2b-6de6de27700a6f6425862, 3–4.

3. Ibid., 5.

4. Ibid.

5. Yuri Kostenko, "Iaderna zbroia Ukraïny: blaho chy zlo? Polityko-pravovyi i ekonomichnyi analiz rozzbroien-nia," *Holos Ukraïny*, no. 164, August 29, 1992, 6; Yuri Kostenko, "Iaderna zbroia Ukraïny: blaho chy zlo? Polityko-pravovyi i ekonomichnyi analiz rozzbroiennia," *Holos Ukraïny*, no. 165, September 1, 1992, 6.

6. Ukraine's mass media provided little coverage of nuclear disarmament process: "Before Parliament's session deliberating on the ratification of START I and the Lisbon Protocol, as well as joining the NPT, the Ukrainian press, radio, and television paid almost no attention to the topic of nuclear weapons. There was no broad discussion of it whatsoever. There was a well-known and highly read article by MP Yuri Kostenko in the newspaper *Holos Ukraïny* that essentially indicated the creation of a group in the Ukrainian parliament that considers it necessary to retain nuclear weapons in Ukraine" (Eduard Lisitsyn, "Suspil´na dumka pro iadernu polityku Ukraïny: Ukraïna, SShA, Rosiis´ka Federatsiia," for the Institute of World Economy and International Relations, Ukrainian Acade-my of Sciences, 11 June 1993. Author's personal archive). Responding to my article, "Nuclear Weapons: Good or Evil?," Ukrainian journalists essentially retransmitted Russia's key messages, discounting Ukraine's need for se-curity guarantees: "Complete nuclear disarmament is our key to the European home. Thus, it is hardly appropriate to continuously bargain for more guarantees in exchange for missile liquidation. This should have been done

before the proclamation of Ukraine's nonnuclear status. Besides, the propositions of a 'nuclear component' in our system of national security are rather controversial. The thing is that strategic weapons do not provide such guarantees. On the contrary, their presence on the territory of the republic threatens the stability of the state. Therefore, we should not build our national security on phantom nuclear foundations (Oleh Strekal, "Iaderni potuhy," *Molod Ukrainy*, September 17, 1992, 2).

7. Videns'ka konventsiia pro ravonastupnytstvo derzhav shchodo dohovoriv no. 2608-XII, 17 September 1992, signed 23 August 1978, effective in Ukraine as of 6 November 1996, http://zakon.rada.gov.ua/laws/show/995_185; http://zakon.rada.gov.ua/laws/show/995_072.

8. The extent of problems inherent in handing the highly toxic missile fuel was explained by the MoD to the Working Group in the fall of 1992: "Liquid-propellant missiles are fueled with highly toxic substance, with its fuel component, unsymmetrical dimethylhydrazine or 'heptyl,' and the oxidizing agent, nitrogen tetroxide or 'amyl.' The threshold limit value (TLV) of these substances' vapors are as follows: heptyl–0,0001 mg/l, approaching the toxicity levels of chemical warfare agents; amyl–0,002 mg/l. One missile contains approximately 21 tons of heptyl, and 108 tons of amyl. The total reserves of fuel and oxidizer comprise nearly 3,000 tons and 14,000 tons, respectively" (The Ministry of Defense of Ukraine, Report prepared and submitted to the Working Group, Fall 1992. Author's personal archive).

9. Dubinin, "Iadernyi dreif Ukrainy."

10. The Treaty of Friendship, Cooperation and Mutual Assistance, commonly known as the Warsaw Pact, was a treaty between the USSR and the socialist countries of Central and Eastern Europe, signed on 14 May 1955 in Warsaw. It was established as a defense and political union to counterbalance NATO. Along with the USSR, its membership included seven other countries: Poland, Romania, Bulgaria, Czechoslovakia, Hungary, Albania (withdrew in 1968), and the German Democratic Republic (withdrew in 1990 after reunification of Germany). The Warsaw Pact was in force until 1 July 1991, when the remaining Pact countries officially terminated it in Prague.

11. It is exceedingly difficult to estimate a US dollar equivalent of this sum in Ukrainian karbovantsi (plural of *karbovanets*) at this time because of sweeping inflation rates of the period discussed.

12. Ministry of Defense of Ukraine, Analytical Report prepared and submitted to the Working Group, "General Conclusions on Scientific, Technical, Financial, and Economic Aspects of Nuclear Disarmament," 1–3. Author's personal archive.

13. Ibid., 1.

14. SS-18, SS-19, SS-24, SS-25 were NATO nomenclature for intercontinental ballistic missiles (ICBMs) in USSR's nuclear arsenal. Two missile types, the SS-18 and the SS-24, were considered the leading edge of modern technology in the early 1990s and were produced and serviced by Ukraine's Pivdenmash. The SS-19 and SS-25 were manufactured in Russia. The SS-18s, made by Pivdenmash and

called Satan by NATO, were liquid-propellant missiles carrying ten independently targetable warheads, each with a 300-kiloton yield; they were deployed on the territory of Russia and Kazakhstan. The Russia-manufactured SS-19 was a liquid-propellant, silo-based ballistic missile with a payload of six independently targetable warheads; it was named the Stiletto by NATO. The Pivdenmash-produced SS-24 (Scalpel in NATO nomenclature), was considered the most powerful missile in the early 1990s; this solid-propellant missile carried 10 individually-targetable warheads and could be both silo-based and rail-mobile; however, in Ukraine, the SS-24 missiles were deployed only in silos. The SS-25 Topol was a solid-propellant road-mobile missile carrying a single 550-kiloton warhead.

15. Aleksandr Konovalov, "Kiev poluchaet garantii," *Moskovskie novosti*, January 16–23, 1993, 13.

16. Serhii Tolstov, "Ukraine's Nuclear Dilemma," *The World Today* 49, no. 6 (June 1993): 104.

17. Mark Shteinberg, "Iadernoe oruzhie–perspektivy nerasprostraneniia," *Dzerkalo tyzhnia*, June 23, 1995, https://zn.ua/POLITICS/yadernoe_oruzhie__perspektivy_nerasprostraneniya.html.

18. William C. Potter, "Ukraine's Nuclear Trigger," *New York Times*, November 10, 1992, Op-Ed, A23.

19. The estimate of 30 bombers is low. It could be that this figure accounted for the number of Tu-160 bombers alone, exclusive of the Tu-95 MS bombers. Shteinberg, "Iadernoe oruzhie" gives an estimate of 43 long-range bombers (14 Tu-95 MSs and 29 supersonic Tu-160s), while expert Western sources give 44 bombers. See Joseph P. Harahan, *With Courage and Persistence: Eliminating and Securing Weapons of Mass Destruction with the Nunn-Lugar Cooperative Threat Reduction Programs* (n.p.: Defense Threat Reduction Agency, U. S. DoD, 2014), 11.

20. Shteinberg, "Iadernoe oruzhie."

21. Hearing before the Subcommittee on European Affairs of the Committee on Foreign Relations, United States Senate, One Hundred Third Congress, First Session, 24 June 1993, U. S. Government Printing Office, Washington, DC, 45–46.

22. Potter, "Ukraine's Nuclear Trigger."

23. Dubinin, "Iadernyi dreif Ukrainy"; Applebaum, "Comfort of Missiles."

24. Shteinberg, "Iadernoe oruzhie," Applebaum, "Comfort of Missiles"; Dubinin, "Iadernyi dreif Ukrainy"; Potter, "Ukraine's Nuclear Trigger." For Western estimates of Ukraine's nuclear arsenal, see Harahan, *With Courage and Persistence, 11; and* Steven Pifer, *The Trilateral Process: The United States, Ukraine, Russia and Nuclear Weapons* Brookings Arms Control Series Paper 6 (Washington, D. C.: Brookings Institution, 2011), 4.

25. The International Atomic Energy Agency [IAEA] is an international organization established in 1957, serving as an intergovernmental platform, reporting to the UN General Assembly and Security Council. As of 2017, the IAEA had 169 member states. Its activities are aimed at peaceful use of nuclear energy and advancement of nuclear energy technologies. One of its main tasks is to reduce

the use of nuclear energy for military purposes. The IAEA has a system of safeguards to ensure peaceful use of nuclear material under the NPT. The IAEA sets standards for nuclear security, provides technical assistance to member states, and fosters scientific and technological exchange in the field of nuclear energy. (www.iaea.org/)

26. Shteinberg, "Iadernoe oruzhie."

27. The Expert Group for the Investigation of Opportunities for Using Nuclear Material Extracted from Nuclear Weapons was comprised of experts from the National Science Center of the Kharkiv Institute of Physics and Technology (KIPT), and was headed by the director of the KIPT, academician Viktor Zelens′kyi. It was tasked with developing recommendations on utilizing nuclear material in preparation for the ratification of START and Ukraine's attainment of nonnuclear-weapon state status.

28. The significant differences in the weight estimates of nuclear material in Ukraine's nuclear weapons were primarily due to the absence of precise data on tactical weapons, as well as to secrecy regarding the construction of nuclear explosive devices.

29. Danilov, "My ne nastol′ko bohaty."

30. Ibid.

31. Iurii Ruban, "Mil′iardni prybutky chy shliakh do sumy?" *Holos Ukrainy*, no. 210, 3 November 1993, 8.

32. Kiselev, "Kogo obogatit obogashchennyi uran?"

33. John W. R. Lepingwell, "How Much Is a Warhead Worth?" *RFE/RL Research Report* 2, no. 8 (February 19, 1993): 62–64.

34. Russia did not intend to disarm at its own expense; with the help of Western aid, it was completing the construction of its own nuclear industry infrastructure. Particularly, it used the US $800 million of US aid to build a plutonium storage facility in Siberia, the so-called Fort Plutonium, to supply equipment for the safe transportation of nuclear warheads, and to manufacture storage cylinders for the plutonium removed from nuclear weapons (John J. Fialka, "U. S. Bid to Help Russia Disarm Moves Slowly," *Wall Street Journal*, March 9, 1993, A14; William J. Broad, "Nuclear Accords Bring New Fears On Arms Disposal," *New York Times*, July 6, 1992, 1).

35. Chrystia Freeland and R. Jeffrey Smith, "Ukrainian Premier Urges Keeping Nuclear Arms," *Washington Post*, June 4, 1993, A1; Lepingwell, "How Much Is a Warhead Worth?"; John W. R. Lepingwell, "Ukraine, Russia, and the Control of Nuclear Weapons," *RFE/RL Research Report* 2, no. 8 (February 19, 1993): 4–20; "Iadernoe toplivo dlia ukrainskikh AES–v polnom ob″eme," *Dzerkalo tyzhnia*, June 21, 1996, https://zn.ua/ECONOMICS/yadernoe_toplivo_dlya_ukrainskih_aes__v_polnom_obeme.html; Mykola Makarevych, first deputy minister of foreign affairs, to Vasyl′ Durdynets′ first deputy speaker of the Parliament of Ukraine, "Note of the US Embassy of 23 April 1993, Roman Popadiuk, US ambassador to Ukraine," no. UKOR23–490, April 30, 1993, author's personal archive; Pierre Goldschmidt, "Corral Plutonium for Peaceful Use," *Wall Street Journal*, Europe Edition, January 14, 1994, 6; Viktor Zelens′kyi, KIPT, to Yuri Kostenko, Working Group, "Report on the Results of Negotiations with the Russian Delegation on Issues of Dismantling, Transportation and

Elimination of Ukraine's Strategic Nuclear Warheads, Moscow, 16–18 February 1993," author's personal archive, English translation of the document is available in appendix; Danilov, "My ne nastol′ko bogaty"; Kiselev, "Kogo obogatit obogashchennyi uran?".

36. In 1992, Ukraine's NPPs were 100-percent dependent on Russia's supply of nuclear fuel. The nuclear energy industry, with five operating NPPs, provided roughly half of Ukraine's energy needs at the time. By the number of power-generating units, Ukraine ranked eighth in the world and fifth in Europe, with its fourteen operating units and six in construction phase. However, Ukraine's entire nuclear power industry was only a segment of the nuclear fuel cycle of the former USSR, key elements of which were located on Russia's territory, particularly, the production of fresh nuclear fuel, reprocessing and storage of spent fuel, disposal of high-level radiation waste, and enrichment, among others. The two problems that made Ukraine particularly vulnerable were its 100-percent dependence on the supply of fresh nuclear fuel from and the storage of spent fuel in Russia. Without an immediate solution to these problems, it was impossible to furnish a steady and secure operation for Ukraine's NPPs. It is in this context that we considered our options for cooperating with the US in the field of nuclear energy.

37. General Atomics (GA) is a US company, established in 1955 as a division General Dynamics for the development and commercialization of nuclear technologies, including nuclear fuel cycle and reactor equipment. GA developed a Generation IV nuclear reactor, the Gas Turbine Modular Helium Reactor (GT-MHR) and, in 2010, the Energy Multiplier Module (EM^2), a gas-cooled reactor powered by spent fuel.

38. The State Committee of Ukraine for Nuclear Power Utilization (Derzhkomatom). In December 1991, Ukraine's nuclear energy enterprises were integrated into a governmental conglomerate, Ukratomenerhoprom, which was later reorganized by the Cabinet of Ministers decree into Derzhkomatom. In addition to operating functions, this governmental structure was also tasked with providing secure operations of Ukraine's NPPs. Overtime, its name and functions have changed several times. In 1996, Derzhkomatom's operational functions were passed onto the newly established National Nuclear Energy Generation Company, Enerhoatom. Year 1997 saw the establishment of the State Department of Nuclear Energy of Ukraine (Derzhatom), with central executive authority status, subordinated to Ukraine's Ministry of Energy.

39. James J. Graham, vice president, General Atomics, to Dr. Yuri I. Kostenko, minister of environment protection, *Via Federal Express*, 15 January 1993, 1. Author's personal archive.

40. Ukrainian government officials referenced information from the Ukrainian press as follows: "V. Shmarov (the vice prime minister) had informed that these 'items' could only be utilized where they were produced, that is, in Russian defense factories" (Valentin Tabuns′kyi, "START-1: ratyfikatsiia iz zasterezhenniamy," *Holos Ukrainy*, November 23, 1993, 3).

41. Correspondence with General Atomics reveals the extent to which Ukraine's highest leadership was misinformed on the volumes of HEU in Ukraine's nuclear weapons. In one of the letters, First Vice Prime Minister Iukhnovs'kyi corrects the American company, stating Ukraine was in possession of 30, and not 60 metric tons of HEU. This would significantly reduce the potential profits for the American company (possibly, the misinformation relied on this fact); however, GA was also prepared to work with these reduced volumes (James J. Graham, president, International Commodities Exchange Corporation, an Affiliate of General Atomics, to Dr. Yuri I. Kostenko, minister of environment protection, *Via Federal Express*, 15 March 1993. Author's personal archive). Later, even the Russian calculations produced by various working groups named a figure of 64 tons of HEU.

42. James J. Graham, General Atomics, to Yuri I. Kostenko, "General Atomics Nuclear Services Proposal for Ukraine," *Via Fax Transmittal (+1.201.301.0068)*, 5 March 1993, 2. Author's personal archive.

43. James J. Graham, vice president, General Atomics, to Bohdan Burachins'kyi, Coordinating Committee to Aid Ukraine (USA), *Via Fax Transmittal (+1.201.301.0068)*, 3 May 1993, 1. Author's personal archive.

44. Here is a more detailed excerpt from GA's proposals regarding the oxidation of HEU: "As an alternative to the redelivery of LEU from Russia or the construction of dilution facilities in Ukraine, GA proposes to provide a simple HEU oxidation facility which can be built in either Ukraine or Russia as the first step in processing HEU. Under this alternative, HEU received from disassembled warheads would be oxidized in a controlled atmosphere furnace and packaged in approved containers for delivery to the United States. GA has made arrangements with Nuclear Fuel Services (NFS) for the purpose of handling the oxidized HEU at its facilities in Erwin, Tennessee. The proposed oxidation system would duplicate the NFS system that is approved by U. S. regulatory agencies and which has been proven to be very efficient. GA/NFS has an experienced team which can provide facility installation, process start up and employee training. The capital cost of this oxidation facility is projected to be less than one million U. S. dollars and the operational and environmental risk of the oxidation process is negligible. GA then proposes to ship the HEU oxide under appropriate government supervision by air to the United States where NFS would proceed to purify, convert to hexafluoride, and dilute the HEU to reactor grade LEU which would be redelivered to Ukraine" (Graham to Kostenko, 1–2).

45. The Shelter Object (Ukrainian: *Ukryttia*), more commonly known as the Sarcophagus, is a structure encasing the destroyed nuclear reactor number four of the Chernobyl Nuclear Power Plant. It was built to contain radioactive contamination resulting from reactor's explosion on 26 April 1986.

46. Westinghouse is a transnational corporation designing nuclear power reactors and manufacturing nuclear fuel. After the collapse of the USSR, it worked to develop opportunities to supply fuel for NPPs in Eastern Europe.

Westinghouse provides crucial technologies for 50 percent of the world's NPPs.

47. Michael H. Jordan, chairman and chief executive officer, Westinghouse Electric Corporation, to His Excellency Leonid Kuchma, president of Ukraine, 28 May 1997. Author's personal archive.

48. Ukraine ultimately received US $0.2 billion from Russia, and not even in actual money, but in lieu of repayment of some questionable energy debts. In general, information on compensation for Ukraine's tactical nuclear weapons looked as follows in the Ukrainian press: "Under the tactical nuclear weapons compensation agreement, Russia agreed to pay nearly $500 million via heretofore unspecified methods (either in lieu of Ukraine's energy debts or in the form of fuel for NPPs). [...] According Ukrainian expert estimates, the agreement would provide Ukraine for three years' worth of fuel for NPP" (Iulia Mostovaia, "K godovshchine iadernoi prem'ery," *Dzerkalo tyzhnia*, December 8, 1995, https://zn.ua/POLITICS/k_godovschine_yadernoy_premiery.html). However, this statement contains a fundamental error: fuel for NPPs was supplied in return for the strategic nuclear weapons already transferred to Russia, the cost of which was estimated at US $40 billion. As to the tactical weapons, President Kuchma lamented the lack of payoffs from Russia to the US Vice President Al Gore in 1997: "Before the meeting with Albert Gore, Leonid Kuchma made an official statement, in which he stressed that the trilateral agreement in which Ukraine stipulated to transferring its nuclear weapons to Russia needed to be adhered by all, including on the provision regarding compensation for tactical nuclear weapons. Exactly three years ago, Ukraine and Russia have executed a protocol on compensation for the transferred tactical weapons. Viktor Chernomyrdin confirmed during his visit to Kyiv that it would be in the sum of $450 million. However, Russia continues to not comply with the agreement" (Iulia Mostovaia, "Nachali za upokoi, zakonchili za zdravie," *Dzerkalo tyzhnia*, May 16,1997, https://zn.ua/ARCHIVE/nachali_za_upokoy,_zakonchili_za_zdravie.html). In conclusion, the official information provided to me by Prime Minister Yulia Tymoshenko in response to an official parliamentary request of 9 September 2008, showed the total compensation for fissile material from tactical nuclear weapons constituted only US $200 million, and these funds, as indicated in her answer, Russia wrote off as Ukraine's repayment of a state debt (accrued under dubious schemes).

49. The document, entitled "Report," was provided at the author's request. It contained projected costs of different options for processing HEU and plutonium; however, as was usual practice at the time, when the official position of the majority of governmental agencies did not officially support the Working Group's position, the document was provided unofficially, and thus has no identifying information. Author's personal archive.

50. Resolution of the Parliament of Ukraine "On Additional Measures to Ensure Ukraine's Attaining a Nonnuclear Weapons State Status." An English translation of text of the document is available in Appendix D.

51. Not only did the Derzhkomatom avoid developing a strategy for utilization of nuclear material for Ukraine's domestic needs, but it also blocked by all means possible the development of Ukraine's nuclear power industry strategy. Even when Ukraine's government appealed to Parliament with a request to lift a moratorium on the completion of two reactors at the Zaporizhzhia and Khmelnytskyi NPPs in June 1993, the Derzhkomatom provided no arguments to the parliament in support of peaceful nuclear power. I reacted to this in my address to Parliament on June 18, 1993 as follows: "I would like to bring to your attention, acknowledging full responsibility for the statement, that in my view, as of today, we do not have data or are in a state of sufficient readiness that could convince the parliament to make the decision one way or the other [regarding lifting moratorium]. To make this decision [on completing the construction of reactors], we first need an overall strategy for nuclear power industry's development, and a strategy for handling radioactive waste. [...] It is in the context of creating Ukraine's own nuclear fuel cycle, and on the basis its economic and environmental benefits, that we should be deciding whether Ukraine can develop its nuclear power industry or not. [...] Additionally, we should consider the problem of utilization of nuclear material extracted in the process of nuclear disarmament, likewise in the context of our own nuclear fuel cycle" (Parliament of Ukraine Session Hall, Session 88, 18 June 1993, 10 am, http://iportal.rada.gov.ua/meeting/stenogr/show/4872.html).

52. The idea that Russia had exclusive capacity to dismantle former USSR's nuclear warheads was reiterated by SBU head Ievhen Marchuk: "[A]ll these munitions were manufactured on Russia's territory. Accordingly, the disassembly of such a complex and classified product, and its dismantling at the end of the lifecycle, could only be done by those who created it. Anyone else would turn into a monkey with a grenade" (Vitalii Kniazhans'kyi, "Odna zustrich i dva uroky," Den, 30 March 2012, https://day.kyiv.ua/uk/article/nota-bene/odna-zustrich-i-dva-uroki).

53. Strekal, "Iaderni potuhy." The article itself, polemicizing against my earlier article "Nuclear Weapon of Ukraine: Good or Evil" is curious because it essentially called for Ukraine's immediate surrender of nuclear weapons and their transfer to Russia. It does not reference any sources other than the SBU. Its key messages, clearly outlined in the article, coincide with those used by Russia in its diplomatic process and during the information warfare with Ukraine. It is possible to infer that the SBU was feeding ideas to the Ukrainian journalist who then presented them as his own.

54. James J. Graham, President, International Commodities Exchange Corporation, to Mr. Ihor R. Iukhnovsky, First Vice Prime Minister of Ukraine, Via Federal Express, 11 December 1992. Author's personal archive.

55. "Report of the Working Group on the Use of Material in Nuclear Warheads: Per Results of the 16–18 February 1993 Moscow Meeting of the Representatives of the Working Group, V. F. Zelens'kyi and V. S. Krasnoruts'kyi, with Representatives of Russia's Working Group," signed by V. F. Zelens'kyi, head of the Working Group; members

of the Working Group: A. A. Afanas´iev, I. I. Kushchin, V. H. Kharytons´kyi, V. S. Krasnoruts´kyi, V. A. Belous, V. A. Rudakov, N. N. Sappa. Author's personal archive.

56. Borys Tarasiuk, chairman of the National Disarmament Committee of Ukraine, deputy minister for Foreign Affairs of Ukraine, to Yuri Kostenko, minister for environmental protection, letter no. UKOR/26–243, 29 March 1993. Author's personal archive.

57. James J. Graham, vice president, General Atomics, to Bohdan Burachins´kyi, Coordinating Committee Aid Ukraine (USA), Via Fax Transmission (+1.303.771.1625), 3 May 1993. Author's personal archive.

58. Yuri Kostenko, Working Group, to Leonid Kravchuk, president of Ukraine, "Certain Remarks on the Implementation of START," working memorandum, 22 April 1992. Author's personal archive.

59. Report of the Academy of Sciences of the Ukrainian SSR, Interdepartmental Commission for Communications with the International Atomic Energy Agency, 6 April 1992, signed by: I. M. Vyshnevs´kyi, chairman of the Commission on Communications with the IAEA and A. P. Trofymenko, scientific secretary of the Commission. Author's personal archive.

60. Danilov, "My ne nastol´ko bogaty."

61. "Rosia no Kaku kaitai, shikin kyouryoku ga fujyou," Nihon Keizai Shimbun, March 27, 1993.

62. KIPT, "Nuclear Material," 1992, 1–5, author's personal archive; Viktor Zelens´kyi, director of the Kharkiv Institute of Physics and Technology, National Academy of Sciences, and Vladimir Krasnoruts´kyi, laboratory head, "Estimated Value of Material in Nuclear Warheads," Kharkiv, February 1992, author's personal archive; "Report of the Working Group on the Use of Material," 1–6.

63. KIPT, "Nuclear Material," 2–3.

64. This is noteworthy, because, according to the US Embassy, the proportion of the value of HEU to the cost of its conversion into fuel for NPPs did not appear as catastrophic (Borys Tarasiuk, head of the National Committee of Ukraine on Disarmament, deputy minister, to Yuri Kostenko, minister of environment and natural resources, Ukrainian Translation of the US Embassy Memorandum, UKOR/26–243, March 1993, 1. Author's personal archive). Furthermore, if transportation and insurance costs were so significant, why were other options not considered to avoid such expenditures? Particularly, GA were offering their services in the disassembly of warheads, along with a cost-effective production of TVELs in Ukraine.

65. KIPT, "Nuclear Material," 3.

66. Ibid., 4.

67. Zelens´kyi to Kostenko, "Report on the Results of Negotiations."

68. Danilov, "My ne nastol´ko bohaty."

69. Zelens´kyi to Kostenko, "Report on the Results of Negotiations," 1.

70. KIPT, "Nuclear Material," 4.

71. Zelens´kyi to Kostenko, "Report on the Results of Negotiations," 1.

72. Ibid., 2.

73. Ibid.

74. Ibid., 3.
75. Ibid., 3–4.
76. Ibid., 6.
77. "Report of the Working Group on Utilization of Fissionable Material from Nuclear Weapons," prepared on the basis of recommendations developed during Working Group Conference at the KIPT, 23–25 February 1993, 1. Author's personal archive.
78. Ibid., 3; see also Viktor Havryliuk, deputy director for scientific research, Institute for Nuclear Research, NANU, "Brief," 23 April 1993. Author's personal archive.
79. Ibid.
80. Moscow made no secret of its use of Ukraine's energy dependence to correct the country's political course. For example, the following was published in the Russian press in June 1993: "Boris Yeltsin essentially admitted the fact that Russia was applying pressure on Kyiv using the energy sector situation when, at a recent press conference, he explained to the journalists how manipulations with the oil pipeline led to 'advances' in Ukraine's position on certain issues. Moscow's hard line on energy issues (which is rather justified: after all, one sovereign state has the right to sell its oil to another sovereign state at global market prices and demand money for it) was superimposed on mistakes, slips, and miscalculations by the Ukrainian leadership, and on the lack of palpable economic reforms in Ukraine. All this led to an eruption of public discontent [e. g., the miners' strikes in Ukraine in June of 1993]" ("Krizis vlasti na Ukraine: Kiev vynuzhden korrektirovat' politicheskii kurs," *Kommersant*, no. 119, June 1993, https://www.kommersant.ru/doc/51857?isSearch=True).
81. The Cassette Scandal, or Kuchmagate, was a political crisis in Ukraine in 2000 named after audio recordings of conversations in the office of the incumbent President Kuchma. Allegedly, the recording was done by the president's bodyguard, Major Mykola Melnychenko, with a Dictaphone placed under the sofa. In the aftermath of the scandal, the European Union and the US immediately curtailed political and economic cooperation with Ukraine, that had intensified after Kuchma's reelection for the second term in 1999, and after his inaugural speech about Ukraine's Euro-Atlantic strategy. Meanwhile, Russia actively used this cooling in a relationship between Ukraine and the West to increase its presence and influence in Ukrainian politics, finance, economics, energy industry, and information domain. The cassette recording was made public by the incumbent Speaker of Parliament, the leader of the Socialist Party, Oleksandr Moroz.
82. Viktor Lytovs'kyi, "Rossiiskaia armiia nachinaet rekonstruktsiiu svoikh strategicheskikh iadernykh sil," *Izvestiia*, 3 November 1992, 1.
83. Ibid.
84. Ibid.
85. Vladimir Trofimov, "My ne dolzhny upuskat' voennye zakazy," *Izvestiia*, May 14, 1992, 4.
86. Volodymyr Tolubko, "Turbota pro bezpeku chy nazad do falangy," *Holos Ukraïny*, 20 November 1992, 7.
87. Ibid., 7.
88. Vladimir Orlov, "Ukraina: 'kvaziiadernaia' superderzhava?" *Moskovskie novosti*, November 2, 1992, 14–15.
89. Yuri Kostenko, Member of the Parliament of Ukraine, to Vasyl' Durdynets' Vice Speaker of the Verkhovna Rada of Ukraine, letter, 7 September 1992. Author's personal archive.
90. Oleksii Ievsikov, "Zadlia harantii bezpeky," *Holos Ukraïny*, November 18, 1992, 1; Ukrinform, "Senatory SShA u Kyievi," *Holos Ukraïny*, November 24, 1992, 2; Oleksii Breus, "Kurs vyznacheno," *Holos Ukraïny*, December 19, 1992, 1; "Rozmova po telefonu L. Kravchuka z Bushom," *Holos Ukraïny*, no. 497, December 26, 1992, 1; "Rozmova po telefonu L. Kravchuka z B. Klintonom," *Holos Ukraïny*, no. 516, January 29, 1993, 1.
91. Ievsikov, "Zadlia harantii bezpeky."
92. Ihor Hrushko, "START nezabarom startuie," *Holos Ukraïny*, 20 November 1992, 1–2.
93. Ukrinform, "Senatory SShA u Kyievi."
94. Ibid.
95. Breus, "Kurs vyznacheno."
96. "Rozmova po telefonu L. Kravchuka z Dzh. Bushom."
97. "Rozmova po telefonu L. Kravchuka z B. Klintonom."
98. Foran, "New Nuclear World."
99. "Polit 'kachky' vid Reiter do Ostankino. Povidomlennia pres-tsentru Verkhovnoï Rady Ukraïny," *Holos Ukraïny*, January 11, 1993, 1.
100. Valerii Pavliukov, deputy minister of Ukraine's Ministry of Machine Industry, and Aleksei Kryzhko, head of the Center for Administrative Management of the Ministry of Defense of Ukraine, "Report on the State of Affairs in Missile Complexes of Strategic Nuclear Forces Deployed in Ukraine, and Suggestions for Ensuring their Safe Operations," 3 March 1993. Author's personal archive.
101. "Iaderna zbroia: kontseptsiia kerivnytstva Ukraïny. Prezydent L. M. Kravchuk vidpovidaie na zapytannia korespondentiv Ukrinformu," *Holos Ukraïny*, December 16, 1992, 1, 3.
102. Ibid., 1.
103. Ibid.
104. For the Declaration of State Sovereignty of Ukraine see *VVR*, no. 31 (1990): 429, http://zakon1.rada.gov.ua/laws/show/55–12; for the Statement of the Parliament of Ukraine on the Nonnuclear Status of Ukraine, see *VVR*, no. 51 (1991): 742, http://zakon.rada.gov.ua/laws/show/1697–12.
105. "Iaderna zbroia," 3.
106. Ibid., emphasis added.
107. Ibid.
108. Ibid.
109. This interview not only failed to demonstrate the presence of a unified official position on the issue of nuclear disarmament, but it also revealed Kravchuk as a weak link in Ukrainian politics. His outright incompetence on the subject of nuclear weapons was all the more regrettable because of the rather comprehensive amalgamation of information that sat atop his presidential desk, with everything that the Working Group was able to amass by then: from cost estimates to issues inherent in the liquidation process, including its technological, financial and social aspects. However, from the onset of these

developments Kravchuk deliberately limited the circle of advisors whose guidance he was willing to accept. These were Zlenko and Marchuk, whose main strategy entailed handing everything over to Russia as quickly as possible. Thus, unable to ignore decisions of Parliament, but leaning towards Zlenko and Marchuk in his decisions, Kravchuk was constantly maneuvering between the two viewpoints: while quoting a decision by Parliament, he would immediately contradict it. This convenient and habitual form of behavior, which some people used to call diplomacy, led to a disastrous outcome in the case of nuclear disarmament: namely, the world could not understand what Ukraine actually wanted to achieve.

Chapter 3

1. START II is a treaty on the further reduction and limitation of strategic weapons, signed between the US and Russia on 3 January 1993, in Moscow. It was intended to establish a new level of strategic parity between Russia and the US, as a basis for preserving strategic stability between the two nuclear states going into the next century. The main aspect of the treaty was to mandate both countries to reduce their nuclear arsenals to 3,000–3,500 units. After eliminating the strategic weapons under the treaty, the number of strategic nuclear missiles in the US and Russia was supposed to be reduced by 75 percent compared to 1990. This new treaty also called for the destruction of all ground-based multiple independently targetable reentry vehicles (MIRVs) on ICBMs, including the SS-18 "Satan." START II was an extension of START I, signed in 1991, and it could only come into effect after Ukraine, Belarus and Kazakhstan acceded to the NPT as nonnuclear states. The US Senate ratified START II in 1996, while the Russian Duma only did so in 2000. The treaty was to be in force for 15 years.

2. Henk Vol′zak [Volsack], "Radians′ka iaderna spadshchyna," paper no. 4 of *Glasnost & Perestroika series for the* Glasnost & Perestroika Institute, Amsterdam. Translated from English to Ukrainian by Dmytro Labuns′kyi with foreword by General Kostiantyn Morozov, Defense Minister of Ukraine, 1992, Author's personal archive. See catalog listing at the Natsional′na Biblioteka Ukraïny im. V. I. Vernads′koho, http://www.irbis-nbuv.gov.ua/cgi-bin/irbis_nbuv/cgiirbis_64.exe? Z21ID=&I21DBN=EC&P21DBN=EC&S21STN=1&S21REF=10&S21FMT=fullwebr&C21COM=S&S21CNR=20&S21P01=0&S21P02=0&S21P03=A=&S21COLORTERMS=1&S21STR=%D0%92%D0%BE%D0%BB%D1%8C%D0%B7%D0%B0%D0%BA%20%D0%A5$.

3. "Zaiava Prezydenta Ukraïny L. M. Kravchuka (z pryvodu pidpysannia rosiis′ko-amerykans′koho Dohovoru SNO-2)," *Holos Ukraïny*, January 5, 1993, 1.

4. Fred Kaplan and Jon Auerbach, "US, Russian Leaders Hail START II Arms Treaty," *Boston Globe*, January 3, 1993, 10.

5. Alexei Portanskii, "Ukraina podtverzhdaet svoe namerenie stat′ beziadernym," *Izvestiia*, December 30, 1992, 1.

6. Department of Information of the Ministry of Foreign Affairs of Ukraine, "MFA data on the leading Western periodicals' reporting of the nuclear disarmament of Ukraine." Forwarded to Pliushch, Speaker of the Parliament of Ukraine, by B. I. Tarasiuk, Chair of the National Committee for Disarmament, via letter UKOR/23–316, 31 March 1993, 1. Author's personal archive.

7. John-Thor Dahlburg and Mary Mycio, "Ukraine Wants More Aid, Holds Up Nuclear Treaties," "Disarmament: Kyiv Has Yet to Ratify START I. It's Virtually a Shakedown, US, Russian Diplomats Say," *Los Angeles Times*, January 8, 1993, http://articles.latimes.com/1993–01–08/news/mn-998_1_nuclear-power; Thomas L. Friedman, "Beyond Start II: A New Level of Instability," *New York Times*, Late Edition, January 10, 1993, A1.

8. Ministry of Foreign Affairs to the Parliamentary Commission on Foreign Affairs, "Issues with Ukraine's Nuclear Disarmament in the Western Press," no. BI/22/104, 10 May 1993, 1–2.

9. Anatolii Zlenko, Minister of Foreign Affairs, to Vasyl′ Durdynets′ First Deputy Parliamentary Speaker, MFA letter regarding the Russian Federation MFA statement, UKOR/23–338, 5 May 1993. Author's personal archive. (Full text of document available in Appendix N.)

10. "Monitoring of Compliance with the Provisions of the Treaty Between the US and the Russian Federation on Further Reduction and Limitation of Strategic Offensive Arms (the START II Treaty) Treaty Doc. 103–1," 24 June 1993, US Senate, Committee on Foreign Relations, Washington, DC. "Statement of Hon. R. James Woolsey, Director, Central Intelligence Agency," 85.

11. Michael R. Gordon, "Aspin Meets Russian in Bid to Take Ukraine's A-Arms," *New York Times*, 6 June 1993, A16.

12. "Starting Over," *Wall Street Journal*, 6 January 1993, A10.

13. Grigorii Kisun′ko, "Razoruzhat′sia, no s umom," *Pravda*, 9 June 1993, 5.

14. Carol Giacomo, "Lack of US Guarantees Makes START Passage Harder, Ukraine Says," Reuters News Agency, June 8, 1993.

15. The security guarantees that the US was prepared to offer were criticized in the Ukrainian press: "If we analyze the previously mentioned official statements (the démarche by then-SecState Jim Baker and a NATO memorandum that was similar in nature), it's hard not to notice the most striking feature in the American position: even if Ukraine were to undertake nuclear disarmament on a voluntary, unilateral basis, the only protection the US commits itself to is in the case of a hypothetical nuclear attack, without any guarantees that it would participate in neutralizing the threat of military pressure on Ukraine involving conventional arms on the part of any country." (Ihor Nediukha, "Iaderne rozzbroiennia Ukraïny: 'za' i 'proty'," *Vechirnii Kyiv*, 5 June 1993, 2).

16. Carol Giacomo, "US Said to Give Ukraine Official Assurances Letter," Reuters News Agency, January 9, 1993.

17. Chrystina Lapychak, "Ukraine Wins US Security Pledge, Official Says: Arms: Top Negotiator Returns from Washington with Letter from Bush. The Ex-Soviet Land Has Balked at Ratifying Nuclear Treaty," *Los Angeles Times*, January 10, 1993, http://articles. latimes. com/1993–01–10/news/mn-1594_1_nuclear-arms.

18. Giacomo, "US Said to Give Ukraine Official Assurances Letter."

19. "Vizyt delehatsii Ukraïny v SShA," *Holos Ukraïny*, January 11, 1993, 1.

20. Anatolii Zlenko, Minister of Foreign Affairs of Ukraine, to Ivan Pliushch, Speaker of the Parliament of Ukraine, "Record of Conversation with RF MFA Special Ambassador M. M. Streltsov, 12 January 1993," UKOR/23–54, 26 January 1993, 2. Author's personal archive.

21. The fact that the MFA had no clear idea of what these "security guarantees" should look like was obvious from Minister Zlenko's speech before the parliament on 3 June 1993: "As to security guarantees for Ukraine, we are talking about the nuclear states drawing up a political and legal document in which they confirm their commitments not to allow the use of any type of force against Ukraine on the part of nuclear states. Of course, such commitments by themselves will not guarantee Ukraine's security but they will have important political significance. We received the texts of such guarantees in the previous plan from the United States, Great Britain, the Russian Federation, France and China." Nevertheless, on the same day, answering a national deputy's question, the Minister defined these documents as "previous political commitments, a sort of guarantees." ("Dopovid' ministra zakordonnykh sprav Ukraïny A. Zlenka na sessiï Verkhovnoï Rady Ukraïny 3 chervnia 1993 r.," *Holos Ukraïny*, June 5, 1993, 3–4).
Until the end of 2014, joining NATO was not a policy priority in Ukraine. The government's position did little to shape a positive attitude towards NATO in the new generation of Ukrainians over 20 years ("Hromads'ka dumka shchodo vstupu do NATO," Ukraine-NATO, https://ukraine-nato. mfa.gov.ua/ua/inform-center/gromadsyka-dumka-shhodo-vstupu-v-nato). In 2005, only 15 percent of Ukrainians favored joining NATO, while 48 percent were against ("Ukraïntsi khochut' vstupaty do ES i ne khochut' do NATO–zahal'nonatsional'ne opytuvannia," Ilko Kucheriv Democratic Initiatives Foundation, 22 February 2005, https://dif.org.ua/article/ukraintsi-khochut-vstupati-do-es-i-ne-khochut-do-nato-zagalnonatsionalne-opituvannya). In April 2008, 21.8 percent were for and 59.6 percent against NATO. "When Ukraine's leadership announced after the New Year's holidays that they wanted to join the Membership Action Plan (MAP) for NATO, Western leaders asked, 'Where's the matching desire in your society?' President [Viktor Yushchenko] and Premier [Yulia Tymoshenko], still in unison at that point, promised to carry out a broad-based public awareness campaign to change anti-NATO attitudes. [...] Last Wednesday [28 May 2008], the Tymoshenko Government quietly approved a resolution 'Confirming the State Targeted Program to inform the public on issues of Ukraine's Euroatlantic integration for 2008–2011.' [...] Over 2008–2011, UAH 40.5 million is planned to be spent on a 'Yes to NATO' campaign, that is, nearly 10 million (US $2 million) annually. Despite the modest spending, the government expects the results of this program to lead to a radical change in the balance of opinion among Ukrainians." ("Iak Tymoshenko z Prutnikom prydumaly ahituvaty za NATO," *Ukraïns'ka Pravda*, June 2, 2008, https://www.pravda.com.ua/articles/2008/06/2/3456321/). Plans were that opinion would rise to 36 percent in 2008, 43 percent in 2009, and over 50 percent by 2010 (ibid.). Yet in 2012, NATO support still amounted to only 13 percent (Anatolii Kotov, "'Pivnichnoatlantychnyi al'ians' chy 'NATO': shcho zminyla informatsiina kampaniia Ukraïny shchodo NATO?" *Ilko Kucheriv Democratic Initiatives Foundation*, 4 April 2016, https://dif.org.ua/article/pivnichnoatlantichniy-alyans-chi-nato-shcho-zminila-informatsiyna-kampani-ya-ukraini-shchodo-nato).

22. Valentyn Labuns'kyi, "START-1: pospishaty, ne kvapliachys," *Holos Ukraïny*, March 10, 1993, 2.

23. This ideology was already defined in the previously mentioned program article (Kostenko, "Iaderna zbroia Ukraïny: blaho chy zlo?").

24. Mykola Makarevych, Deputy Foreign Minister, to Ivan Pliushch, Parliamentary Speaker, "From the diary of O. Bilorus, Washington, 7 December 1992. Record of a conversation with US Under Secretary of State F. Wisner," transmitted by telefax no. 15/49–25057, 21 December 1992. Author's personal archive.

25. "Don't Stop START," *The Economist*, Vol. 326, Issue 7793, January 9, 1993, 14.

26. Jim Hoagland, "Race Between Nightmares East of the Urals," *Washington Post*, May 30, 1993, C07.

27. Chrystia Freeland and R. Jeffrey Smith, "Ukrainian Premier Urges Keeping Nuclear Arms," *Washington Post Foreign Service*, June 4, 1993, A1.

28. R. Jeffrey Smith, "US Fears Ukrainian-Russian Clash; Steps Weighed to Head Off Confrontation Over Nuclear Weapons," *Washington Post*, June 6, 1993, A32.

29. After Ukraine's "capitulation" in Budapest with the signing of the Budapest Memorandum in December 1994, and the removal of the last nuclear warhead in June 1996, Ukraine faded from the world's attention altogether. The West did not hear the messages about Ukraine's priority policy of Euroatlantic integration announced in Leonid Kuchma's inaugural address when he was re-elected president in 1999. Neither did it hear nor see the desire of Ukrainians and the Ukrainian government to be part of Europe even after the 2004 Orange Revolution. Washington's foreign policy orientation to benefit Russia on the matter of Ukraine continued even under the pro-Western presidency of Yushchenko. On 3 July 2008, Henry Kissinger, advisor to several US presidents, called on the US to stop the expansion of NATO to the east to avoid worsening relations with Russia. At the Bucharest NATO summit in April 2008, Ukraine's and Georgia's applications for the MAP had been deferred until December. Kissinger wanted this issue removed

from the agenda. Barely a month later, on 7 August, Russia attacked Georgia.

30. "Obhovoriat' vsi kliuchovi pytannia," *Holos Ukraïny*, January 12, 1993, 1.

31. Dubinin, "Iadernyi dreif Ukrainy."

32. "Nashoii kraïny bezpeka," *Holos Ukraïny*, December 12, 1992, 6.

33. "Spil'ne komiunike pro zustrich derzhavnykh delehatsii Rosiis'koii Federatsii i Ukraïny, ocholiuvanykh Prezydentom RF B. M. Yeltsynym i prezydentom Ukraïny L. M. Kravchukom," *Holos Ukraïny*, January 12, 1993, 2.

34. "Agreement on the principles for forming the Navy of Ukraine and the Naval Fleet of the RF on the basis of Black Sea Fleet (BSF) of the former USSR," or the Yalta Treaty on the Black Sea Fleet, was signed by the presidents of Ukraine and the RF on 3 August 1992 in Mukholatka, outside Yalta. It signaled Ukraine's intentions to transfer to Russia part of the assets that were on Ukrainian territory when Ukraine declared independence, in this case the Black Sea Fleet. The agreement violated the Law on Enterprises, Institutions and Organizations of the Former Soviet Union Located on Ukraine's Territory, dated 10 September 1991 (*VVR, no. 46 (1991): 615,* http://zakon.rada.gov.ua/laws/show/1540–12), under which all assets that were on Ukrainian territory at the time that the Declaration of Independence was passed were the property of Ukraine. Thus, by signing the Yalta Treaty, Kravchuk overstepped his authority. The agreement established the basis for foreign military units to be deployed on Ukrainian territory, which contradicted the intention of being a neutral state, written into Ukraine's Declaration of State Sovereignty. The key articles of this agreement provided that: (1) "the BSF shall be divided among the parties to the Agreement in order for the Navy of Ukraine and the Naval Fleet of the RF to be established;" (2) the period through 1995 shall constitute a transition period;" (3) during this period, the presidents of Ukraine and Russia shall appoint the Joint Command of the BSF on a consensual basis; (4) during this time "the parties shall jointly use the existing system of bases and infrastructure" (meaning that Russia not only got Ukraine's part of the fleet but also got to use all the infrastructure and territory). The further procedures for basing the Ukrainian Naval Forces and the RF Naval Fleet was supposed to be designated in a separate agreement. The outcome of this "joint" use was less than a year after the treaty was signed, 80 percent of all vessels had raised the Russian military flag and most of the infrastructure of the BSF was taken over by Russia without any payment of rent.

35. "Spil'ne komiunike," *Holos Ukraïny*.

36. Essentially, with this document Russia aimed at reserving for itself monopoly rights to the entire package of disarmament services for Ukraine, from manufacturer's oversight over the warheads to their dismantling, transporting, and utilization, and producing the fuel rods for Ukraine's NPPs. Although Kravchuk's signature on the communiqué did not obligate Ukraine to work with Russia, should Moscow's terms and conditions for this cooperation prove unacceptable, Russian diplomats continued to actively use it as an instrument during negotiations, as they did with everything Kravchuk signed.

37. "Protocol to the Treaty Between the Union of Soviet Socialist Republics and the United States of America on the Reduction and Limitation of Strategic Offensive Arms," *Ofitsiinyi visnyk Ukraïny*, 2007, *no. 1*, 299, Article 58, Act Code 38332/2007, http://zakon.rada.gov.ua/laws/card/998_070.

38. John-Thor Dahlburg, "Russia Offers Ukraine Its Nuclear Protection: Arms: The guarantee makes Kyiv more likely to follow through on transferring its remaining atomic weapons to the Kremlin," *Los Angeles Times*, Collections, January 16, 1993, http://articles. latimes.com/1993–01–16/news/mn-1336_1_nuclear-weapons.

39. Juliet O'Neill, "Yeltsin Promises Ukraine Oil and Nuclear Protection," *The Gazette* (Montreal, Quebec), January 16, 1993, H6.

40. Dahlburg, "Russia Offers Ukraine Its Nuclear Protection."

41. Bogdan Turek, "Ukraine May Oppose New Structure of CIS," *United Press International*, January 19, 1993, *https*://www.upi.com/Archives/1993/01/19/Ukraine-may-oppose-new-structure-of-CIS/1590727419600/.

42. Rostislav Khotin, "CIS Split Opens as Summit Looms," Reuters News Agency, January 20, 1993.

43. "Ukraine Needs Reassurance from West; Real Russian Threat," *New York Times*, January 21, 1993, 24 (*The New York Times Archives*: https://www.nytimes.com/1993/01/21/opinion/l-ukraine-needs-reassurance-from-west-real-russian-threat-476693.html).

44. In theory, I understood that negotiations would be difficult. Still, this was my first experience with Russian diplomacy, which meant that I had to even meet this phenomenon face-to-face. The open disregard for Ukrainian legislation, condescending attitude towards their counterparts that was, of course, covered over with a polite smile, the brazen twisting of basic facts, the crude use of force, this signature that I felt on my skin in 1993 remained unchanged 20 years later. The speeches of Russia's representative in the UN on matters related to Ukraine over 2014–2015 gave an approximate idea of these "traditions:" the entire world established as fact the presence of Russian forces on Ukrainian soil, but he kept repeating that they were not there and even blaming Ukraine for the human casualties.

45. It was not only during the first years of independence that members of Ukrainian state and Government delegations were not given materials to conduct negotiations. In 1997, I was in a delegation that flew to Washington for talks as part of the Kuchma-Gore Commission. Opposite us sat US Vice President Albert Gore, with a thick book of recommendations with tabs sticking out of it. Whenever a question arose, he would open one of the tabs and could immediately voice the official position on this specific issue, prepared by executive staff. Before President Kuchma lay just a clean sheet of paper. After a question was addressed to Kuchma, he either answered in general terms or turned to other ministers who had accompanied him. The MFA had not prepared any specific information even for presidents, just as had been the case with

Kravchuk in Bishkek in 1992 at talks among the heads of CIS states and in 1997 when Kuchma was talking to the US Vice President.

46. Dubinin, "Iadernyi dreif Ukrainy." In this article, Dubinin in clear detail outlines the logic and order of Russia's actions aimed at Ukraine's nuclear disarmament over 1993–1994. Events are depicted in the article from the point-of-view of Russia's national interests, on behalf of which Russian diplomats were working. Dubinin uses idiosyncratic interpretations of international and Ukrainian documents related to Ukraine's commitments regarding nuclear disarmament, his masterful manipulation of the letter of the documents, individual words and emphases leading to fundamental changes in their meaning in many cases. Indeed, Dubinin demonstrates just how Russian diplomacy, using pressure, shifting concepts, the press, and influence in the highest echelons of the Ukrainian government, achieved its goal: getting Ukraine's entire nuclear arsenal transferred to Russia and vastly expanding its arsenal at Ukraine's expense. However, the way Russian diplomats presented it, this went down in world history as a success story in the global disarmament process.

47. Ibid.
48. Ibid.
49. "Agreement between the Government of Ukraine and the Government of the Russian Federation about the status of locations (facilities) "S" deployed in Ukraine," Article 9, proposed law with handwritten note: "Negotiations" and "The last version from the Ukrainian side," 24 June 1993, 4. Author's personal archive.
50. Dubinin, "Iadernyi dreif Ukrainy."
51. Ibid. Such obvious desire of the Russian colleagues to knock a powerful trump out of Ukraine's hand looked less like a negotiating process than a fool's game. Combined with their aggressive behavior, the proposition, presented in the form of a demand, yet again illustrated the lack of respect the Russians had for international rules and ethics.
52. Ibid.
53. Danilov, "My ne nastol'ko bogaty."
54. "Rosia no Kaku kaitai, shikin kyouryoku ga fujyou," Nihon Keizai Shimbun, March 27, 1993.
55. The politics of double standards that Russia played in relation to the nuclear inheritance of the Soviet Union was growing more and more obvious. Russia tried to shift the burden of the colossal cost of eliminating this arsenal squarely on Ukraine's miserly budget. Ukraine was offered US $175mn to eliminate the entire strategic arms infrastructure, while Russia was given US $5 billion merely for transporting the warheads to its factories, the warheads that contained what was the most valuable, uranium and plutonium. That was the political reality of our negotiations.
56. Session of the National Defense Council with President Kravchuk participating, 2 April 1992, Author's Personal Notes, 1–5. Author's personal archive.
57. Yuri Kostenko to President Leonid Kravchuk, "Some Observations Regarding the Implementation of START," an analytical brief, 22 April 1992. Author's personal archive.

58. Oleksii Kryzhko, minister of defense, and Valerii Pavliukov, Minmashprom, to the working group, "Report on the State of Missile Complexes of the Strategic Nuclear Forces Deployed in Ukraine and Suggestions for Ensuring Their Safe Exploitation," 3 March 1993. Author's personal archive.
59. Ibid., 3.
60. Ibid., 4.
61. On 3 January 1992, the defense minister issued a decree requiring the service personnel to swear allegiance to the Ukrainian people. By 7 January, some 8,000 had done so and by the end of January, 350,000 had. With the exception of the Black Sea Fleet, the 43rd Strategic Rocket Army, and the Bolgrad Airborne Division, which belonged to the Strategic Containment Forces during the transition period, this amounted to nearly two thirds of the service personnel in Ukraine's military formations. The total number of personnel in the Ukrainian Armed Forces at the beginning of 1992 was 726,536 (Petro Kostiuk, "Tak tvorylys' suchasni Zbroini syly Ukraïny," Portal Spilky Ukrains'kykh Ofitseriv, December 4, 2012, http://port-sou. at.ua/publ/istorichni_statti/istorichni_statti/tak_tvorilis_suchasni_zbrojni_sili_ukrajini/2-1-0-199).
62. At that point, information began to come to our working group about agitation on the part of "Russian specialists" in Ukraine's Strategic Nuclear Forces to have them subordinated to the Russian Armed Forces. Things had already begun moving in that direction with the Strategic Forces of Belarus and Kazakhstan.
63. "Issues of Nuclear Safety," a seminar organized by the Foreign Economic Relations Committee at the Ministry of Foreign Affairs of the Supreme Soviet of the Russian Federation, in collaboration with the Science and Technology Committee of the NATO Parliamentary Assembly, Moscow, 29 June to 1 July 1992.
64. Dubinin, "Iadernyi dreif Ukrainy."
65. "B. Iel'tsin potreboval uskorit' formirovaniie voiennoi doktriny," Izvestiia, November 25, 1992, 1.
66. Kisun'ko, "Razoruzhat'sia, no s umom."
67. Vadym Dolhanov, counselor for information and the press, Embassy of Ukraine in RF, to Borys Tarasiuk, MFA of Ukraine, letter no. 8/112, 6 April 1993. Author's personal archive.
68. Dubinin, "Iadernyi dreif Ukrainy."
69. Iulia Mostova, "V kiievskom glazu Moskva nashla iadernuiu sorinku, a v svoem i raketu ne zametila," Kievskie vedomosti, March 10, 1993, 1–2.
70. Ibid.
71. Yuri Kostenko, speech at parliamentary hearings on the subject of "Political and Legal Aspects of Nuclear Disarmament," 10 March 1993. Author's personal archive.
72. Ibid.
73. Petro Martynenko, Chair of the Department of Comparative Law, Ukrainian Institute of International Relations, Shevchenko National University in Kyiv, "Main Conclusions in Assessing the Legal Ramifications of Ukraine's Ratification of START I and the NPT," materials from the first public parliamentary hearings on the topic of "Preparing the Treaty on Strategic Offensive Weapons,

START I, for Ratification," 5 March 1993, 1. Author's personal archive.

74. Ibid., 4.

75. Vasyl' Repets'kyi, chair of the Department of International Law at Franko University in Lviv, "Political and Legal Justification for the Reduction and Elimination of Nuclear Weapons and Issues of Succession," materials from the closed session of the working group, 12 April 1993, 4. Author's personal archive.

76. Ibid., 4–5.

77. Vitalii Kriukov, senior researcher at the Institute for State and Law, National Academy of Sciences of Ukraine, "Ukraine's Nuclear Status: Legal and Political Issues," materials from the first public parliamentary hearings on the topic of "Preparing the Treaty on Strategic Offensive Weapons, START I, for Ratification," 5 March 1993, 1. Author's personal archive.

78. "Deputaty vymahaiut' dlia Ukraïny status iadernoï derzhavy," *Svoboda*, March 19, 1993, 1.

79. Ibid.

80. "Political and Legal Aspects of Nuclear Disarmament," hearing in the parliament, 10 March 1993, author's personal notes. Author's personal archive.

81. Destroying missile silos was a very expensive idea: "According to calculations by Ukrainian experts, [...] the elimination of the ICBMs and the missile silos [was likely to cost] US $3 billion" (Sergei Klimovich, "Proshchai, oruzhie," *Podrobnosti*, January 16, 2007, http://podrobnosti.ua/386204-proschaj-oruzhie-2.html). With assistance from Western countries capped at US $175mn, "Ukraine was still paying out of its own budget for disarmament even in the early 2000s, including for the destruction of the missile silos, which the working group had categorically argued against" ("Kostenko schitaet, chto Ukraina sdelala bol'shuiu oshibku," *UNIAN*, January 12, 2007, https://www.unian.net/politics/29436-kostenko-schitaet-chto-ukraina-sdelala-bolshuyu-glupost.html). At the official level, it was always stated that financing for the projects to destroy the silos came from US grants that ended in 2001. However, the state budgets of 2009–2012 funded a program called "Utilizing ammunition and liquid components of missile fuel, ensuring the active state of the Armed Forces of Ukraine's arsenals, bases and warehouses while securing them against combustion and explosions." To finance the program, Ukrainian budget has allocated UAH 35.8mn, UAH 114.8mn, UAH 229.2mn, and UAH 392.4mn, respectively, or, combined, nearly US $100mn at the exchange rate of the time. From an environmental viewpoint, the detonating the silos led to significant tectonic disruptions that could have unforeseen consequences, including the depletion of underground waters in densely populated areas. In the end, the silos that the state failed to use for peaceful purposes led to looting at the supposedly destroyed sites: "When the military abandoned the missile silos in Pervomaisk and Kryvo-ozersk Counties in Mykolaïv Oblast at the beginning of the 2000s, after they had been brought to the public's attention several times because of their illegal and dangerous dismantling, they were given over to private individuals on the basis of sales agreements in which they were referred to as 'underground vertical cylindrical reinforced concrete constructions of a diameter of nearly 9 meters.' [...] A silo on the outskirts of the village of Kamianka Balka in Pervomaisk County went to a resident of Kirovohrad Oblasts on 6 September 2011 for the symbolic amount of UAH 988 [roughly US $125]. The seller, incidentally, was a local who had been given ownership only the day before. [...] It was later established that the sale of the components of a single missile site, which has about 1,000 tons of metal, including specialty steel, stainless steel, aluminum, bronze, specialty iron, titanium, and so on, as well as the sale of other elements of the structure, brought close to UAH 25mn [US $3 million]" ("Na Nikolaevshchine za kopeiki prodaiutsia raketnyie shakhty," *PRESTUPNOSTI.net*, November 29, 2011, https://news.pn/ru/politics/49524). The press reported that in May and June 2011 that "illegal work using heavy machinery was taking place simultaneously at eleven sites (Oleksandr Danyliuk, "Iaderni raketni shakhty prodaiut' pryvatnym osobam," *Korrespondent*, November 30, 2011, https://blogs. korrespondent.net/blog/celebrities/3248689-yaderni-raketni-shakhty-prodauit-pryvatnym-osobam).

82. "On Urgent Measures to Carry out Ukraine's Nuclear Policies," 15 March 1993, confidential document. Author's personal archive.

83. Ibid.

84. John J. Fialka, "US Bid to Help Russia Disarm Moves Slowly," *Wall Street Journal*, March 9, 1993, A14.

85. "Issues of Nuclear Safety," a seminar organized by the Foreign Economic Relations Committee at the Ministry of Foreign Affairs of the Supreme Soviet of the Russian Federation, in collaboration with the Science and Technology Committee of the NATO Parliamentary Assembly, Moscow, 29 June –1 July 1992.

86. Borys Tarasiuk, chair of the National Committee of Ukraine for Disarmament and deputy foreign minister, to Yuri Kostenko, Minister of the Environment, "Memorandum of the US Embassy Regarding the Problem of Sale and Utilization of HEU and Plutonium Extracted From Nuclear Weapons (translation from English)," UKOR/26–243, 29 March 1993. Author's personal archive.

87. Fialka, "US Bid to Help Russia."

88. Ibid.

89. Henk Vol'zak [Wolsack], "Rosiia: iadernyi shchyt kraïny roz'idaie irzha khaosu i kryzy," *Holos Ukraïny*, December 3, 1993, 9.

90. Ibid.

91. Anatolii Zlenko, minister of foreign affairs of Ukraine, to Vasyl' Durdynets' first deputy parliamentary speaker, letter from the MFA with information about a statement from the MFA of the Russian Federation, UKOR/23–338, 5 May 1993. Author's personal archive. (Full text of document available in the appendix.)

92. Ibid.

93. "Zaiavlenie pravitel'stva RF po povodu iadernoho oruzhiia na Ukraine," *Izvestiia*, April 6, 1993, 2.

94. Ibid.

95. Liliia Hryhoriieva, "Hra tryvaie," *Holos Ukraïny*, April 7, 1993, 10.

96. In April 1993, the NAS Institute of Sociology ran a survey of public opinion on the issue of nuclear weapons. Thirty-one percent of respondents were in favor of keeping nuclear weapons in Ukraine, 52 percent were more in favor of destroying them, and 17 percent had no opinion. If Ukraine decided to give the nuclear weapons to Russia, 48 percent thought Ukraine should be compensated, seven percent thought that no compensation was necessary, 25 percent were against handing the weapons to anyone, and 20 percent declined to answer. This meant that nearly one third of Ukrainians favored Ukraine gaining nuclear status, and at least 48 percent understood that nuclear weapons were not a burden but a valuable asset ("Shcho my dumaiemo pro iadernu zbroiu," *Holos Ukraïny*, April 23, 1993, 4)

97. The tone of the Ukrainian Government's response to Russia's aggressive informational attack can be seen in the following passage of the Cabinet of Minister's press statement, which was issued a day after the press conference in Moscow: "Ukraine welcomes Russia's readiness to provide additional security guarantees. Right now, what matters is to ensure the proper form, timing, level and extent of these guarantees. We are also pleased to accept Russia's readiness to resolve the issue of compensation for nuclear material which is property of Ukraine. We understand this readiness as extending to the components of the tactical nuclear weapons that were previously removed" ("Zayava pres-sluzhby Kabinetu Ministriv Ukraïny," *Holos Ukraïny*, April 7, 1993, 10).

98. Ibid.

99. In reaction to the brazen statements being issued by the RF MFA, Ukraine's Foreign Ministry kept justifying itself: "Although the operational control of the strategic forces on Ukrainian territory remains under the Joint Command of the Strategic Forces of the CIS, their administrative control has been turned over to the Ministry of Defense of Ukraine, which in no way contradicts the objective of ensuring a single, reliable control over nuclear weapons. Nothing in the concept of administrative control of strategic nuclear forces deployed in Ukraine offers grounds to see intent to take over direct control of nuclear weapons" (ibid).

100. Hryhoriieva, "Hra tryvaie."

101. "Pro problem iadernoï zbroï v Ukraïni," *Holos Ukraïny*, April 7, 1993, 10.

102. The Massandra Accords were three agreements signed by the Heads of Government of Ukraine and Russia, Leonid Kuchma and Viktor Chernomyrdin on 3 September 1993, in the town of Masandra, near Yalta: "On the Utilization of Nuclear Warheads" (https://zakon.rada.gov.ua/laws/show/643_132), "The Basic Principles of Utilizing Nuclear Warheads of the Strategic Nuclear Forces Deployed in Ukraine" (https://zakon.rada.gov.ua/laws/show/643_133), and "On the Provision of Warranty and Manufacturer Oversight for the Operation of the Strategic Forces' Strategic Missile Systems Deployed on Their Territories" (https://zakon.rada.gov.ua/laws/show/643_131). They contradicted the decisions of the parliament regarding the elimination of nuclear weapons and instead stipulated to all the nuclear warheads of the strategic nuclear forces to be transferred to the Russian Federation without any legal commitments on the part of the RF regarding their destruction and no international oversight procedures over the conversion of nuclear materials into "peaceful atoms." These agreements also finally knocked the trump out of Ukraine's hand. Suggestions that they be signed started with the first round of negotiations. Ukraine was interested in having Russia service its warheads, and Russia in having Ukraine service its carriers. Ukraine should have signed a package of agreements, but with the Massandra Accords, only Ukraine took on any obligations. This allowed Moscow to blackmail Kyiv later with talk of potential explosions of nuclear warheads.

103. "Statement of National Deputies of Ukraine Regarding Ukraine's Nuclear Status," Parliament of Ukraine, Kyiv, April 1993. Author's personal archive.

104. Ibid., 1–2.

105. Natalia A. Feduschak, "US Stance on Ukraine Pushes It Closer to Declaring Itself a Nuclear Nation," *Wall Street Journal*, May 3, 1993, A10.

106. "Record of a Conversation with the First Secretary of the US Embassy in Ukraine W. Sulzynsky" from the diary of Vladyslav Demianenko, MFA Ukraine; forwarded via a letter to Anatolii Zlenko, foreign minister, to Leonid Kuchma, prime minister, UKOR/ 21–495, 5 May 1993, 1. Kuchma's note: "FYI Kostenko." Author's personal archive.

107. "Record of a Conversation with the First Secretary of the US Embassy in Ukraine W. Sulzynsky" from the diary of Vladyslav Demianenko, MFA Ukraine; forwarded via a letter to Anatolii Zlenko, foreign minister, to Leonid Kuchma, prime minister, UKOR/ 21–495, 5 May 1993, 1. Kuchma's note: "FYI Kostenko." Author's personal archive.

108. "Zaiava pres-tsentru MZS Ukraïny," *Holos Ukraïny*, April 29, 1993, 2.

109. Ibid.

110. Vadym Dolhanov, counselor for information and the press, Embassy of Ukraine in Russia, to Borys Tarasiuk, MFA of Ukraine, "Information About Responses to the Statement Issued by the MFA of Ukraine," Letter no. 8–281, 4 May 1993, 1. Author's personal archive.

111. Ibid.

112. "Nashoï kraïny bezpeka."

113. Volodymyr Tolubko, "Ukraine's Security and Military Doctrine–Formula and Philosophy," 1992, 10. Author's personal archive. SBU head Ievhen Marchuk considered those who were against immediate transfer of nuclear weapons to Russia "dilettante and philistine"; he stated in a 2012 interview: "We had to think very carefully whether to leave all this nuclear potential in Ukraine or not. However, we are now often being upbraided that Ukraine gave up its nuclear weapons, and that is why no one in the world takes us into consideration anymore. Utter nonsense! Germany and Japan are also nonnuclear countries and were completely destroyed during the Second World War [...]. Yet the world certainly takes them into account [...]. Right now, many are thinking: if we had nuclear weapons, everyone would be afraid

of us and would seriously respect us. That's the view of the dilettante and philistine. Thinking that the world would respect you more because you are capable of a first strike and can use nuclear weapons with impunity, and that a strike in response won't just wipe us off the face of the earth, is lower-brain thinking" (Vitalii Kniazhans'kyi, "Odna zustrich i dva uroky," *Den'*, March 30, 2012. https://day. kyiv.ua/uk/article/nota-bene/odna-zustrich-i-dva-uroki).

114. Ministry of Defense of Ukraine to the working group "Cost Assessment of Select Types of Armament and Military Equipment (in 1990–1991 prices)," Analytical Brief, 1992, 2. Author's personal archive.

115. Tolubko, "Ukraine's Security and Military Doctrine," 17.

116. Viktor Tymoshenko, "Philosophical, Theoretical, Legal, Military and Political Issues Concerning the Destruction of Nuclear Weapons and Nuclear Strategic Armament in Ukraine," materials to the first public parliamentary hearings on "Preparing the Treaty on Strategic Offensive Weapons, START I, for Ratification," March 1, 1993, 5, submitted to Yuri Kostenko as head of the working group, to prepare for parliamentary hearings. Author's personal archive.

117. Vitalii Lazorkin and Valerii Kokhno, "Additions and Observations to Ukrainian MFA's Brief, 'Possible Consequences of Alternate Approaches to Ukraine's Implementation of Nuclear Policy,'" no. 1699/02, 4 February 1993, submitted to the head of the working group. Author's personal archive.

118. "An Estimate of Ukraine's Possible Quantity of Weapons," Analytical Brief, 1. Author's personal archive.

119. The Russian press confirmed the strengthening of hawkish positions in Ukraine's political circles, who saw Russia as the main threat: "According to information leaked at the conference, there is no unity in the Ukrainian leadership on the country's nuclear future: the Defense and Foreign Affairs Ministries and the Presidential Administration have not been coordinating their steps. Many here believe that owning nuclear weapons is a good 'vaccination' against the disintegration of a country that, if not today, then tomorrow will be on the verge of collapse." However, Moscow also understood that Ukraine really only needed to defend itself against Russia: "Its defense policy is based on a double standard. For the future: neutrality and nonnuclear state status. For today: maintaining status quo because of the 'unpredictable behavior of neighbors.' It's no secret to anyone that although it's pluralized, what's really meant is Russia" ("Ukraina: 'kvaziiadernaia' superderzhava?" 14–15).

120. Stanislav Koniukhov, "Ukraine's Attitude to START I (An Assessment by Pivdenne Design Bureau), Dnipropetrovsk: Pivdenne Design Bureau, 1993. Author's personal archive.

121. Iurii Pakhomov, Director of Institute for Economics and International Relations, National Academy of Sciences, to Borys Tarasiuk, Head of Ukraine's National Committee for Disarmament and Deputy Foreign Minister, "Letter/Expert analysis no. 7" (in response to Tarasiuk's inquiry UKOR/2912 of 11 January 1993), 26 February 1993. Author's personal archive.

122. Anatolii Zlenko, Foreign Minister of Ukraine, to Leonid Kuchma, Prime Minster of Ukraine, Letter UKOR/21–492, 5 April 1993, with Ukrainian translation of the US Embassy Memorandum attached. Author's personal archive.

123. On 1 February 2005, the parliament ended the force of the military doctrine adopted on 19 October 1993, as a new military doctrine had been brought into force by Decree of President Kuchma on June 15, 2004. Among the new provisions that looked somewhat theoretical at the time but proved realistic in 2014 were Art. 33: "Actions that Ukraine sees as military aggression are defined in the Law of Ukraine 'On the Defense of Ukraine.'"

124. *VVR*, no. 43 (1993): 409, http://zakon.rada.gov.ua/laws/show/3529–12.

125. Pakhomov, "Letter/Expert Analysis No. 7."

126. Ibid.

127. Ibid.

128. The civic position of the academic whose conclusions were used by Ukraine's MFA on the issue of nuclear disarmament became more clearly evident in 2010, with the publication under the title, "NASU Academic Iurii Pakhomov: 'Ukrainians Work Ideally Only When There Is a Russian Overlord.'" The article states: "Ukrainians are mentally unprepared to build an independent state and to ensure its success. These were the words of professional economist and academic with the National Academy of Sciences of Ukraine Iurii Pakhomov. 'The country has not happened because it never was a country,' he stated. The academic believes that Ukraine is missing a link mentally, which is why it cannot find its way in politics or economics. [...] According to Pakhomov, Ukraine's prospects for economic success directly depend on its level of cooperation with Russia. 'The solution is closer interaction with Russia. Ukraine only works well when it is being supervised by Russia. Why was Ukraine the best republic in the USSR? Because when there is an overlord, the Ukrainian works ideally, whereas on his own, without the overlord, he will not work,' said Pakhomov ("Akademik NANU Iurii Pakhomov: Ukrainets rabotaet ideal'no tol'ko kogda nad nim est' russkii khoziain," *censor.net.ua*, October 2010, https://censor.net.ua/forum/2545282/akademik_nanu_yuriyi_pahomov_ukrainets_rabotaet_idealno_tolko_kogda_nad_nim_est_russkiyi_hozyain).

129. Ibid.

130. Iurii Romanov, MP, to Ivan Pliushch, Parliamentary Speaker, "Conclusions and Recommendations from the Seminar on 'Issues of International Environmental Safety in the Process of Rocket Forces' Conversion in the Context of Political and Socioeconomic Changes in CIS Countries,' 14–16 April 1993," Letter *no*. 02–3/610, 22 April 1993. Author's personal archive.

131. Ibid.

132. "Za neiadernu pryntsypovist'," *Holos Ukrainy*, May 26, 1993, 1.

133. Jonathan Eyal, "Still Hugged by the Russian Bear: Jonathan Eyal Accuses the West of Betraying the Former Soviet Republics," *Independent*, May 11, 1993, https://www.independent.co.uk/voices/still-hugged-by-the-russian-bear-jonathan-eyal-accuses-the-west-of-betraying-the-former-soviet-2322228.html.

134. "Ukraine: Barrier to Nuclear Peace," *New York Times*, January 11, 1993, A16; Warren Strobel, "US Presses Ukraine to Ratify Nuclear Arms Pact," *Washington Times*, January 8, 1993, A7; David White, "No Further Nuclear Arms Cuts in View," *Financial Times*, January 8, 1993, 2; "Little Russia's Chip," *The Times*, May 10, 1993; "Washington Wire: Nuclear Missiles," *Wall Street Journal*, May 14, 1993, A1.

135. Potter, "Ukraine's Nuclear Trigger."

136. "General'nyi director NPO 'Khartron,' razrabatyvaiushchego sistemy upravleniia dlia iadernykh raket: Nikto v Ukraine ne znaiet, kuda natseleny rakety, raspolozhennye na territorii," *Interfax-Ukraine*, November 26, 1993.

137. Hoagland, "Race Between Nightmares."

138. Steve Coll and R. Jeffrey Smith, "Ukraine Could Seize Control Over Nuclear Arms," *Washington Post*, June 3, 1993, A01.

139. Smith, "US Fears Ukrainian-Russian Clash."

140. Hoagland, "Race Between Nightmares."

141. Ibid.

142. Peter B. Martin, Letters to the Editor, *International Herald Tribune*, Opinion, November 26, 1992.

143. "Little Russia's Chip.".

144. Smith, "US Fears Ukrainian-Russian Clash."

145. Ibid.

146. Don Oberdorfer, "Administration Rejects Ukrainian Appeal on START I Ratification," *Washington Post*, January 7, 1993, A27.

147. Hoagland, "Race Between Nightmares."
Unlike Western countries, Ukraine did little to gain support for its position abroad, let alone resist Moscow's propaganda. This passivity on the informational front was one of the main reasons for Ukraine's foreign policy failures during the entire independence period. In 2014, Moscow waged an especially active informational war in order to present its military incursion into the Donbas as Ukraine's "internal conflict," with which Russia supposedly had no connection, and to represent reactions in the West as "interference in the domestic affairs of Ukraine." The West began to acknowledge that Russia would beat Europe in the information war: "The Russians are prevailing noticeably in terms of their media strength," said John Whittingdale, chair of the Culture, Media and Sport Committee of the British House of Commons. "This is what I have already told the BBC management. It's frightening to even imagine the degree to which we might lose this information war" ("BBC prohraie informatsiinu viinu propahandi Kremlia," *Radio Svoboda* [RFE/RL], December 22, 2014, https://www.radiosvoboda.org/a/26756881.html).

148. "Posly Ukraïny v nyztsi kraïn ES ihnoruiut' ZMI, popry informatsiinu viinu," *Ievropeis'ka pravda*, November 19, 2014, https://www.eurointegration.com.ua/news/2014/11/19/7027972/.

149. Christian Faber-Rod, Danish ambassador to Ukraine, on behalf of member states of the European Union, to the Ministry of Foreign Affairs of Ukraine, "Démarche in Regard to the Situation with Ukraine's Ratification START I and Accession to the NPT," 17 February 1993. Forwarded by Anatolii Zlenko, MFA, to Ivan Pliushch, parliamentary Speaker, letter UKOR/23–144, 19 February 1993. Author's personal archive.

150. Zlenko to Kuchma, UKOR/21–492.

151. Ibid.

152. Feduschak, "US Stance on Ukraine."

153. "Little Russia's Chip"; Coll and Smith, "Ukraine Could Seize Control Over Nuclear Arms"; Hoagland, "Race Between Nightmares;" "Washington Wire: Nuclear Missiles"; Steven Erlanger, "Ukraine Asks Russia for Talks on Black Sea Fleet," *New York Times*, May 31, 1993, 1, 4.

154. At the time, Canada, like the US, was giving Russia priority over other former Soviet republics. According to information sent by Counsellor Andrii Veselovs'kyi to the MFA in 1993, Canada had maintained the position for the preceding eighteen months that, after the collapse of the Soviet Union the only nuclear state among the successor states was supposed to be Russia. Other republics were supposed to sign the NPT as nonnuclear. At the same time, the Ukrainian Embassy in Canada began to note "a fundamental shift in tone in the Canadian press. On numerous occasions, there have been the so-called balanced commentaries, where Ukraine, Russia and even the US were equally criticized, while in some of the articles Ukrainian positions were being unequivocally justified. [...] Finally, conversations with Canadian citizens, officials and foreign diplomats showed that Canadians were recognizing the fairness of Ukraine's demands and were of the opinion that pressures from the US were not reasonable. Canadian diplomats expressed this same line in private conversations, but some of them expressed concern about the "eastern shift." Doubts were also expressed about the possibility of changing the deep alliance between the US and Russia in the unconditional removal of nuclear weapons from Ukraine without providing the country with firm, unambiguous guarantees. For internal policy reasons, Russia was not prepared to guarantee, while Washington could not offer unilateral ones and risk losing Yeltsin" (Andrii Veselovs'kyi, Counsellor of the Ukrainian Embassy in Canada, "Nuclear Weapons in Ukraine: Canadian Position. A Political Letter," 18 May 1993. Author's personal archive).

155. Here is an example of attitudes in the Ukrainian military towards disarmament in the run-up to the closed session in the parliament of 3 June 1993: "The Ovruch Society of Officers of Ukraine (SOU), Unit 18876, declares that certain super-peaceloving deputies are prepared to disarm Ukraine, give over its nuclear weapons, in return for a pittance of US $176m, even as nearly all the country's neighbors are making claims on its territory. Only if we have nuclear weapons will we be able to freely develop. There's no point to having peace on our land by making the lives of the people subject to trading [...]. Ratifying the treaty on nonproliferation of nuclear weapons will be the last, tragic mistake [...]. SOU Head for Unit 18876 Holub" (Ovruch Society of Officers of Ukraine [SOU] to parliamentary Speaker Ivan Pliushch, Telegram *no.* 02–3/874, 3 June 1993. Author's personal archive).

156. Erlanger, "Ukraine Asks Russia for Talks."

157. Leonid Kuchma, prime minister of Ukraine, to Ivan Pliushch, parliamentary Speaker, letter no. 29–1041/4, forwarding analytical note from Ukraine's MFA, "Possible

Consequences of Not Acceding to the NPT," 17 May 1993. Author's personal archive.

158. "Little Russia's Chip."

159. Volodymyr Knysh, "Start-1 ne dolzhen stat' finishem...: Na kakikh usloviiakh Ukraine ratifitsirovat' dogovor o sokrashchenii strategicheskikh iadernykh vooruzhenii?" *Narodna armiia*, April 14, 1993, 1.

160. Hoagland, "Race Between Nightmares."

161. Anatolii Zlenko, foreign minister, to Leonid Kuchma, prime minister, letter UKOR/21–482 forwarding Ukrainian translation of Systems Planning and Analysis, Inc. to the Ministry of Foreign Affairs of Ukraine, "Demilitarizing Project for the Nuclear Missile Arms Deployed on Ukraine's Territory, and Ensuring Their Reliable and Safe Storage until Their Complete Elimination," 28 April 1993, 1 (Kuchma's note: "FYI Kostenko," 10 May 1993). Author's personal archive.

162. Ibid., 3.

163. Ibid., 4.

164. Ibid., 5.

165. Ibid., 8.

166. Ibid., 2.

167. Ibid.

Chapter 4

1. The National Disarmament Committee (NDC) was an interagency state body for according and coordinating Ukraine's policy on disarmament and arms control. It was established by presidential decree by Leonid Kravchuk on 15 June 1992, and lost effect on 9 August 2002. The NDC functioned under the general direction of the MFA and reported to the president. It consisted of the deputy directors of all enforcement, defense and industrial agencies, as well as the Academy of Science. On matters that related to arms control and disarmament, the NDC was expected to organize research, analytical, forecasting, communication, coordination, and oversight activities, and to take part in developing regulations and proposals for defining and adjusting the direction of Ukraine's foreign policy activities regarding disarmament. The committee organized and coordinated the work of the delegations during negotiations and approved draft directives for Ukraine's delegations. The NDC had its own executive bodies, whose functions were carried out by the Office of Arms Control and Disarmament (OACD) under the Foreign Ministry and the Verification Center of the Armed Forces of Ukraine (AFU).

2. "Ne prostyi shliakh do bez'iadernoho svitu," *Holos Ukrainy*, January 5, 1993, 12. For comparison purposes, the next four footnotes present texts from the Borys Tarasiuk interview published in this article, which are almost identical to the text of Anatolii Zlenko's speech at the 50th session of the parliament, 3 June 1993.

3. "Experts know that nuclear warheads are extremely complex mechanisms that teams of thousands of highly qualified scientists and technologists have worked to develop in nuclear states [...]. And even if we could still manage to produce the first three elements through many years of extremely exhausting effort on the part of the entire Ukrainian nation at the cost of billions, the nuclear testing on Ukrainian territory is simply out of the question. The same kind of infrastructure is necessary to support nuclear warheads that Ukraine inherited from the former USSR under safe conditions. Each of them has a guaranteed lifespan, after which it needs to be dismantled and completely remade. The vast majority of warheads that are deployed on the territory of Ukraine have already exhausted a good part of their guaranteed term. And yet, at a certain point after their shelf life ends, even those enterprises that produced them in the first place will not be prepared to take on dismantling these nuclear warheads. This simply becomes dangerous" ("Ne prostyi shliakh do bez'iadernoho svitu").

4. "Ukraine's declared intention to gain status as a nonnuclear weapons state in the future became a kind of pass to the international community of civilized nations. This is one of the conditions without which Ukraine cannot expect to take part, as an equal, in international economic or technological relations. On the contrary, if Ukraine were to make the opposite choice, all the levers of political, economic and psychological pressure would be brought to bear against the country–our Western partners have been pretty open about this" (Ibid).

5. "As to security guarantees for Ukraine, we're talking about the nuclear states making a political commitment in the form of an appropriate document that would confirm their commitment to treat any kind of use or threat of use of force against Ukraine by a nuclear state, as unacceptable. Of course, these kinds of commitments, in and of themselves, do not guarantee Ukraine's security, but they will have considerable political and legal significance" (Ibid).

6. "For Ukraine, maintaining normal, friendly relations with all its neighbors, developing political, economic and technological relations with the US and other Western countries as a priority, participating in Europe-wide security systems, getting security guarantees from the nuclear states, establishing the necessary conditions for foreign investment, undertaking deep economic reforms based on a market economy, further democratizing Ukrainian society–all these are the components of our national security, and they will only be possible provided that Ukraine acquires the status of a nonnuclear weapons state" (Ibid).

7. Ukraine presented mutually contradictory figures about the funds needed for nuclear disarmament. The figure mentioned by Foreign Minister Zlenko, US $2.6 billion, was based on the calculations made by our Working Group. The ministry did not come up with it on its own. But Zlenko left out the fact that this referred to dismantling only of that part of the weapons whose reduction

was required under the Lisbon Protocol: 36 percent of carriers and 42 percent of warheads. Obviously, the minister had no other numbers but did not dare admit that he was talking about partial reduction at this point. By the beginning of 1993, various figures began to be bandied about in the press, that Ukraine was supposedly demanding to finance its disarmament. In January, the *Washington Post* wrote: "K[yiv] has also radically increased its financial demand for denuclearization to $1.5 billion" (Virginia I. Foran, "Ukrainian Holdout: The Real Problem with the Treaty," *Washington Post*, January 3, 1993, C03), but three month later, with reference to a meeting between Zlenko and officials at the Administration in Washington, the paper reports about US $3 billion (Don Oberdorfer, "Ukraine Is Loath to Yield Nuclear Arms; Foreign Minister, Here for Talks, Notes Current Crisis in Russia," *Washington Post*, March 25, 1993, A29). Then, two months later, a US $2 billion figure was quoted as necessary for dismantling in the *Wall Street Journal* ("Washington Wire: Nuclear Missiles," *Wall Street Journal*, May 14, 1993, A1). On the surface, the impression was that Ukraine did not have solid figures and was just making them up as it went along. This situation also testified that Ukraine's position had not been convincingly presented through MFA channels to the negotiating parties as the single, official and necessary one to start talking about dismantling.

8. Session 50, transcript of the plenary session, 3 June 1993, 10:00, http://static.rada.gov.ua/zakon/skl1/BUL17/030693_50.htm.

9. Ibid.

10. Session 50, closed part of the session, 3 June 1993. Author's personal archive.

11. Ibid.

12. Ibid.

13. Each SS-19 carried six nuclear warheads.

14. Ibid.

15. Ivan Pliushch, parliamentary Speaker, "Excerpt of Protocol no. 51 of the Parliament of Ukraine, "On Authorization with Regard to Reviewing the Question of Ratifying the Strategic Arms Reduction Treaty and Acceding to the Treaty on the Nonproliferation of Nuclear Weapons. Protocolary," Parliament of Ukraine, 3 June 1993. Author's personal archive.

16. A week after the parliament session in which deputies rejected the Russian scenario for Ukraine's nuclear disarmament, a widespread miners' strike began in the Donbas: "The strikers' political demands–a snap election, regional autonomy, and the restoration of ties with Russia–were undoubtedly beneficial to Russia. Russia's foreign policy concept approved by Boris Yeltsin notes that the prevalent actions of Ukraine's current government show 'a tendency to distance itself from the Russian Federation and from the CIS as a whole.' At the same time, the concept states, 'actions intended to weaken ties between the Russian Federation and former USSR republics' represent a threat to Russia's vital interests. From this, it follows logically that Russia is interested to see either a change in the current government in Ukraine or an adjustment in the current leadership's foreign policy course to reflect Russia's geopolitical interests to a greater degree and to show more activity in integrating within the framework of the CIS. Ukraine's leadership was able to feel this once again during a recent visit to Moscow. *Kommersant* picked up a phrase dropped by Deputy Premier Vasyl' Ievtukhov that was clearly not intended for the press: 'What will be, will be [...] toeing the line'" ("Kiev vynuzhden korrektirovat' politicheskii kurs," *Kommersant*, June 25, 1993, https://www.kommersant.ru/doc/51857?isSearch=True).

17. Chrystia Freel and R. Jeffrey Smith, "Ukrainian Premier Urges Keeping Nuclear Arms," *Washington Post*, June 4, 1993, A01.

18. Freel and Smith, "Ukrainian Premier Urges Keeping Nuclear Arms."

19. Ibid.

20. Ibid.

21. Stanislav Kondrashov, "Neozhidannyi iadernyi treugol'nik," *Izvestiia*, June 26, 1993, 7.

22. Chrystia Freel, "Ukrainian Leader Says No Change on A-Arms; President Contradicts Premier's Statement," *Washington Post*, June 5, 1993, A17.

23. Chrystia Freel, "Ukrainian Leader Says No Change on A-Arms; President Contradicts Premier's Statement," *Washington Post*, June 5, 1993, A17

24. International observers believed that Ukraine's MFA was suffering because it had no position of its own: "During the second half of 1992 and the early months of 1993, Ukrainian diplomacy encountered a number of difficulties and obstacles. Some of these were domestic in origin–the Ukrainian government failed to redraw a reasoned and well-thought-out concept of foreign policy in the new and rapidly changing geopolitical situation of Eastern Europe, where newly independent states frequently found their national interests in conflict with those of their neighbors. The lack of a clear doctrine and the absence of allies–and also, in part Russian reluctance to abandon direct influence throughout former Soviet space–led to a foreign-policy crisis and, to some extent, the isolation of Ukraine from the mainstream of world affairs. In the United States, the Bush Administration supported the concept of Russia as a guarantor of strategic stability in the former Soviet geopolitical space [...]" (Serhii Tolstov, "Ukraine's Nuclear Dilemma," *The World Today* 49, no. 6 [June 1993]: 104).

25. Zlenko, "Review of American Press."

26. "Dopovid' ministra zakordonnykh sprav Ukraïny A. M. Zlenka na sesiï Verkhovnoï Rady Ukraïny 3 chervnia 1993 roku," *Holos Ukraïny*, June 5, 1993, 3–4.

27. Freel and Smith, "Ukrainian Premier Urges Keeping Nuclear Arms."

28. The Parliamentary Commission on Budget Planning, Finance and Pricing sided with the nuclear hawks: "The world community should look at the question of succession fully, not only at issues that affect individual states or groups of states. [...] This should be the gist of Verkhovna Rada statement. We can't blindly join treaties that were drawn up under completely different political conditions in Ukraine's transition to nonnuclear weapons status. Here we need to consider an entire gamut of

issues: returning Ukraine its share of Soviet assets, gold reserves, diamond fund, cultural artifacts, and property abroad; security against territorial incursions, against military, political and economic blackmail; financial and technical support for disarmament; the destruction of tactical nuclear weapons that are moved from Ukraine to Russia under international supervision with the participation of Ukraine. [...] There can be no talk about destroying rockets and tearing down missile silos [...]. We never said we intended to become the defenseless victim of any aggressor. No one can take away our right to properly take care of our level of defense. If we're talking about destroying nuclear warheads, this should not be delegated to Russia. That country's actions have shown pretty convincingly that it cannot be trusted." ("Proposals from the Members of the Parliamentary Budget Planning, Finance and Pricing Commission on the Ratification of Nuclear Treaties," 23 July 1993. Author's personal archive).

29. Les Aspin's proposition regarding the dismantling and storing of warheads in Ukraine demonstrated the shift in the US position towards Ukraine: "Defense Secretary Les Aspin went to Kyiv last week to 'turn the page' in US-Ukrainian relations. The US offered to help separate Ukraine's nuclear warheads from the missiles they're deployed on and store them in Ukraine under international supervision. Mr. Aspin is also seeking to form closer military relations between the US and Ukraine. [...] [H]is trip to Ukraine was a large step in the right direction. The existence of the disputed ICBMs and operational control over them are not small matters, and little is to be gained by an extended argument between Russia and Ukraine. Ukraine's responsiveness to the Aspin mission at least suggests that the US can indeed play a useful, active and evenhanded role in the future of these two important nations" ("Whose Nukes Are They, Anyway?" *Wall Street Journal*, June 17, 1993, A10).

30. In the US, American pressure on Ukraine was recognized: "Some US officials admit candidly to having botched the Ukrainian situation, at least until the most recent Clinton administration initiatives. A few even plead guilty to Ukraine's central charge–that the United States has so far signaled, in effect, that it will not respond vigorously if Russia seeks to retake Ukraine" (Steve Coll, "Reborn Ukraine Faces Growing Pains in Its Quest for Global Respect," *Washington Post*, June 20, 1993, A24). This was also mentioned by the Center for Security Policy in response to a vote in the Russian legislature "for a resolution proclaiming the Ukrainian city of Sevastopol to be Russian territory" ("Russian Parliament's Grab for Sevastopol Demands Immediate Shift in US Stance on Ukrainian Nukes," *Center for Security Policy*, July 9, 1993, https://www.centerforsecuritypolicy.org/1993/07/09/russian-parliaments-grab-for-sevastopol-demand-immediate-shift-in-u-s-stance-on-ukrainian-nukes-2/).

31. Hearing before the Subcommittee on European Affairs of the Committee on Foreign Relations, United States Senate, One Hundred Third Congress, First Session, 24 June 1993, *US Government Printing Office*, Washinghton, DC, 45–46.

32. Ibid., 43–45.

33. "Monitoring of Compliance with the Provisions of the Treaty Between the US and the Russian Federation on Further Reduction and Limitation of Strategic Offensive Arms (the START II Treaty) Treaty Doc. 103–1," 24 June 1993, US Senate, Committee on Foreign Relations, Washington, DC. "Statement of Hon. R. James Woolsey, Director, Central Intelligence Agency," 85.

34. Hearing before the Subcommittee on European Affairs, 19.

35. Russia's aggression against Ukraine in 2014 shows that, back then, American officials understood the situation better than many Ukrainian ones.

36. Hearing before the Subcommittee on European Affairs, 47.

37. Ibid., 46–47.

38. Ibid., 46.

39. Ibid., 46.

40. The Center for Security Policy is a neo-conservative think tank in Washington. In 1993, its board included a large number of officials from the Reagan Administration. Its president and founder was Frank Gaffney, a one-time executive at the Pentagon. Among the other board members was Jeane Kirkpatrick, who was US ambassador to the UN under Reagan. In June 1993, the Center clearly stated its opinion that Russia's goal was to return Ukraine to it sphere of influence, which made Ukraine's insistence on security guarantees justified and such as required US support.

41. "What Strobe Talbott Won't Tell the Senate Today: Insisting On A Nuclear-Free Ukraine Is Folly," *Center for Security Policy*, 24 June 1993, https://www.centerforsecuritypolicy.org/1993/06/24/what-strobe-talbott-wont-tell-the-senate-today-insisting-on-a-nuclear-free-ukraine-is-folly-2/.

42. Mearsheimer, "Debate: The Case for a Ukrainian Nuclear Deterrent," 50.

43. Ibid., 61.

44. Ibid., 65.

45. Ibid., 65–66.

46. Miller, "Debate: The Case Against a Ukrainian Nuclear Deterrent," 67–80.

47. Fareed Zakaria, managing editor, *Foreign Affairs*, to Yuri Kostenko, minister of the environment, facsimile transmission (212) 861–1849, 19 July 1993. Author's personal archive.

48. Yuri I. Kostenko, "Kiev & the Bomb: Ukrainians Reply," *Foreign Affairs*, Letters to the Editor, vol. 72, no. 4 (September/October 1993): 183–184.

49. Ibid., 183.

50. Ibid.

51. Ibid., 184.

52. Tolstov, "Ukraine's Nuclear Dilemma," 104.

53. Kondrashov, "Neozhidannyi iadernyi treuhol'nik."

54. Ibid.

55. "What Strobe Talbott Won't Tell the Senate Today," *Center for Security Policy*.

56. "Russian Parliament's Grab for Sevastopol Demand Immediate Shift In U. S. Stance On Ukrainian Nukes," *Center for Security Policy*, July 9, 1993, https://www.centerforsecuritypolicy.org/1993/07/09/

russian-parliaments-grab-for-sevastopol-demand-imme-diate-shift-in-u-s-stance-on-ukrainian-nukes-2/.

57. George Melloan, "Global view: Clinton is Being Drawn Into a Russian Web," *Wall Street Journal*, January 10, 1994, A13.

58. Ibid.

59. *VVR*, no. 37 (1993): 379, http://zakon.rada.gov.ua/laws/show/3360–12.

60. Session 72, transcript of plenary session, 2 July 1993, 10:00, 92, http://iportal.rada.gov.ua/meeting/stenogr/show/4886.html.

61. Ibid., 90.

62. Ibid.

63. *VVR*, no. 43 (1993), 409, http://zakon.rada.gov.ua/laws/show/3529–12.

64. Parliament Session 72, 83.

65. Anatolii Zlenko, minister of foreign affairs, to Vasyl′ Durdynets′, first deputy speaker of Parliament, UKOR23–779 dated 7 July 1993; letter FYI from Vasyl′ Durdynets′ to Yuri Kostenko no. I0207/02–3/1193, 7 July 1993. Author's personal archive.

66. Ibid.

67. Ibid., emphasis added.

68. Dubinin, "Iadernyi dreif Ukrainy."

69. "On the question of nuclear weapons in Ukraine," *Holos Ukrainy*, April 7, 1993, 10.

70. Ukraine did not hand over its Objects S because the Russian Federation would have then removed everything, as it had done with the tactical nuclear weapons: "Ukraine does not have a single nuclear warhead whose guaranteed lifespan has ended or is coming to an end. The lifespan of the carriers is coming to an end. Indeed, there's a problem in that the warheads taken off the rockets are being kept in units at missile bases, and not in the Objects S that are designed for this purpose. Ukraine can't hand over its Objects S to Russian jurisdiction because this is an unresolved legal matter. Russia continues to argue with us over the issue of who owns the warheads. By handing over the right to ownership of our Objects S to Russia, we would lose legal control over the warheads. Then Russia will be in a position to move these warheads out of Ukraine under the guise of routine maintenance. This will happen if Moscow has the right to dispose of the Objects S. In no time at all, we would lose all our warheads. We've already had the unpleasant experience of Russia removing our tactical weapons. At the time, we had not signed an agreement that the weapons were Ukrainian property and that Russia was supposed to compensate us for the loss of material assets. As soon as the tactical weapons crossed our border, Russia announced that all discussions on this matter were closed and we would not be compensated." (Viktor Tkachuk, "Iaderna zbroia–tse zasib strymuvannia ahresiï," *Visti z Ukraïny*, no. 33, February 1994, 8).

71. Dubinin, "Iadernyi dreif Ukrainy."

72. Ibid.

73. Ibid.

74. Russia saw the Massandra Accords as a definitive victory: "The issue was of such fundamental significance that they decided that agreement would be formed as a special protocol that was not for publication. The moment of truth was approaching. Obviously, this was why the deep differences that existed in Ukraine over the issue of nuclear weapons broke to the surface. Defense Minister Kostiantyn Morozov criticized the agreements at a plenary session, speaking against his own president. And although Kravchuk did not agree with him, when they prepared the final documents the Ukrainian side did everything it could to distort the essence of what had been agreed. At one point the diplomatic contretemps reminded me of hand-to-hand combat. In the end, our Ukrainian colleagues were unable to change anything in the three draft agreements. The Protocol on removing all the nuclear warheads was short and to the point" (Ibid).

75. Ibid.

76. Ibid.

77. "Protokoly, kotorye potriasli mir," *Kievskie vedomosti*, September 9, 1993, 2.

78. Dubinin, "Iadernyi dreif Ukrainy."

79. The domestic press was actively reporting on the political situation in Ukraine during the period of negotiations on nuclear disarmament: "In Kyiv, yesterday [26 August 1993], the Parliament of Ukraine opened its plenary session. The presidium proposed placing three items on the agenda: voting on the government's latest program for overcoming the crisis, opinions on the upcoming 26 September referendum, and the issue of measures for social protection in connection with a nearly fivefold rise in prices expected by the beginning of September. Undoubtedly, the central issue of this session is the referendum. On 17 June, under pressure from widespread strikes that have paralyzed the economy in the industrial east of the republic, the parliament decided to hold a national referendum on confidence in the Rada and the president on September 26" (Alla Kovtun, "Parlament poka ne prinial reshenie o samorospuske," *Kommersant* #163, 27 August 1993, http://www.kommersant.ru/doc/57838).

80. Agenda of Working Group sessions 9 June, 25 June, and 9 July 1993. Author's personal archive.

81. "Working Group propositions regarding the ratification of START I," 16 November 1993. Author's personal archive.

82. "On ratifying the Treaty between the USSR and the USA on the reduction and limitation of strategic arms, signed in Moscow on July 31, 1991, and the attached protocol, signed in Lisbon on behalf of Ukraine on 23 May 1992." Author's personal archive.

83. Ibid.

84. "Pozytsiia Ukraïny chitka i iasna."

85. "Povidomlennia pres-sluzhby prezydenta Ukraïny i Kabinetu ministriv Ukraïny," *Holos Ukraïny*, November 23, 1993, 3.

86. "START-1: ratyfikatsiia iz zasterezhenniamy," *Holos Ukraïny*, November 23, 1993, 3.

87. Ibid.

88. Ibid.

89. Ibid.

90. Ibid.

91. "Pravitel′stvo Rossii schitaet ratifikatsiiu Ukrainoiu dogovora SNV-1 s ogovorkami grubym narusheniem

mezhdunarodnykh obiazatel′stv po iadernomu oruzhiiu," *Interfax-Ukraine*, News on Ukraine, November 26, 1993.

92. Ibid.

93. Dubinin, "Iadernyi dreif Ukrainy."

94. "O ratifikatsii dogovora mezhdu Rossiiskoi Federatsiei i Soedinennymi Shtatami Ameriki o dal′neishem sokrashchenii i ogranichenii strateicheskikh nastupal′nykh vooruzhenii," Federal Law no. 56-FZ, dated 4 May 2000, *Rossiiskaia gazeta*, May 6, 2000, 6.

95. Ibid.

96. Ibid.

97. The fact that Russia basically was not destroying its nuclear arsenals according to START I is confirmed by these figures: "At the beginning of 1992, Russia had a strategic nuclear force of 627 ICBMs with 3,727 warheads. [...] The last day of 1999, President Yeltsin transferred to his successor, Vladimir Putin, the following nuclear forces: 756 ICBMs with 3,540 warheads [...]." In short, all reductions of nuclear arms under this treaty were at the expense of Ukraine, Kazakhstan and Belarus. All Russia did was make money by removing all nuclear warheads and the best strategic carriers its own territory. On the other hand, "Between 2000 and 2007, the Strategic Nuclear Forces lost 405 carriers and 2,498 warheads" (Nikolai Poroskov, "V plenu mifa: Rossiiskaia armiia: vozrozhdenie ili degradatsiia," *Flot.com*, December 17, 2010, https://flot.com/nowadays/structure/revive.htm). These figures show that in reality Russia began the disarmament process only after ratifying START II in May 2000.

98. A Research and Development enterprise, called Khartron-Arkos Ltd. today.

99. "'Heneral′nyi director NPO 'Khartron'" (General Director of Khartron), *Interfax-Ukraine.*

100. Democratic practice of ratifying interstate treaties allows for various kinds of amendments, reservations and interpretations to be included prior to having the documents signed by the country's top officials. These are added by lawmakers, based on national interests. Thus, in line with the rules of the US Senate, "the resolution of ratification when pending shall be open to amendment in the form of reservations, declarations, statements, or understandings" (Rule XXX (c), Executive Session–Proceedings on Treaties, United States Senate Manual, 112th Congress, S. Doc. 112–I, Standing Rules of the Senate, US Government Publishing Office, https://www.govinfo.gov/content/pkg/SMAN-112/html/SMAN-112-pg56.htm, 56). Based on this rule, "an 'amendment' changes the actual language of the agreement; a 'reservation' does not actually alter the language of the agreement itself, but it changes or restricts its legal implications; 'understandings' do not change the agreement and do not restrict its legal implications." ("Agreements in the US Senate," *Zakhid: Window in Ukraine*, Pylyp Orlyk Institute of Democracy, no. 41, 3 June 1993. Author's personal archive). In this way, Moscow's claims that "Ukraine's ratification of START I with reservations [is] a gross violation of international commitments regarding nuclear arms" showed a gross–albeit standard practice for Russia–interference in the domestic affairs of Ukraine. In fact, the parliament made amendments and reservations to START I and the Lisbon Protocol in line with democratic practice.

101. "Pozytsiia Ukraïny chitka i iasna."

102. This quote can be found verbatim in a slew of publications, without any source indicated, which suggests that this opinion had become axiomatic for many Ukrainian writers. Moreover, it coincides with the arguments presented by the MFA, which disagreed with the parliament resolution of 18 November 1993 (Oleksandr Zadorozhniy, *Mizhnarodne parvo v mizhnarodnykh vidnosunakh Ukraïny i Rosiis′koi Federatsii 1991–2014*, [Kyiv: KIS, 2017], 82; Tamila Shutak, 2009, "Iadernyi chynnyk y zovnishnii politytsi administratsii prezydenta SShA B. Klintona (1993–2000 rr.)," *Vernads′kyi National Library of Ukraine*, http://dspace. nbuv.gov.ua/bitstream/handle/123456789/12983/13-Shutak.pdf; C. I. Namoniuk, 2017, "Vtrata iadernoho statusu iak zovnishn′opolitychnyi prorakhunok Ukraïny," *Dnipropetrovsk University Bulletin* no. 2, https://visnukpfs. dp.ua/index. php/PFS/article/view/886/0, 72).

103. "Komporomis bude znaideno. Ale chy skoro?" *Holos Ukraïny*, January 11, 1994, 1; "Trilateral Statement by the Presidents of Ukraine, the United States and Russia," the Parliament of Ukraine, Legislation of Ukraine, http://zakon.rada.gov.ua/laws/show/998_300.

104. Launch facilities (LF), or missile silos, were powerful fortifications built underground to up to 30 meters deep, 9 meters in diameter, and clad in 1.25 meters of reinforced concrete and 40 milimeters of metal. At the beginning of the 1990s, we were told by specialists at the Defense Ministry that there were several known ways to dispose of them. Russian technology involved exploding a 3-ton explosive in the shaft of the silo at a depth of 6 m, which meant irreversible destruction. Other technologies were less barbarous, but costly: RUB3 million, which was then nearly US $3 million (as per the exchange rate in the last days of the USSR) per facility. The Paton Institute under the Academy of Sciences in Ukraine had developed a technology that was an order of magnitude cheaper and 5–6 less time-consuming. This made it possible to eliminate costly lifting equipment and to preserve the shaft of the silo, which could further be used in the agricultural sector. The main engineer at the Pivdenne Construction Bureau, Stanislav Koniukhov, wrote in a policy brief called "Ukraine's Position Regarding START I (an evaluation of Pivdenne CB)," "The launch facilities that are to be disposed of, 36 percent, will be disposed of [...] by dismantling the upper section of the silo. The LFs that are part of the quota will be maintained with the purpose of further utilization." The LF were then disposed of using two methods: small explosions involving only 3–4 kilograms of explosives if the LF was no closer than 3 km to the nearest settlement, and also by abrasive cutting, for which the German company Alba was engaged. All the sites were reclaimed by 1 September 2002.

105. Borys Tarasiuk, chair, National Committee for Disarmament, to Vasyl′ Durdynets′ first deputy speaker of the Parliament of Ukraine, letter no. UKOR/23–786, 13 July 1993, forwarded by Durdynets′ to Kostenko "FYI," letter

no. 10507/02–3/1212, 15 July 1993. Author's personal archive.

106. Ibid.

107. "Iaderna zbroia–tse zasib strymuvannia," *Ukraïns'ka hazeta*, January 20, 1994, 8.

108. Ibid.

109. Aleksandr Shalnev, "Senatory gotovy utverdit' dogovor ob SNV," *Izvestiia*, October 2, 1992, 6.

110. William J. Broad, "Nuclear Accords Bring New Fears on Arms Disposal," *New York Times*, July 6, 1992, 1, https://www.nytimes.com/1992/07/06/world/nuclear-accords-bring-new-fears-on-arms-disposal.html.

111. It was standard practice for Russia to misinterpret the commitments Ukraine had taken upon itself: "However, this problem was already considered resolved several times, starting with the adoption in October 1991 of the parliament's Declaration 'On the Nuclear-Free Status of Ukraine' and ending with the signing of the Lisbon Protocol in May 1992 by the United States, Russia, Belarus, Ukraine, and Kazakhstan. The protocol provided for the elimination of all nuclear weapons on Ukrainian territory within seven years and Ukraine's accession to the Non-Proliferation Treaty as a nonnuclear weapon state. Yet these obligations did not prevent Ukraine's leadership from including, in 1992, the 43rd Rocket Army (176 ICBM silos) and the 46th Air Force Division (strategic bombers and 670 nuclear warheads for ALCMs) in the Ukrainian Armed Forces. And in July 1993, the parliament proclaimed Ukraine 'the owner of nuclear weapons due to historical circumstances.' This situation led to inevitable consequences. Firstly, Ukraine's nuclear warheads were left without absolutely necessary maintenance. Secondly, this created a direct threat to the treaty and the entire non-proliferation program […]." (Aleksandr Konovalov, "Kiev poluchaet garantii," *Moskovskie novosti*, January 16–23, 1993, 13).

112. *VVR*, no. 31 (1990): 429, http://zakon1.rada.gov.ua/laws/show/55–12.

113. *VVR*, no. 14 (1991): 168, http://zakon2.rada.gov.ua/laws/show/698–12.

114. "Vienna Convention on Succession of States in Respect of State Property, Archives and Debts," Vienna, 8 April 1983, *United Nations Treaty Collection*. https://treaties.un.org/Pages/ViewDetails.aspx?src=IND&mtdsg_no=III-12&chapter=3&clang=_en.

115. *VVR*, no. 51 (1991): 742, http://zakon.rada.gov.ua/laws/show/1697–12

116. *VVR*, no. 29 (1992): 405, http://zakon.rada.gov.ua/laws/show/2264–12.

117. *VVR*, no. 49 (1993): 464, http://zakon.rada.gov.ua/laws/show/3624–12.

118. Ibid.

Chapter 5

1. "Partnership for Peace: Framework Document," Ministerial Meeting of the North Atlantic Council/North Atlantic Cooperation Council, *NATO Headquarters*, Brussels, 10–11 January 1994, https://www.nato. int/docu/comm/49–95/c940110b.htm.

2. Aleksei Pushkov, "Rossiia i NATO: chto dal'she? (Russia and the NATO: What's Next?), *Moskovskie novosti*, no. 3, 16–23 January 1994, 13.

3. Ukraine's economic crisis demanded a pursuit for new resources, wrote Steve H. Henke and Sir Alan Walters in *Forbes*: "To finance moneylosing state-owned enterprises and meet its payroll, the government has simply ordered its central bank to print money. Inevitably, hyperinflation has surged, so that it now stands at 100 percent a month–five times the level in Russia. Since mid-August the currency, the karbovanets, has depreciated from 6,000 per dollar to more than 31,000 per dollar. Inflation is a fatal disease. If not checked, it destroys a society and paves the way for revolution. Revolution invites foreign interference. Russia is ready. […] In all of Central Europe and the former Soviet Union, Ukraine is the only powerful balancing force to the military might of Russia. It is important for the whole region that Ukraine overcome its economic crisis and be able to survive independent of Russia" (Steve H. Henke and Sir Alan Walters, "Emerging Hopes for the Russian and Eastern European Economies May be Dashed Unless Ukraine Can Straighten out Its Economic Mess: Powder Keg," *Forbes*, As We See It, 3 January 1994, 64).

4. George Melloan, "Clinton is Being Drawn into a Russian Web," *Wall Street Journal*, Global View, 10 January 1994, A13.

5. George Melloan, Deputy Editor of the *Wall Street Journal*, reported thusly on KGB's interference in Georgia: "Last week, the news from the Republic of Georgia was that its deposed president, Zviad Gamsakhurdia, had committed suicide on Dec. 31 after many months on the run. Georgian intrigues are complex, but the war in that little country bears all the fingerprints of an old Soviet KGB technique for keeping the republics in line. When a regional leader didn't behave you caused trouble on his patch and ended it only when the victim repented. Georgia's current president, former Soviet foreign minister Eduard Shevardnadze, repented by bringing Georgia back into the Commonwealth of Independent States (CIS). Lo and behold, the Abkhazian gunnery subsided" (Ibid).

6. Ibid.

7. Ibid.

8. Ihor Hrushko, "Iakshcho tuman ne zavadyt'" (If We Can See through the Fog), *Holos Ukraïny*, 12 January 1994, 1, 8.

9. Ihor Hrushko, "My ne stanemo na dorozi iadernoho rozzbroiennia" (We Will Not Stand in the Way of Nuclear Disarmament), *Holos Ukraïny*, 14 January 1994, 1–2.

10. Ibid.

11. As we already know, the war of the nuclear Russia against the nonnuclear Ukraine in 2014 refuted this claim by the US president entirely.

12. Ibid. This promise, on the other hand, was fulfilled by the United States in its entirety: billions of dollars associated with elimination of the third largest nuclear potential in the world fell on Ukraine's shoulders.

13. Melloan, "Clinton is Being Drawn into a Russian Web."

14. As it turned out later, Ukrainian diplomats took part in the negotiations of the Trilateral Statement. Ukraine's MPs, and even 's leadership, knew nothing about these negotiations. Borys Tarasiuk stated during a live broadcast of "Shuster LIVE" talk show on December 5, 2014: "The whole world at the time–from the US and its NATO allies to Russia–categorically insisted that Ukraine should give up its nuclear weapons. Essentially, this is when the negotiations commenced, in 1992–1993, and I took part in these negotiations, that led to the Trilateral Statement signed in Moscow on January 1994, and that was subsequently reworked into the Memorandum on Security Guarantees for Ukraine. Was this document good or bad? I can say this as a participant in the negotiations, at the time, Ukraine was the only country pursuing the idea of this document to be legally-binding, for it to put legally-binding obligations onto guarantor countries. Both the US and Russia were categorically against it. [...] We had no choice but to sign the Budapest Memorandum at the time" ("Shuster LIVE," *UA: Pershyi*, National Public Broadcasting Company of Ukraine, 5 December 2014, https://www.youtube.com/watch?v=vapbFNz3fQ4).

15. "'START'–Tse nadto ser'izno, shchob pospishaty" (START is too Serious to be Rushed), *Holos Ukraïny*.

16. Russian military analysts were rejecting a possibility that, some day, Russia would bring to power politicians with invading intentions: "After all, if we let our policy be influenced by those yearning to wash the boots of Russian soldiers in the warm waters of the Indian Ocean, then tomorrow they will want to quench their cavalry horses' thirst in the Dnieper river. In this case, it will be very difficult to convince Ukraine of the benefits of getting rid of nuclear weapons" (Konovalov, "Kiev poluchaet harantii" [Kyiv Receives Guarantees]).

17. George Melloan also wrote on the Russian threat and KGB's influence in the countries of the former socialist bloc: "The shrewd operatives of the former KGB still have many links to behind-the-scenes power brokers in the former Soviet republics and former Warsaw Pact nations. They are most likely not yet under effective presidential control. It is primarily for this reason that Poland, the Czech and Slovak republics, Hungary and Bulgaria have pleaded with Mr. Clinton to grant them NATO admission. They know from long experience that the principle tool of Russian imperialism was subversion. They are more concerned about internal stability than any external threat" (Melloan, "Clinton is Being Drawn into a Russian Web").

18. "Trilateral Statement by the Presidents of Ukraine, the US and Russia," signed in Moscow, 14 January 1994, http://zakon.rada.gov.ua/laws/show/998_300.

19. *VVR*, no. 29 (1992): 405, http://zakon.rada.gov.ua/laws/show/2264-12.

20. *VVR*, no. 49 (1993): 464, http://zakon.rada.gov.ua/laws/show/3624-12.

21. Ibid.

22. The US knew that the Budapest Memorandum did not guarantee protection from the use of military force. Explaining the restrained response by the US to Russia's military aggression against Ukraine, US Ambassador to Ukraine Geoffrey Pyatt stated: "The Budapest Memorandum was not an agreement that gave security guarantees. I asked Ambassador Steven Pifer and some of the other experts who also took part in the negotiations of this Memorandum, and, according to what I had heard from them, all parties then clearly understood that the essence of this document was in the signatories' committing to respect the sovereignty and territorial integrity of Ukraine," quoted from "Posol SShA: Budapeshts'kyi memorandum ne buv dohovorom pro harantiï bezpeky," *Dzerkalo tyzhnia: Ukraïne*, May 30, 2014, https://dt.ua/POLITICS/posol-ssha-budapeshtskiy-memorandum-ne-buv-dogovorom-pro-garantiyi-bezpeki-144263_.html).

23. Later, the US increased this sum to US $350 million; in other words, Kravchuk agreed to an aid that constituted 1/10 of the realistic expenditures, with the remaining 9/10 to be financed from Ukraine's national budget.

24. Resolution of the Parliament of Ukraine "On Ratification of the Treaty," 464.

25. Ibid.

26. Ibid.

27. "Basic Principles of Utilization of Nuclear Warheads in Strategic Nuclear Forces Stationed in Ukraine," signed in Yalta, 3 September 1993, http://zakon.rada.gov.ua/laws/show/643_133.

28. "Trilateral Statement by the Presidents of Ukraine, the US and Russia."

29. Konovalov, "Kiev poluchaet harantii" (Kyiv Receives Guarantees).

30. Konovalov, "Kiev poluchaet harantii" (Kyiv Receives Guarantees).

31. However, in December 2014, former US Ambassador to Ukraine Steven Pifer was forced to admit that, in 1994, Washington simply issued "a check to Kyiv, avowing US support under the Budapest Memorandum," instead of providing security guarantees, hoping that "it would never be cashed" ("SShA maiut' zrobyty dlia Ukraïny bil'she nizh zaraz" (The US Should Do More for Ukraine than What It Is Doing Now), *Ukrinform*, 10 December 2014).

32. "A Non-Nuclear Ukraine," *Washington Post*, January 11, 1994, A18.

33. Konovalov, "Kiev poluchaet harantii."

34. Ibid.

35. Ibid.

36. Leonid Kravchuk, president of Ukraine, to the members of Parliament, "On Implementation of Recommendations in Provision 11 of Parliament's Ratification Resolution by the President and the Government of Ukraine," No. 1-14/23, 24 January 1994. Author's personal archive.

37. "Conclusion of the of the Parliamentary Working Group on START I Treaty Implementation Regarding Adherence of the Trilateral Statement on Ukraine's Nuclear Disarmament, signed 14 January 1994 in Moscow, to the Provisions of Parliament's Resolution Regarding Ratification of START I of 18 November 1992," signed by Yuri Kostenko

and members of the Working Group, 25 January 1994. Author's personal archive.

38. "Abstract of the Protocol no. 99 of the Parliamentary Rada Commission on Foreign Affairs," 26 January 1994. Author's personal archive.

39. Fedir Burchak and Anatolii Matsiuk, "On the Issue of Assessment of the Trilateral Statement by Presidents of Ukraine, Russia and the US," 24 January 1994. Materials distributed to members of the Parliament of Ukraine for discussion. Author's personal archive.

40. "Abstract of the Protocol no. 50 of the Parliamentary Defense and National Security Commission," January 26, 1994, "Abstract of the Protocol no. 321 of the Meeting of the Parliamentary Law and Constitutional Compliance Commission," 25 January 1994. Author's personal archive.

41. Ministry of Foreign Affairs, "Exploratory Brief Regarding the Treaty on Nonproliferation of Nuclear Weapons," distributed to Members of the Parliament of Ukraine for discussion, 2 February 1994. Author's personal archive.

42. Ian Brzezinski, director of the Parliamentary Advisory Council's National Security Program, "Treaty on the Nonproliferation of Nuclear Weapons, and Ukraine's Accession to the Treaty (Ukrainian version)," analytical brief, 1 November 1994. Author's personal archive.

43. Draft Resolution "On Trilateral Statement by Presidents of Ukraine, USA and Russia," 2 February 1994. Materials distributed to Members of the Parliament of Ukraine for discussion. Author's personal archive.

44. In March 1994 the snap elections to Parliament took place. Parliament's political composition essentially remained the same; the left were in the majority, and a socialist, Oleksandr Moroz, was elected Speaker.

45. "Treaty on the Nonproliferation of Nuclear Weapons (NPT)," adopted at United Nations, New York, 1 July 1968, entered into force 5 March 1970, extended indefinitely 11 May 1995, *United Nations Office for Disarmament Affairs*, https://www.un.org/disarmament/wmd/nuclear/npt/text/.

46. NPT's definition of a nuclear-weapons state did not cover Ukraine's situation after the dissolution of the USSR: on the one hand, as the successor to the former USSR, it inherited both the nuclear weapons and the property rights to them; on the other hand, Ukraine did not have its own production of nuclear weapons. The discrepancy between the letter of the law and these new circumstances allowed Ukraine to bring up the issue of amending NPT's definitions by adding a third category of states, in addition to nuclear and nonnuclear. Although Ian Brzezinski noted in his afore-mentioned brief that "[c]onsidering the amendment procedure, this would be very difficult to achieve politically," it still did not prevented Ukraine from bringing up the issue in order to ease international pressure and gain time to prepare for a more sensible disarmament process (Brzezinski, "Treaty on the Nonproliferation of Nuclear Weapons").

47. General Major Oleksandr Kovtunenko, Doctor of Technical Sciences, Professor, Deputy Head of Research, Kharkiv Military University, "Forecasts for Nuclear Weapons Proliferation, and Prospects of Transitioning to Nonnuclear Weapons. Condition in 1993." Forwarded by Volodymyr Tolubko, head of Kharkiv Military University, to General Lieutenant Vasyl' Durdynets', first deputy speaker of the parliament of Ukraine, via letter no. 14, April 14, 1993, forwarded by Vasyl' Durdynets', to Yuri Kostenko and Valerii Semenets', members of the Working Group, no. 5464/02–3/566, April 19, 1993. Author's personal archive.

48. Ibid.

49. Ibid.

50. Ibid.

51. The MFA, on the other hand, kept assuring the parliament that the NPT was as an effective mechanism: "In over 20 of its existence, the NPT has proven its effectiveness [...]. Over 150 countries are signatory to the NPT. The Treaty's founding principles of nonproliferation of nuclear weapons are widely recognized by the international community [...]. Attempts by some countries, such as Israel, India, Pakistan, Iraq, North Korea, South Africa and others, to violate the status quo established by the Treaty have sparked only a negative reaction from members of the NPT [...]. Thus, it is an undeniable fact that, thanks to the NPT, there exists an effective international mechanism for the nonproliferation of nuclear weapons" (Cabinet of Ministers of Ukraine "Possible Consequences of Ukraine's Refusal to Accede to the NPT," analytical brief, 17 May 1993. Author's personal archive).

52. Shteinberg, "Iadernoe oruzhie."

53. Ibid.

54. Ibid.

55. Ibid.

56. Ibid.

57. Viktor Tkachuk, "Iaderna zbroia–tse zasib strymuvannia ahresii," *Visti z Ukraïny*, no. 33, November 1993, 8.

58. Parliament of Ukraine Session Hall, Session 15, 3 February 1994. Transcript of the plenary session, http://iportal.rada.gov.ua/meeting/stenogr/show/4261.html; http://iportal.rada.gov.ua/meeting/stenogr/show/4262.html.

59. Ibid.

60. Ibid.

61. Defense Minister was supposed to know that the design of a nuclear weapon contained a powerful defense system. A contemporary publication by the Head of the 12th Chief Directorate [for nuclear provision and safety] of Russia's Ministry of Defense, Iurii Sych, provides testimony to the information our Working Group had supplied to Ukraine's leadership on the topic in the early 1990s: "In the years since the first nuclear test, nuclear specialists have worked with tens of thousands of nuclear weapons, carrying out millions of various technological operations with these weapons–all without a single case of nuclear or radioactive incident that would trigger a nuclear process or radioactive fallout. The technical measures are incorporated in the very design of a nuclear weapon. They prevent any possibility of an unsanctioned nuclear detonation or release of nuclear energy of a considerable output in case of incidental detonation of an explosive" (Iurii Sych, "Istoki iadernoi moshchi strany," *Nezavisimoe voienne obozrenie*, March 5, 2014, http://nvo.ng.ru/nvo/2014–09–05/1_nuclear.html).

62. Parliament of Ukraine Session Hall, Session 15, 3 February 1994, 4:00 pm. Transcript of the plenary session, http://iportal.rada.gov.ua/meeting/stenogr/show/4262.html.
63. Ibid.
64. Ibid.
65. Ibid.
66. This is how Western aid to Ukraine for the liquidation of nuclear weapons looked like at the end of 1994, from the interview with the 43rd Missile Army Commander Volodymyr Mykhtiuk: "[Correspondent:] Does the army receive aid from the West for the liquidation process? [Volodymyr Mykhtiuk:] The aid has not been very tangible thus far. In terms of dollar amounts, we received roughly US $4 million as of today. In terms of materials, we received engine oil, three cranes, several utility vehicles, and some emergency equipment. However, in all practicality, we have not really been able to use any of it. The cranes, which we need, arrived late and also defective. We had to order specialists from abroad to train our crane operators, since the cranes were produced in the Netherlands. [Correspondent:] Why did you order these cranes in the first place? [Mykhtiuk:] That is the thing, we were not the ones ordering them. The orders were made without us, and were unprofessionally done, and this was the main mistake. I believe it is still not too late to fix it (Anatolii Murakhovs'kyi, "Den' rozhdeniia–grustnyi prazdnik u raketchikov 43-i raketnoi armii," *Dzerkalo tyzhnia*, 16 December 1994, https://zn.ua/POLITICS/den_rozhdeniya_-_grustnyy_prazdnik_u_raketchikov_43-y_raketnoy_armii.html)
67. Parliament of Ukraine, Session 15, 4:00pm.
68. Ibid.
69. Ibid.

Chapter 6

1. Leonid Kuchma unexpectedly won the snap presidential election, campaigning on the pro-Russian slogan, "Fewer walls, more bridges." In the early parliamentary elections, the political shape of the parliament barely changed: leftists were again in the majority and Socialist Party leader Oleksandr Moroz was elected Speaker.
2. Leonid Kuchma, president of Ukraine, to Oleksandr Moroz, Speaker of the Parliament of Ukraine, Letter no. 1–14/54–1, 2 November 1994. Materials distributed to MPs of Ukraine for familiarization. Author's personal archive.
3. Ibid.
4. Ibid.
5. Draft Parliament Resolution submitted by MPs Oleh Vitovych, Iurii Tyma, Stepan Khmara, Iaroslav Iliasevych,
and Bohdan Iaroshyns'kyi. Materials distributed to MPs for familiarization, 15 November 1994. Author's personal archive.
6. Amendment to a draft resolution submitted by Andrii Mostys'kyi. Materials distributed to MPs for familiarization, 15 November 1994. Author's personal archive.
7. "On Ukraine's Accession to the Treaty on the Nonproliferation of Nuclear Weapons of 1 July 1968," submitted by the Iednist' coalition in Parliament. Materials distributed to MPs for familiarization, 16 November 1994. Author's personal archive.
8. "To the Bill on Ukraine's Accession to the Treaty on the Nonproliferation of Nuclear Weapons of 1 July 1968," submitted by the parliamentary commissions on Foreign Affairs, and Defense and National Security. Materials distributed to MPs of Ukraine for familiarization, 16 November 1994. Author's personal archive.
9. Main Chamber of the Parliament of Ukraine, Session 32, 16 November 1994, 12:30, Transcript of plenary session, Parliament of Ukraine, official web portal. http://iportal.rada.gov.ua/meeting/stenogr/show/3573.html
10. *Parliament of Ukraine*, Session 32.
11. Ibid.
12. Ibid.
13. Ibid.
14. Ibid.
15. Ibid.
16. Memorandum on Security Assurances in connection with Ukraine's accession to the Treaty on the Non-Proliferation of Nuclear Weapons, an international document signed on 5 December 1994, by Ukraine and three nuclear states: the US, Russia, and Great Britain. The signatories committed themselves to "respect the independence and sovereignty and current borders of Ukraine," that "none of their weapons would ever be used against Ukraine," and also to "refrain from using economic pressure aimed at subjugating Ukraine's exercise of the rights inherent to its sovereignty to its own interests." The MFA presented this to Parliament as a "unique" document that represented, in fact, the "security guarantees" that would protect Ukraine in case of need. FM Udovenko said that the mechanisms for exercising it were "by holding consultations between the three nuclear states and Ukraine in case the question of one of the sides upholding its commitments under the Memorandum were to arise." All parties, including the Ukrainian side, acknowledged that the document was not legally binding.
17. *Parliament of Ukraine*, Session 32.
18. Ibid.
19. Ibid.
20. Ibid.
21. Law of Ukraine "On Ukraine's Accession to the Treaty on the Non-Proliferation of Nuclear Weapons of 1 July 1968," no. 248/94-VR, Kyiv, dated 16 November 1994, *Vidomosti Verkhovnoi Rady Ukraïny*, 1994, no. 47, 421. http://zakon.rada.gov.ua/laws/show/248/94-%D0%B2%D1%80.
22. "Law of Ukraine 'On Ukraine's Accession to the Treaty on the Non-Proliferation of Nuclear Weapons of 1 July 1968,'" *Holos Ukraïny*, 19 November 1994, 3.

23. *VVR*, no. 29 (1992): 405, http://zakon.rada.gov.ua/laws/show/2264-12.

24. "Written Statement by the Ukrainian Side Regarding the Signing of the Protocol to the Treaty between the USSR and the US on the Reduction and Limitation of Strategic Offensive Arms," Ministry of Foreign Affairs, 23 May 1992, 2. Author's personal archive.

25. Ambassador Oleh Bilorus, "From the Journal of O. Bilorus. Washington, 7 December 1992. Recording of Talk between US Secretary of State F. Wisner," forwarded via Facsimile transmission no. 15/49-25057 by M. P. Makarevych, Deputy Foreign Minister of Ukraine, to I. S. Pliushch, Speaker of Parliament, 21 December 1992. Author's personal archive.

26. Ukrinform, "Senatory SShA u Kyievi."

27. Carol Giacomo, "US Said to Give Ukraine Official Assurances Letter," Reuters News Agency, January 9, 1993.

28. Giacomo, "US Said to Give Ukraine Official Assurances Letter."

29. Anatolii Zlenko, Foreign Minister, to Ivan Pliushch, Speaker of Parliament, Letter UKOR/23-54, "Recording of a Conversation between Deputy Foreign Minister Borys Tarasiuk and Special Envoy of the Russian Federation's Foreign Ministry Mikhail Strieltsov, 12 January 1993," 26 January 1993, 2.

30. Conference for Security and Cooperation in Europe (CSCE), renamed to Organization for Security and Co-operation in Europe (OSCE) on 1 January 1995.

31. "Iak zapobihty rospovsiudzhennia iadernoï zbroï. Vystup Prezydenta Ukraïny L. M. Kravchuka na Vsesvitniomu ekonomichnomu forumi v Davosi," *Holos Ukraïny*, February 4, 1993, 2-3.

32. Ibid.

33. Among the guarantees Russia proposed giving Ukraine after nuclear disarmament, it offered: "3. The Russian Federation will support its commitments in relation to the principles of the Concluding Acts of the OSCE to respect the independence and sovereignty and existing state borders of members of the OSCE, and recognizing that borders may only be changed in a peaceful and consensual manner, and recognizes its duty to refrain from threats or the application of force against the territorial integrity or political independence of any country and that none of its weapons will ever be used, other than in self-defense or any other form, in line with the Charter of the United Nations. 4. The Russian Federation confirms its commitment to Ukraine in line with the Agreement Establishing the Commonwealth of Independent States to recognize and respect Ukraine's territorial integrity and the inviolability of existing borders within the framework of the Commonwealth. The Russian Federation also reaffirms its adherence to the provisions stipulated in the Statement of CIS Heads of State of 14 February 1992, that any and all disputes arising among member CIS states shall be resolved exclusively by peaceful means, through negotiations, and notes that this provision applies fully to its relations with Ukraine" (MFA of the Russian Federation to the MFA of Ukraine, Draft "Declaration on Security Guarantees," sent by letter No. 1024/

DRK to the Ukrainian Embassy in the Russian Federation, 25 February 1993. Author's personal archive).

34. Ambassador Michel Peissik, the French Embassy in Ukraine, to Anatolii Zlenko, minister of foreign affairs, "Projet de Declaration," Kyiv, 6 May 1993. Forwarded by B. I. Tarasiuk, Chair of the National Committee of Ukraine for Disarmament and Deputy Foreign Minister of Ukraine, to V. V. Durdynets', first deputy speaker of the Parliament of Ukraine, 15 June 1993, via Letter UKOR/23-625. Author's personal archive.

35. Main Chamber of the Parliament of Ukraine, Session 15, 3 February 1994, 10:00, official portal, http://static.rada.gov.ua/zakon/skl1/BUL19/030294_15.htm.

36. Parliament of Ukraine, Session 32, 16 November 1994, 12:30.

37. MFA of Ukraine, brief "On Ukraine's Accession to the NPT: Progress with Implementing the Decisions of Parliament dated 18 November 1993 and 3 February 1994," 15 November 1994. Materials distributed among MPs prior to discussing the agenda. Author's personal archive.

38. Ibid.

39. Ibid.

40. Ianina Korniienko, "Stiven Paifer: 'Iakby Ukraïna zalyshyla iadernu zbroiu, ii vidnosyny z Rosiieiu bylu b shche skladnishymy," *Platforma*, 16 July 2014, https://platfor.ma/magazine/text-sq/experience/stiven-paifer/.

41. In 2014, Russian Prime Minister Dmitry Medvedev stated in an interview with *Bloomberg* that Russia was not prepared to guarantee Ukraine's territorial integrity: "Medvedev also refused to guarantee that Russia won't incorporate the Ukrainian regions of Donetsk and Luhansk, which voted to break away in disputed referendums this month, saying that the U. S. and the EU must pledge not to interfere in Ukraine and push it into NATO. 'We don't have to guarantee anything to anyone because we never undertook any obligations on this matter,' Medvedev said. 'The most important task is to calm down the situation in Ukraine'" (Ryan Chilcote, Henry Meyer & Olga Tanas, "New Cold War May Emerge in Ukraine Crisis, Medvedev Says," *Bloomberg*, Business, May 20, 2014, https://www.bloomberg.com/news/articles/2014-05-20/second-cold-war-may-emerge-in-ukraine-medvedev-says).

42. In 2012, Ievhen Marchuk considered the Memorandum a reliable protection for Ukraine: "Today, some are saying that more could have been bargained for. Maybe it could have, he agreed skeptically, and easily brought listeners to the next section of his speech, In addition to compensation, Ukraine was able to also get the Budapest Memorandum adopted in 1994, that is, to get guarantees from the nuclear states. [...] The guarantee meant that these countries would not use nuclear weapons to blackmail Ukraine in any form. Marchuk explained, and made a commitment to defend Ukraine" (Kniazhans'kyi, "Odna zistrich i dva uroky"). Still, just before the 20th anniversary of the Budapest Memorandum, Leonid Kravchuk admitted that he felt conned because of the failure of the signatories to uphold the Budapest Memorandum: "'No one came through on their commitments to Ukraine,' said the country's first president. He calls on the new legislature to immediately appeal to the countries, which

once gave the state guarantees of its territorial integrity. He meant all the countries that had signed the Budapest Memorandum in an item on TSN, 19:30. This was Great Britain, the United States of America and the Russian Federation. Kravchuk thinks that they should get together and come up with a mechanism for implementing this agreement. Kravchuk reminded his audience that Ukraine did not renounce nuclear weapons of its own accord at the time but was pressured to do so by the international community. Then, they promised to protect Ukraine's territorial integrity and now the guarantor countries aren't even prepared to provide ordinary weapons. 'Promised? They tricked Ukraine, in fact! They conned us. I personally feel like I was conned. We lay down our nuclear weapons and now they're debating whether or not to provide Ukraine with rifles!' Kravchuk said with outrage" ("Kravchuk pochuvaietsia oburenym cherez 'Budapeshts′kyi Memorandum'," *TSN.ua*, December 2, 2014, https://tsn.ua/politika/kravchuk-pochuvayetsya-obdurenim-cherez-nedotrimannya-krayinami-svitu-budapeshtskogo-memorandumu-394962.html).

43. Dubinin, "Iadernyi dreif Ukrainy."
44. Ibid.
45. Ibid.
46. Iulia Mostova, "K hodovshchine iadernoi premiery" (On the Anniversary of the Nuclear Debut), *Dzerkalo tyzhnia*, 8 December 1995. https://zn.ua/POLITICS/k_godovschine_yadernoy_premiery.html.
47. Although Russia was ill-prepared, the last nuclear warheads were moved from Ukraine already by 1 June 1996.
48. Russia moved nuclear assets from the territories of former republics at a blistering pace: "The first task is to disassemble and ship to Russia all ICBM combat units deployed in Kazakhstan, which will thus give up its entire strategic nuclear potential. Of the two divisions of Belarus's rocket army, the material and weapons of one division have been moved to Russia and now there are no more than 42 warheads on missile carriers left in Belarus. According to a report by Russia's Foreign Intelligence Service, by April of this year, around 600 warheads from ICBM combat blocks and cruise missile were moved from Ukraine to Russia. This has cut Ukraine's nuclear potential by one third." (Mark Shteinberh, "Iadernoe oruzhie" [Nuclear Weapons]).
49. Leonid Petrovs′kyi, "Vzryv na troikh" (Explosion for Three), *Dzerkalo tyzhnia*, 5 January 1996, https://zn.ua/POLITICS/vzryv_na_troih.html.
50. Ibid.
51. Ibid.
52. Rostyslav Khotin, "Saliut poslednei boeholovke" (Farewell to the Last Warhead), *Dzerkalo tyzhnia*, 7 June 1996, https://zn.ua/POLITICS/salyut_posledney_boegolovke.html.
53. Ibid.
54. "'Patriarkh' amerikanskoi politiki predlagaet ostavit′ Ukrainu na zadvorkakh NATO," *Ukraïns′ka pravda*, 3 July 2008. https://www.pravda.com.ua/rus/news/2008/07/3/4444295/.
55. On 16 March 2014, Russia took over a part of Ukraine's territory, the Autonomous Republic of Crimea, violating treaties on the inviolability of state borders. Russian officials refused to comply with the Budapest Memorandum on the basis that they did not recognize the Ukrainian government that came to power after the 2013–2014 Revolution of Dignity. The world community recognized this as the first precedent of a change in borders in post-WWII Europe. The EU and the US limited themselves to economic sanctions against Russia and economic assistance to Ukraine. The Ukrainian Armed Forces and internal forces proved unable to localize the threat and the situation was destabilized across all of eastern Ukraine.
56. Stockholm International Peace Research Institute (SIPRI), "Nuclear forces reduced while modernizations continue, says SIPRI," *SIPRI for the media*, 16 June 2014, https://www.sipri.org/media/press-release/2014/nuclear-forces-reduced-while-modernizations-continue-says-sipri.
57. "So viele Atomwaffen sind weltweit in Umlauf" (So Many Nuclear Weapons in Circulation Worldwide), *Spiegel Online*, 16 June 2014, http://www.spiegel.de/politik/ausland/atomwaffen-zahl-der-nuklearsprengkoepfe-sinkt-laut-sipri-langsamer-a-975368.html.
58. "Vitse prem′ier Rosiï: onovymo iadernyi arsenal i vsikh zdyvuiemo zbroieiu," *Ukraïns′ka pravda*, 22 September 2014, https://www.pravda.com.ua/news/2014/09/22/7038484/. Additionally, in March 2015, NATO Secretary General Jens Stoltenberg acknowledged that there was a nuclear threat coming from Russia: "We are not interested in a confrontation with Russia. As before, we continue to look for constructive relations. Transparency and reliability are key to understanding, but right now the conditions aren't there. [...] Russia is also activating its nuclear training. It is developing new nuclear weapons and testing new rockets that can carry a nuclear payload. This will do little to build trust or reduce tensions." ("Iaderni pohrozy Rosiï ne spryiaiut porozuminniu z Ievropoiu–hensek NATO," *Ievropeis′ka pravda*, 30 March 2015, https://www.euro-integration.com.ua/news/2015/03/30/7032423/).
59. Kier A. Lieber and Daryl G. Press, "The Rise of US Nuclear Primacy," *Foreign Affairs*, March/April 2006, https://www.foreignaffairs.com/articles/united-states/2006-03-01/rise-us-nuclear-primacy.
60. Ibid.
61. Ibid.
62. On the contrary, NATO considers Russia under President Putin more dangerous than the USSR. This is the opinion of former NATO Secretary General, Anders Fogh Rasmussen: "Even in soviet times, they did not dare to speak about a nuclear war. Now we see open discussions. In this sense, Russia today is more dangerous than under the Soviet Union. The USSR was more predictable than the current leadership of the Russian Federation" ("Rosiia nebezpechnisha za Radians′kyi soiuz–eks-hensek NATO" (Russia is more dangerous than the Soviet Union–NATO ex-Secretary General," *Siohodni*, 16 April 2015, https://ukr.segodnya.ua/politics/nyneshnyaya-rossiya-opasnee-sovetskogo-soyuza-eks-gensek-nato-608398.html).
63. By March 2014, the West was calling Russia's actions the first instance of an attempt to change the borders of Europe since World War II, and Vladimir Putin was being

compared to Adolf Hitler: "'We should be talking about the danger threatening Europe. Post-war borders are being redrawn. And that is the signal Russia is sending us,' stated Lithuanian President Dalia Grybauskaitė" ("Prezydent Lytvy: Rosiia khoche peredilu pisliavoiennykh kordoniv," *Ukraïns'ka pravda*, March 6, 2014, https://www.pravda.com.ua/news/2014/03/6/'47017832/). "'There is a good deal of truth in such a comparison because Putin is using the same ethnic issue in his foreign policy. The Second World War began under the pretext that Hitler wanted to protect the German-speaking population of Sudetenland in Poland and other parts of Eastern Europe. He believed that this gave him the right to attack these countries. And in that sense, Putin is doing exactly the same as Hitler did,' said Francis Fukuyama." (Mustafa Naiem, "Fransis Fukuiama: Putin robyt rivno te same, shcho robyv Hitler" [Francis Fukuyama: Putin is Doing Exactly What Hitler Did], *Ukraïns'ka pravda*, 27 August 2014, https://www.pravda.com.ua/articles/2014/08/27/7035903/). Former Secretary of State Hillary Clinton comparing "actions by Russian President Vladimir Putin in Ukraine," commented: "'Now if this sounds familiar, it's what Hitler did back in the 30s,' she said. 'All [...] the ethnic Germans, the Germans by ancestry who were in places like Czechoslovakia and Romania and other places, Hitler kept saying they're not being treated right'" (Karen Robes Meeks, "Hillary Clinton compares Vladimir Putin's actions in Ukraine to Adolf Hitler's in Nazi Germany," *Press-Telegram*, 5 March 2014, https://www.presstelegram.com/2014/03/05/hillary-clinton-compares-vladimir-putins-actions-in-ukraine-to-adolf-hitlers-in-nazi-germany/).

64. However, as of 1993, the Foreign Ministry declared Ukraine's battle-readiness adequate: "I want to emphasize one more point that is forgotten too often: even after all the nuclear weapons are removed from our territory and weapons are reduced in line with the Treaty on Conventional Armed Forces in Europe, Ukraine will have the third most numerous army in Europe–and if we don't count Russia and Turkey, who are not entirely European states, then the biggest." ("Neprostyi shliakh do bez'iadernoho svitu," 12).

65. In 2014, as in 1991, the border between Ukraine and Russia was completely open. Russian military men and materiel freely crossed into Ukraine: "The OSCE Special Monitoring Mission officials observing Russia's Gukovo checkpoint reported in the special commission's report published on the OSCE site that the border was being crossed by people in uniform every single day" ("Na kordoni vidbulysia zbroini sutychky z rosiins'kymy dyversantamy–RNBO," *Ukraïns'ka pravda*, August 14, 2014, https://www.pravda.com.ua/news/2014/08/14/7034806/); "Journalists witnessed how a column of Russian armored vehicles crossed the border into Ukraine. Moscow correspondent for *The Guardian* Shaun Walker wrote about this in his Twitter microblog, reporting that at least 28 APCs had violated the state border of Ukraine. This number was confirmed by *The Telegraph*'s Roland Oliphant. The *New York Times* correspondent who was based in Kamensk-Shakhtinsk in Russia also reported Ukraine's border being crossed from the Russian side, where he stood next to a 'humanitarian convoy' of KamAZes. The *New York Times* special correspondent reported that the column of armored vehicles included a few dozen infantry fighting vehicles, APCs, and military trucks pulling artillery, TSN reports." ("Inozemni zhurnalisty vpershe pobachyly, iak cherez rosiis'kyi kordon v Ukraïnu ide viis'kova tekhnika," *Telekrytyka*, August 15, 2014, http://ru. telekritika.ua/education/2014–08–15/97026?theme_page=830&). In September 2014, the Ukrainian government announced the building of a wall to protect the Ukrainian-Russian border: "The State Border Service, which initiated the plan, has drafted a massive engineering project called "The Wall," which will cover nearly 2,300 km of hitherto undemarcated border between Ukraine and the Russian Federation. The government has approved it." (V'iacheslav Shramovych, "Proekt 'Stina': chy potribna ukraïns'ka liniia Mannerheima u XXI stolitti?" *BBC Ukraine*, September 15, 2014, http://www.bbc.com/ukrainian/politics/2014/09/140912_experts_on_ukrainian_wall_project_vs); "Without building concrete barriers along the border sections with Russia, it will be impossible to carry out any peace plan says Yatseniuk." ("Uriad vydilyv shche shist milioniv ievro na proekt 'Stina' (video)," *Podrobnosti*, October 8, 2014, http://podrobnosti.ua/997068-urjad-vidliv-sche-shst-mljonv-vro-na-proekt-stna-vdeo.html).

66. Right from the beginning of the 1990s, Russia began applying economic and political pressure on Ukraine. Its policy did not change after the signing of the Budapest Memorandum, nor after all of Ukraine's nuclear warheads were moved to Russian territory. Moscow began to squeeze Ukrainian businesses out of the Russian market, first by suddenly withdrawing its defense procurements from Ukrainian MIC enterprises, for whom such procurements were up to 75 percent of production, and after Yeltsin's secret decree in 1994, the Russian Federation began to set up its own defense production facilities. In 1993, for the first time Russia issued a territorial claim on the city of Sevastopol. Energy was actively used as leverage to resolve political issues. Among others, there was the famous phrase by Yeltsin during negotiations over the Black Sea Fleet, when the Ukrainian delegation did not accept the Russian proposition: "That is how the tap can be turned off, too." (Author's notes from negotiations between the Ukrainian and Russian Federation delegations regarding the division of the Black Sea Fleet. Author's personal archive). Thanks to its gas blackmail, Russia not only got part of the Black Sea Fleet which, based on the principles, on which Soviet assets were split, should have been 100 percent Ukraine's, but also a military base in Crimea, which gave it the foothold for the eventual annexation of the peninsula in 2014.

67. Kier A. Lieber and Daryl G. Press, "The End of MAD? The Nuclear Dimension of US Primacy," *International Security* 30, no. 4 (Spring 2006): 7–44.

68. Lieber and Press, "The End of MAD?"

69. Ibid.

70. Ibid.

71. "Little Russia's Chip."

72. According to the National Security and Defense Council of Ukraine (Rada Natsionalnoï Bezpeky ta Oborony Ukraïny [RNBO]), "During the last five years, the war [in Donbas] has claimed thirteen thousand Ukrainian lives, three thousand of them service persons" ("O. Turchynov: Mitsna sim'ia staie kliuchovym faktorom natsionalnoï bezpeky Ukraïny," http://www.rnbo.gov.ua/news/3221.html). Meanwhile, *UNIAN* information agency reported the following, referring to official information provided by the UN Monitoring Mission on Human Rights: "From April 2014 to the end of 2018, some 12,800 to 13,000 people were killed in Donbas hostilities," with further "30,000 people [...] wounded and injured as a result of hostilities" ("Donbas War Death Toll Rises up to Nearly 13,000–UN," *UNIAN*, January 22, 2019, https://www.unian.info/war/10416549-donbas-war-death-toll-rises-up-to-nearly-13–000-un.html).

Illustration Credits

All images are reproduced in this volume by permission from the copyright holders. Authors and sources of the digital files for images available in the public domain or with a creative commons license are credited where such information was available.

1. Main chamber of the Parliament of Ukraine (Verkhovna Rada), the first convocation, September 1994. Volodymyr Samokhots′kyi / UkrInform   19
2. Ivan Drach, 1996. Pavlo Zdorovylo / UkrInform   23
3. Volodymyr Vasylenko, 2011. Iurii Il′ienko / UkrInform   25
4. Georgii Arbatov, 1983. Rob Bogaerts / Anefo. Source: Nationaal Archief / WikiMedia   26
5. Boris Yeltsin and George H. W. Bush. Dmitrii Donskoi / Boris Yeltsin Presidential Center   30
6. The Parliament of Ukraine. Shotshop GmbH / Alamy Stock Photo   32
7. Serhii Holovatyi (at the microphone), 1991. UkrInform   34
8. Leonid Kravchuk (left). UkrInform   35
9. The signing of the Minsk Agreement, 8 December 1991. U. Ivanov. Source: WikiMedia   36
10. Alma-Ata Protocol, 21 December 1991. Dmitrii Donskoi / RIA Novosti. Source: WikiMedia   38
11. Boris Yeltsin. Dmitrii Sokolov / Boris Yeltsin Presidential Center   39
12. Andrei Sakharov, 1989. R. C. Croes / Anefo. Source: WikiMedia   43
13. Iurii Ryzhov. Sputnik / Alamy Stock Photo   44
14. Ievhen Marchuk, 1995. UkrInform   45
15. Ivan Pliushch, 1990. UkrInform   46
16. Zbigniew Brzezinski, 1991. Anatolii Kolesnyk / UkrInform   53
17. Dmytro Pavlychko, 1993. UkrInform   57
18. Anton Buteiko, 2005. Anatolii Mykhailov / UkrInform   58
19. Anatolii Zlenko, 1991. Volodymyr Samokhots′kyi / UkrInform   59
20. Pivdenmash. Courtesy of Mykhailo Markiv   60
21. Ihor Iukhnovs′kyi, 1991. Ihor Demchuk / UkrInform   62
22. Signing of the Lisbon Protocol. Fernando Ricardo / AP Photo   67
23. Signatures of the parties to the Lisbon Protocol. Photo by the author   69
24. Ministry of Foreign Affairs. Vimoculars / WikiMedia   70

25. Viacheslav Kebich. Sputnik / Alamy Stock Photo   72
26. Nursultan Nazarbayev, 1991. Egemen Media / WikiMedia   73
27. Andrei Kozyrev. Mikhail Evstafiev / WikiMedia   75
28. Stanislav Koniukhov (center) and Leonid Kuchma (right). State Space Agency of Ukraine   76
29. Protective gear for working with heptyl. Exhibition at the Strategic Missile Forces Museum of the National Military Historical Museum of Ukraine. Photo courtesy of Tetiana Tkachenko   80
30. Leonid Kravchuk (center, seated) visiting NATO Headquarters, 1992. UkrInform   83
31. Ivan Bizhan (left), 2000. Oleksandr Sieryi / UkrInform   84
32. Volodymyr Mykhtiuk, 2001. Volodymyr Samokhotsʹkyi / UkrInform   87
33. SS-24 Scalpel. Exhibition at the Strategic Missile Forces Museum of the National Military Historical Museum of Ukraine. Photo courtesy of Tetiana Tkachenko   91
34. SS-19 engine case. Exhibition at the Strategic Missile Forces Museum of the National Military Historical Museum of Ukraine. Photo courtesy of Tetiana Tkachenko   93
35. Types of missiles used by the USSR (1). Exhibition at the Strategic Missile Forces Museum of the National Military Historical Museum of Ukraine. Photo courtesy of Tetiana Tkachenko   95
36. Types of missiles used by the USSR (2). Exhibition at the Strategic Missile Forces Museum of the National Military Historical Museum of Ukraine. Photo courtesy of Tetiana Tkachenko   95
37. Soviet air defense rockets. Exhibition at the Strategic Missile Forces Museum of the National Military Historical Museum of Ukraine. Photo courtesy of Tetiana Tkachenko   96
38. R-12 Dvina. Exhibition at the Strategic Missile Forces Museum of the National Military Historical Museum of Ukraine. Photo courtesy of Tetiana Tkachenko   98
39. Kh-22NA cruise missile. Exhibition at the Strategic Missile Forces Museum of the National Military Historical Museum of Ukraine. Photo courtesy of Tetiana Tkachenko   99
40. Model of a warhead. Exhibition at the Strategic Missile Forces Museum of the National Military Historical Museum of Ukraine. Photo courtesy of Tetiana Tkachenko   100
41. Missile silo for the SS-24 Scalpel. Exhibition at the Strategic Missile Forces Museum of the National Military Historical Museum of Ukraine. Photo courtesy of Tetiana Tkachenko   103
42. Viktor Zelensʹkyi. "80 let so dnia rozhdeniia V. F. Zelenskogo," *Pytannia atomnoï nauky i tekhniky*, part 2, no. 4–2 (2009): 302, https://vant.kipt.kharkov.ua/   105
43. Transporting the SS-24. Exhibition at the Strategic Missile Forces Museum of the National Military Historical Museum of Ukraine. Photo courtesy of Tetiana Tkachenko   106
44. Mock-up of a Unified Command Post (UCP). Exhibition at the Strategic Missile Forces Museum of the National Military Historical Museum of Ukraine. Photo courtesy of Tetiana Tkachenko   108

45. Passageway to the UCP. Exhibition at the Strategic Missile Forces Museum of the National Military Historical Museum of Ukraine. Photo courtesy of Tetiana Tkachenko   110

46. Map of 43rd Rocket Army unit locations. Exhibition at the Strategic Missile Forces Museum of the National Military Historical Museum of Ukraine. Photo courtesy of Tetiana Tkachenko   111

47. Residential quarters of UCP personnel. Exhibition at the Strategic Missile Forces Museum of the National Military Historical Museum of Ukraine. Photo courtesy of Tetiana Tkachenko   112

48. Control panel of the security and defense shift operator in the central missile launch complex. Exhibition at the Strategic Missile Forces Museum of the National Military Historical Museum of Ukraine. Photo courtesy of Tetiana Tkachenko   113

49. Transport and trailer with container for SS-24 missile. Exhibition at the Strategic Missile Forces Museum of the National Military Historical Museum of Ukraine. Photo courtesy of Tetiana Tkachenko   114

50. Viktor Chernomyrdin (right), 1994. Iurii Abramochkin / RIA Novosti. Source: WikiMedia   116

51. Protests against President Kuchma, 2001. Ihor Huz´. Source: WikiMedia   117

52. Yevgeny Shaposhnikov. Sueddeutsche Zeitung Photo / Alamy Stock Photo   119

53. Vasyl´ Durdynets´, 1994   120

54. Borys Tarasiuk, 2003. Oleh Dykan´ / UkrInform   130

55. Boris Yeltsin and Leonid Kravchuk at the Yalta Agreement of 1992. Aleksandr Sentsov / Boris Yeltsin Presidential Center   134

56. Iurii Dubinin, 1986. Dennis Brack / Alamy Stock Photo   137

57. Stanislav Koniukhov, 2003. Oleh Dykan´ / UkrInform   142

58. Kostiantyn Hryshchenko, 2000. UkrInform   144

59. Boris Gromov, 2003. Volodymyr Tarasov / UkrInform   145

60. Administration of the military oath to Ukraine, in Soviet uniforms, 1992. UkrInform   146

61. Viktor Bar'iakhtar, 2005. Oleh Dykan´ / UkrInform   148

62. Valerii Kukhar. National Academy of Sciences of Ukraine   149

63. Volodymyr Tolubko (left) and Kostenko (right), 1992. Photo by the author   163

64. TASS. Stanislav Kozlovskii. Source: WikiMedia   169

65. President Leonid Kuchma, 1999. UkrInform   186

66. Prime Minister Leonid Kuchma speaking in Parliament, 1993. UkrInform   187

67. Les Aspin. Robert Ward. Source: WikiMedia   192

68. Senator Joe Biden, 1991. Dennis Brack / Alamy Stock Photo   193

69. Jim Woolsey. Central Intelligence Agency. Source: WikiMedia   194

70. Stepan Khmara, 1996. Valerii Soloviov / UkrInform   199

71. Bohdan Horyn´, 1996. Volodymyr Strumkovs´kyi / UkrInform   200

72. Valerii Shmarov. Volodymyr Samokhots´kyi / UkrInform   203

73. Boris Yeltsin and Leonid Kravchuk in Masandra, 3 September 1993. Aleksandr Sentsov / Boris Yeltsin Presidential Center   204

74. Inside a missile silo. Exhibition at the Strategic Missile Forces Museum of the National Military Historical Museum of Ukraine. Photo courtesy of Tetiana Tkachenko   216

75. Missile silo being decommissioned. Exhibition at the Strategic Missile Forces Museum of the National Military Historical Museum of Ukraine. Photo courtesy of Tetiana Tkachenko   220

76. Vladimir Zhirinovsky. A. Sdobnikov. Source: WikiMedia   222

77. Bill Clinton and Leonid Kravchuk, 1993. Volodymyr Repik / UkrInform   224

78. Trilateral Statement, 14 January 1994. Dmitrii Sokolov / Boris Yeltsin Presidential Center   227

79. Vitalii Radets'kyi, 2005. Oleh Dykan´ / UkrInform   237

80. Leonid Kuchma and Leonid Kravchuk, 1994. UkrInform   246

81. Oleksandr Moroz (left, at the microphone), 1995. Anatolii Piddubnyi / UkrInform   247

82. Vitalii Masol, 1985. Volodymyr Repik / UkrInform   251

83. The loading of a nuclear warhead into a container for transport (1), 1992. Valerii Soloviov / UkrInform   257

84. Signing the Budapest Memorandum, 5 December 1994. Marcy Nighswander / AP Photo   258

85. The loading of a nuclear warhead into a container for transport (2), 1992. Valerii Soloviov / UkrInform   260

86. "The Value of the Budapest Memorandum," cartoon by Oleh Smal´. Courtesy of the artist   262

87. The destruction of the twenty-first ICBM silo, Khmelnytskyi Oblast, 1996. Valerii Soloviov / UkrInform   263

88. The last RS-22 ICBM launcher is destroyed, 2001. Volodymyr Strumkovs´kyi / UkrInform   264

89. Major General Robert P. Bongiovi (left) and Lieutenant General Leonid Fursa at the opening ceremony for a nuclear weapon processing facility at the Ozerne Airfield, 2002. Volodymyr Tarasov / UkrInform   266

Index

Page numbers in italics refer to illustrations and captions;
page numbers in boldface refer to tables.

Academy of Sciences of Ukraine, 17, 57, 61, 69, 103, 104, 105, 115
Afanasiev, Iurii, 44
Aizenberh, Iakiv, 56, 58, 125, 170
Alma-Ata Protocols, 37, 38, 64, 102, 271–72
Altunian, Henrikh, 46
amyl, 58, 80, 307n8
Anti-Ballistic Missile Treaty, 301n12
Applebaum, Anne, 28, 92, 302n16, 305n69
Arbatov, Georgii, 26, *26*, 27, 302n14
Arzamas-16 Nuclear Center, 156
Aspin, Les, 31, 133, *192*, 192–193, 197, 322n29
Attali, Jacques, 110
August Putsch of 1991, 2, 39, 43, 45, 59, 119, 120, 301n7

Balandiuk, Mykola, 91
Baltic countries, 20
Baltin, Eduard, 135
Bar'iakhtar, Viktor, 56, 125, 148, *148*
Belarus: position on nuclear disarmament, 68; ratification of nuclear treaties, 11, 74, 127, 305n73; surrender of nuclear weapons, 29, 31, 38, 72, 74, 118, 306n86, 324n97, 330n48
Belavezha Accords, 29, 36, 37
Biden, Joe, 193, *193*, 194
Bilinsky, Yaroslav, 195
Bilorus, Oleh, 8, 131, 132, 253
Bizhan, Ivan, 84, *84*, 86, 139, 143
Black Sea Fleet: Crimean bases of, 264, 304nn55–56; dispute over, 55; division of, 46–47, 259, 314n34, 331n66; nuclear weapons on, 92; Russia's claim of, 60, 175; Yalta Treaty on, 4, 314n34
black soil: contamination of, 49
Blair, Bruce G., 193–6
Blix, Hans, 122
Bongiovi, Robert P., *266*
Borzykh, Oleksandr, 76
Broad, William J., 36, 217
Brzezinski, Ian, 232
Brzezinski, Zbigniew, 9, 53, *53*, 55

Budapest CSCE summit, 256
Budapest Memorandum: cartoon about, *262*; consequences of, 257; failure of, 260–61, 329n42; limitations of, 326n22; original idea of, 255; provision on security assurance, 130, 255; Russia's foreign policy and, 263–65; signing of, 4, 255, 256–57, 258, *258*; terms of, 257–258; text of, 292–293; US approach to, 326n22, 326n31; violation of, 330n55
Budjeryn, Mariana, 6, 299n1
Burbulis, Gennadii, 36
Burchak, Fedir, 231
Burns, William F., 41
Bush, George H. W.: address to Parliament of the Ukrainian SSR, 29; "Chicken Kiev" speech, 2, 30; foreign policy of, 2, 67; Kravchuk's letters to, 68; official visit to Ukraine, 29; Yeltsin and, *30*
Buteiko, Anton, 18, 56, 58, *58*, 86–88, 205–6

Canada's foreign policy, 319n154
Cassette Scandal (Kuchmagate), 10, 117, 311n81
Celec, Fred S., 155
Center for Security Policy, 196, 198, 322n40
Cheney, Dick, 53
Chernenko, Vitalii, 76
Cherniak, Volodymyr, 44
Chernobyl disaster, 100, 304n61
Chernobyl Nuclear Power Plant: Shelter Object (Sarcophagus), 309n45
Chernomyrdin, Viktor, 116, *116*, 117, 156, 205, 309n48
China's nuclear arsenal, 21, 22
Chornovil, V'iacheslav, 183, 301n1
Chuchuk, Markian, 46
Clinton, Bill: at Budapest summit, 256, *258*; foreign policy, 198, 202, 222, 223, 224; Kravchuk and, 4, 122; negotiations at Boryspil Airport, 223–24, *224*
Clinton, Hillary, 331n63
Collective Security Treaty (Tashkent Pact), 306n87
Collective Security Treaty Organization (CSTO), 74, 306n87
Commonwealth of Independent States (CIS): creation of, 29; division of military assets, 118; Minsk summit of 1993,

133; Russia's policy toward, 3, 28, 134; Strategic Forces, 154, 269–70

Coordinating Committee for Multilateral Export Controls (CoCom), 82

Crimea: Russia's annexation of, 1, 261, 304n55; Russia's military bases in, 264, 304n55, 304n56

Davos forum (February 1993), 253

Declaration of State Sovereignty of Ukraine, 23, 25, 124, 217–18, 300n4, 300n5

Declaration "On Ukraine's Nonnuclear Status," 267–68

Decree on Urgent Measures for the Construction of the Armed Forces of Ukraine, 305n62

Defense Council of Ukraine (DCU), 56–61, 303n48

Dem'ianenko, Vladyslav, 160

Democratic Party of Ukraine, 301n1

Demydov, Hryhorii, 46

Derkach, Leonid, 304n54

Derzhkomatom (Ukrainian State Committee for Nuclear Power Utilization), 97, 100, 103, 104, 115, 308n38, 310n51. See also Ukratomenerhoprom

Dolhanov, Vadym, 158, 160

Donbas miners' strike, 206, 321n16

Donbas war, 14, 174, 261, 266, 319n147, 332n72

Drach, Ivan, 23, 23, 23–24, 301n1

Dubinin, Iurii: career of, 137; estimate of number of nuclear warheads, 92, 305n69; negotiations on Ukraine's nuclear disarmament, 82, 133, 136, 137–39, 141, 145, 157, 315n46; on November 10 parliamentary resolution, 212; photograph of, 137; on Russian-Ukrainian draft agreements, 204–5; on security guarantees for Ukraine, 256

Durdynets', Vasyl': debates on nuclear disarmament, 33, 120, 147, 148, 158, 180; "hawks' declaration" and, 179, 183; on legality of the Trilateral Statement, 240; photograph of, 120; political career of, 120–21; position on nuclear treaty ratification, 166, 192

economic security, 46, 48, 49

Egorov, Nikolai, 41

electronic warhead simulators ("equivalents"), 176

environmental security, 46, 49

European Bank for Reconstruction and Development (EBRD), 64

European security, 9, 132, 221

Expert Group for the Investigation of Opportunities for Using Nuclear Material Extracted from Nuclear Weapons, 308n27

Eyal, Jonathan, 31

Filonenko, Vitalii, 155

Fokin, Vitold, 36, 47

foreign ministers meeting in Moscow, 62

Fort Plutonium, 156, 308n34

43rd Rocket Army: map of unit locations, 111; missile launch facilities, 90; nuclear arsenal of, 86, 201; oath of allegiance, 157, 315n61

France's nuclear arsenal, 21, 22

fuel rod assembly. See TVEL

Gamsakhurdia, Zviad, 222

Garnett, Sherman, 9

General Atomics (GA): expertise of, 97–8; meetings with Ukrainian officials, 96–97; nuclear technologies of, 308n37; proposal on repurposing of nuclear materials, 86, 96–97, 98, 99, 309n41, 309n44

Global Learning and Observations to Benefit the Environment (GLOBE), 53

global security system, 144, 266

Gorbachev, Mikhail, 2, 8, 301n7

Gore, Albert, 173, 309n48

government negotiations on nuclear deal: American involvement in, 202; delegations, 143; discussion of compensation for the tactical weapons, 144; first round of, 136–42; Kostenko's opening speech at, 137; results of, 145, 146–47; Russia's position at, 139–40, 141, 144–45, 202, 203–4; second round of, 144–47; Shmarov's role in, 202–4; Ukrainian position at, 140–41, 202–3. See also Massandra summit

Grachev, Pavel, 120, 146

Graham, James J., 96, 99

Gromov, Boris, 144, 145, 145, 157, 158

Havryliuk, Viktor, 114, 148

heavy bombers, 90, 91

Heiets', Valerii, 148

heptyl, 58, 80, 307n8

highly enriched uranium (HEU): conversion program, 96, 109, 113, 309n44; cost of processing, 93, 109, 309n49, 310n64; Derzhkomatom strategy for utilization of, 310n51; as energy source, 80; estimated amount of, 114, 309n41; prospect of selling on the global market, 110, 112; recommendations on handling, 109–10; storing options, 110; US-Russia agreement on, 141

Holovatyi, Serhii, 23, 24, 33, 34, 241

Horyn', Bohdan, 43, 76, 151, 200, 207, 208, 301n1

Hryshchenko, Kostiantyn, 143–44, 144

Iakheieva, Tetiana, 33, 183

Iakymenko, Oleksandr, 16

Ianaiev, Gennadii, 301n7

Iaroshyns'kyi, Bohdan, 247

Iazov, Dmitrii, 301n7

Ievtukhov, Vasyl', 100

Iliasevych, Iaroslav, 247, 251

India: Nonproliferation Treaty and, 233, 327n51; nuclear weapon tests, 21, 235

information wars: American media in, 170–71; nuclear disarmament and, 168–69; Russia's methods of, 171–72, 319n147; Ukrainian officials and, 172–73

Institute for Nuclear Research, 105, 114

Institute for US and Canadian Studies (Russia), 26

Institute of Government and Law, National Academy of Sciences of Ukraine, 69

intercontinental ballistic missiles (ICBMs): in Belarus, 72; destruction of silo of, 263; limits under SALT I and SALT II Treaties, 2; types of, 90, 307n14; in Ukraine, 90, **92,** 125, 184

Intermediate-Range Nuclear Forces (INF) Treaty, 2, 235, 301n12

International Atomic Energy Agency (IAEA), 92, 307n25

Inter-Regional Deputy Group, 44

Israel: Nonproliferation Treaty and, 233, 327n51; as nuclear weapon state, 21, 22, 195, 233–34
"Issues of Nuclear Safety" seminar, 84, 155, 315n63, 316n85
Iukhnovs'kyi, Ihor, 35, 61, 62, *62*, 95, 100, 309n41
Izmalkov, Valerii, 46, 65

Japan: as member of G8 group, 94; as nonnuclear country, 14, 317n113; nuclear technology, 57, 80, 98, 108; scientific and technical capability, 50

Kazakhstan: four-way talks between Ukraine, Russia, Belarus and, 206, 207, 215; nonnuclear-weapon status of, 73; nuclear treaties signed by, 11, 64, 66, 69, 71, 159, 183, 256, 305n73, 312n1, 325n111; position on nuclear disarmament, 68, 72, 73; Soviet nuclear weapon in, 26, 63, 142, 172; surrender of nuclear weapons, 29, 31, 37–38, 74, 127, 235, 264, 265, 306n86, 324n97, 330n48
Kebich, Viacheslav, 36, 72, 73
Kendzior, Ia. M., 183
"Key Directions of Ukraine's Foreign Policy" (document), 199–201
KGB: First Directorate of, 17; influence in the former Soviet bloc, 303n46, 326n17; interference in Georgia affairs, 222, 325n5
Kh-22NA cruise missile, *99*
Kharkiv Research and Development Association Khartron, 58, 138, 143, 160, 170
Khmara, Stepan, 43, 66, 181, 199, *199*, 247
Kirimov, Ivan, 251
Kiselev, Vladimir, 94
Kissinger, Henry, 260
Kisun'ko, Grigorii, 129
Klerk, F. W. de, 77
Kokhno, Valerii, 164
Kokoshin, Andrei, 118
Koniukhov, Stanislav: career of, 142; at Defense Council meeting, 56, 58–59, 142; on disposal of launch facilities, 324n104; at government negotiations, 143; Leonid Kuchma and, *76*; nuclear expertise of, 59; on operation of nuclear facilities, 59–60; photograph of, *142*; proposal on nonnuclear deterrent systems, 164
Konovalov, Vitalii, 114, 229
Korzh, Anatolii, 46
Kostenko, Hryhorii, 47
Kostenko, Yuri: as chair of Working Group, 76; conversations with Brzezinski, 54–55; conversations with Pliushch, 76, 179–80; on creation of Working Group, 75–76, 306n6; at Defense Council meeting, 60; on destruction of nuclear arsenal, 216; experience with Russian diplomacy, 314n44; "Kiev & the Bomb: Ukrainians Reply," 196–7; letter to Kravchuk on plutonium, 108; memo on national security, 46–47; negotiations on Ukraine's nuclear disarmament and, 15, 63–64, 146–47, 151; personal archive of, 16–17; photograph of, *163*; on ratification of START I, 151; on right of ownership of nuclear weapons, 137, 242–43; "Some Observations Regarding the Implementation of START," 142; speech at the Irpin sanatorium meeting, 137; trip to the US, 99; visit to Russian nuclear plant, 88–89; visit to the 35rd Rocket Army, 86
Kotsiuba, Oleksandr, 231

Kovtunenko, Oleksandr, 151, 233–34
Kozhushko, Oleksandr, 250
Kozyrev, Andrei, 75, *75*, 127, 191
Kravchuk, Leonid: at Alma-Ata summit, 38; approach to security guarantee, 121–22, 253–54; on Budapest Memorandum, 329n42; on Commonwealth of Independent States, 134; on compensation for nuclear material, 95; conversation with nuclear weapons specialists, 56; criticism of, 229; at Davos forum, 253–54; decrees of, 61, 157, 305n62; diplomacy of, 62, 68, 102, 121, 122, 124, 158, 191, 253; interview for *Holos Ukraïny* on 16 December 1992, 123–24; Kostenko's analytical memo to, 46–47; Kuchma and, *246*; at Massandra summit, 11, *204*; at Minsk summit, 36; at Moscow summit of 1993, 105, 135; at National Defense Council meeting, 151; national security policy of, 47–48, 303n48; at NATO headquarters, 83; negotiations at Boryspil Airport, 223–24, *224*; nuclear disarmament concept of, 10, 11, 40, 56–57, 90, 122, 123–26, 202; opposition to, 136; Parliament vs., 229–32; photographs of, *35*, *187*; political career of, 35, 173, 311n109; on ratification of nuclear treaties, 87–88, 210–11; Resolution no. 33 (Res. 33), 47; at Russian-Ukrainian meeting, 133; Russia's pressure on, 4, 147; as signatory of Minsk Agreement, 36; on START ratification, 121; testimony before the Parliament on 3 February 1994, 237, 238, 239, 240, 244; understanding of politics, 172, 173; visit to Moscow on 14 January 1994, 225; at Yalta summit, *134*
Kravets, Nadiya, 7, 299n12
Kriuchkov, Vladimir, 301n7, 303n46
Kriukov, Vitalii, 150
Kryzhanivs'kyi, Volodymyr, 128
Kryzhko, Oleksii, 143
Kuchma, Leonid: at Budapest summit, 256, *258*; criticism of Nunn-Lugar Act, 246; diplomacy of, 173, 259; draft treaties of Massandra Accords, 205; election as president, 245, 328n1; foreign policy of, 311n81; General Atomics representatives' meeting with, 96–7; Koniukhov and, *76*; Kravchuk and, *246*; letter to the Parliament on 2 November 1994, 245–47; on Lisbon Protocol, 243; nuclear disarmament strategy of, 4, 10, 243, 254, 303n48, 304n54, 309n48; photograph of, *186*; political career of, 100, 117, 186, 202, 207, 222; presidential campaign of, 328n1; protests against, *117*; ratification of NPT and, 245, 247, 248, 249–50; resignation as prime minister, 222; Russia's criticism of, 190; statement on 3 June, 190, 192; technical expertise of, 76, 179; Western media on, 190
Kuchmagate. *See* Cassette Scandal
Kuchma-Gore Commission, 100, 314n45
Kukhar, Valerii, 148, 149, *149*
Kyrylenko, V'iacheslav, 300n1

launch facilities. *See* missile silos
Law "On Entrepreneurship," 218
Law "On Ukraine's Accession to the Treaty on the Nonproliferation of Nuclear Weapons," 290–91
Lazorkin, Vitalii, 164
Lemish, Valentyn, 46, 231
Liberal Democratic Party of Russia (LDPR), 222

liquid-propellant missiles, 80, 91, 93, 138, 307n8, 307n14. *See also* SS-18 Satan missiles; SS-19 Stiletto missiles

Lisbon Protocol: Article II of, 68; Article V of, 69, 70, 231; contradictions of, 72, 185, 189; distribution of quotas under, 306n80; execution of, 52; goal of, 66, 68; legal analysis of, 69; media coverage of, 72; outcome of, 74; ratification process of, 15, 67, 303n37, 305n73; Russia's official statement on, 71–72; signatures of the parties to, 69, 138; signing ceremony, *67*, 68; terms of, 3, 11, 68, 73, 124; text of, 275–76; timeframes of, 166; Ukraine and, 67–68; violation of, 135; Zlenko's statement on, 70–71

low enriched uranium (LEU): conversion program, 41–42, 96, 98, 104; cost of transportation and insurance, 109, 277, 280; revenues from the sale of, 109, 112

Lugar, Richard, 121

Luk'ianenko, Levko, 43

Luk'ianov, Anatolii, 303n46

Lytvyn, Volodymyr, 300n1

Major, John, 256, *258*

Makarevych, Mykola, 136

Manhattan Project, 21

Marchuk, Ievhen: on Budapest Memorandum, 329n42; career of, 45; development of security policy, 47; photograph of, *45*; position on nuclear disarmament, 134, 163, 223, 310n52, 312n109, 317n113; statement on foreign policy, 162

Martynenko, Petro, 148–49

Masol, Vitalii, 247, 251, *251*

Massandra Accords: Buteiko's amendment to, 205–6; criticism of, 203; Russia's view of, 323n74; signing of, 158, 203, 206, 221, 227, 317n102; terms of, 140, 221, 227, 317n102; text of, 277–84

Matsiuk, Anatolii, 231

Mazepa, Ivan, 23

McCurry, Mike, 223

Mearsheimer, John, 5, 28, 29, 196

Medvedev, Dmitry, 329n41

Melloan, George, 325n5, 326n17

Melnychenko, Mykola, 117, 311n80

Memorandum on Security Assurances in connection with Ukraine's accession to the Treaty on the Non-Proliferation of Nuclear Weapons, 256, 328n16

Mikhailov, Viktor, 41

"Military and Political Aspects of Arms Reduction in Ukraine, The" (report), 150–51

Miller, Steven E., 5, 196

Ministry of Defense (Ukraine), 64–65, 104, 163

Ministry of Foreign Affairs (Ukrainian MFA): disagreement with Ukrainian parliament, 175; headquarters of, *70*; information war and, 172, 173; Non-Proliferation Treaty ratification and, 120, 231, 327n51; nuclear disarmament strategy, 35, 72, 85–86, 128, 146, 166, 177–78, 202, 217, 305n68, 317n99; search for security guarantee, 130; START ratification and, 120; statement on Ukraine's battle-readiness, 331n64

Minsk Agreement (1983): ratification of, 39; signing of, *36*, 102, 302n32; terms of, 302n32

Minsk Agreement on Strategic Forces, 38

missile fuel utilization, 46, 80, 314n36, 316n81, 324n104

missile silos: destruction of, 216–17, 316n81, 324n104; views of, *216*, *220*

mixed-fuel technology, 57, 108

Mokin, Borys, 76

Moroz, Oleksandr, 43, 157, 245, *247*, 247, 251, 311n81

Moroz, Viktor, 47

Morozov, Kostiantyn, 88, 205, 323n74

Mostys'kyi, Andrii, 181, 182, 247

Motiuk, Myroslav, 46

Movchan, P. M., 242

Mukhin, Volodymyr, 245

Mykhtiuk, Volodymyr, 86–87, *87*, 201, 258, 328n66

Mysnyk, Pavlo, 76

Mytrakhovych, Mykhailo, 47

Nahulko, Taras, 46

National Committee on Disarmament and the State Administration Committee on Nuclear Policy, 125, 126

National Defense Council of Ukraine (NDC), 151–52, 153–54

National Disarmament Committee, 180, 186, 320n1

National Nuclear Energy Generation Company (Enerhoatom), 308n38

National Science Center of the Kharkiv Institute of Physics and Technology (KIPT), 105–6, 109, 113, 308n27

national security: concept of, 44, 45–48; Kostenko's memo on, 46–47; military interpretation of, 46, 48; parliamentary working group on, 45–46, 48

National Security and Defense Council of Ukraine (NSDCU), 46, 303n48

National Security Council of Ukraine (NSCU), 46, 47

NATO: conceptual change of, 121; eastward expansion of, 223, 225; "Issues of Nuclear Security" seminar, 84–85; Russia's relations with, 221–22, 330n62; security summit in Brussels, 222; Ukraine's perception of, 132

Navy of Ukraine: formation of, 314n34

Nazarbayev, Nursultan, 37, *38*, 73, 73–74

Nazarenko, Arnold, 44, 45, 46

9 May celebrations, 16

nongovernmental organizations (NGOs), 166

Non-Proliferation Treaty (NPT): arms race and, 234; Belarus's accession to, 74; definition of nuclear-weapons in, 327n46; goal of, 233; Kazakhstan's accession to, 74; parliamentary debates on ratification, 188–89; participants of, 233; Review and Extension Conferences, 234; sanctions for violation of, 83; signatories of, 21–22, 233; terms of, 21–22; vote on extension of, 22

Non-Proliferation Treaty ratification: debates over, 87, 245–52; draft Resolutions on, 247–48; Iliasevych's appeal, 250–51; Kuchma's arguments, 245, 248, 249–50, 251; security guarantees and, 252; Udovenko's speech on, 248–49, 250; Ukraine's position on, 74, 120, 152–53, 201; vote on, 251

NPT Review Conference of May 1995, 188, 189, 234, 236

North Korea's nuclear arsenal, 21, 22

nuclear arms race, 21

nuclear button issue, 58, 170

nuclear club: expansion of, 13, 233–34, 235

Nuclear Fuel Services (NFS), 309n44

nuclear materials: compensation for, 140; cost of production, **94**; estimated quantities of, **94**; General Atomics' proposal on repurposing of, 96–7; for power generation, repurposing of, 80–81, 95–1; transportation and storage of, 114; utilization of, 105–8, 110–12, 115, 116, 155, 209, 218–19, 221, 227, 281, 310n51; value of, 93–4; weight estimates, 308n28. *See also* highly enriched uranium (HEU); low enriched uranium (LEU)

nuclear power plants (NPP) in Eastern Europe, 309n46

nuclear weapons: compensation for the cost of, 81, 259; defense system against incidental explosion of, 327n61; detonation tests, 234–35; distribution of, 22; of the former Soviet republics, 67, 79; fuel for, 95, 177; global arsenal of, 22, 234, 261; loading into container, *257, 260*; modernization of, 261; photograph of warheads, *100*; protective gear, *80*; public opinion poll on, 317n96; security measures, 235; as seed capital, 34; shipment to Russia, 102, 105, 258, 259, 306n86; treaties on control of, 2; types of, *95*; value of, 80, 299n1; weight characteristics of, 88, 89

Nunn, Sam, 121

Nunn-Lugar Act, 246

Objects S (storage facilities), 203, 323n70

Oliinyk, Borys, 245

Oliinyk, Ivan, 85

Omelchenko, Mykola, 46

"On Possible Options to Reducing and Limiting Strategic Arms in Line with START I," 215

"On the Nonnuclear Status of Ukraine" (Parliamentary Statement), 218

Orange Revolution, 35, 247, 313n29

Organization of the Ukrainian Nationalists (OUN), 16, 300n3

Ozerne Airfield nuclear weapon processing facility, *266*

Pakhomov, Iurii, 166, 167, 318n128

Parliamentary Commission on Budget Planning, Finance and Pricing, 321n28

Parliamentary debates of 3 June 1993: closed session, 186; doves' move, 179–83; Durdynets' arguments, 183–84; hawks' move, 183–88; Kuchma's speech, 186–88, 192; outcome of, 188–89; Pliushch's summary of, 189; on right to ownership of nuclear weapons, 184–85; Tarasiuk's speech, 186; Western media on, 191–192; Zlenko's speech, 180–83, 191

Parliamentary session of 3 February 1994: debate on Article V of the Lisbon Protocol, 236; debate on Ukraine's nuclear status, 236–37, 238; Durdynets's statement, 240–41; idea of effective disarmament, 241, 243; Kravchuk's testimony, 237, 238, 239, 240, 243–44; MPs deliberations, 241–44; Radets'kyi's testimony, 237–39, 240; resolution, 244; Zlenko's testimony, 237, 238, 239, 240

Parliament Commission on Foreign Affairs, 128

Parliament of the Ukrainian SSR, 23–26, 44

Parliament of Ukraine (Verkhovna Rada): analysis of the provisions of the NPT, 232; assessment of the Trilateral Statement, 229–32; Committee on nuclear disarmament, 33–34; debate on Lisbon Protocol, 236; debate on Ukraine's accession to Non-Proliferation Treaty, 236–37;

Defense and National Security Commission, 51; expertise in nuclear weapons, 75; first convocation of, *19*; hearing on nuclear disarmament, 152; hearings on ratification of START I, 147–48; March 1994 elections, 232, 245; photograph of the building of, *32*; statement "On the Nonnuclear Status of Ukraine," 35, 65, 124, 218; statement "On Ukraine's Right of Ownership of Nuclear Weapons," 159; Working Group for Addressing the Issues Related to Refining Ukraine's System of National Security, 302n29

Parliament of Ukraine's resolutions: On Additional Measures to Ensure that Ukraine Gains Nonnuclear Status, 102–3, 218, 252, 273–74; On Additional Measures to Ensure Ukraine's Attainment of Nonnuclear Weapon State Status, 61, 65; On the Implementation by the President of Ukraine and Government of Ukraine of Recommendations in Point 11 of the Resolution of the Parliament of Ukraine, 288–89; on ratification of START I (18 November 1993), 208, 210–14, 229, 232, 238, 240, 243, 244, 254; On Ratifying the Treaty Between the USSR and USA on the Reduction and Limitation of Strategic Nuclear Arms, 218–19; On the Reduction and Limitation of Strategic Offensive Arms, 285–87; On the Trilateral Statement by the Presidents of Ukraine, the US and Russia, 232

Partial Test Ban Treaty of 1963, 2

Partnership for Peace, 221–22

Pavliukov, Valerii, 143

Pavlychko, Dmytro, 33, 56–57, *57*, 122, 174, 199–0, 211, 301n1

People's Council (Narodna Rada), 19, 300n5

People's Movement of Ukraine for Reconstruction (Narodnyi Rukh), 23, 300n5, 301n1

Perry, William, 259

Pifer, Steven, 6, 255, 326n31

Pivdenmash (Yuzhmash, Southern Machine-Building Plant): chief designer of, 56, 58, 76, 142, 162, 164, 240; competition in the global market, 59; director of, 4, 186, *186*; interior view of, *60*; Kostenko's visit to, 304n54; production of nuclear missiles, 63, 73, 90, 138, 152, 179, 304n58, 307n14

Pliushch, Ivan: background and career of, 18, 46; diplomacy of, 121, 122; Kostenko's conversations with, 179–80; at National Defense Council, 151; order to assess the Trilateral Statement, 229; personality of, 179; photograph of, *46, 187*; ratification of START I and, 123, 147; support of the Working Group, 45, 75–76, 303n37

plutonium: conversion to nuclear fuel, 104, 107–8; as energy source, 80, 99; estimated value of, 98, 107, 229; Japan's accumulation of, 108; storage of, 107, 108; utilization of, 80–81, 227

political prisoners, 43

Popov, Gavriil, 44

Porovskyi, Mykola, 66

Potter, William C., 91

Powell, Robert, 13

precision-guided weapons, munitions (PGMs), 59, 119, 164, 165, 167, 174, 185, 215, 239–40

Pressler, Larry, 216

Prykhodko, Viktor, 46

Pugo, Boris, 301n7

Putin, Vladimir, 212–13, 328–31n63
Pyatt, Geoffrey, 326n22
Pylyp Orlyk Institute for Democracy, 169

R-12 Dvina, *98*
Radets'kyi, Vitalii, 211, 223, *237*, 237–39, 240
Rasmussen, Fogh, 330n62
ratification of interstate treaties: democratic practice of, 324n100
Ratushnyi, Mykhailo, 251
"Relationship between Society and Military in Central and Eastern Europe, The" seminar, 84
Repets'kyi, Vasyl', 150
Revolution of Dignity, 15, 29, 73, 204, 265, 300n2, 330n55
Riabchenko, Serhii, 44
Rogozin, Dmitry, 261
Romanov, Iurii, 167
RS-22 ICBM launcher, *264*
Rudia, Kostiantyn, 97
Russia: control over CIS's nuclear forces, 60, 154; double standards politics, 315n55; economic crisis, 41; foreign policy, 222, 326n16; imperialistic ambitions, 222–223; international arms trade, 119; as major threat to world peace, 266, 330n62; media attacks on Ukraine, 62, 139, 147, 197, 228–29, 317n97, 318n119; military doctrine of, 120, 145; military-industrial complex, 302n14; non-compliance with START I, 324n97; nuclear industry infrastructure, 156, 308n34; nuclear power of, 22, 261, 264–65, 324n97; proposal on Soviet weaponry, 27; revision of defense system, 118; sanctions against, 14, 330n55; secret talks with G8 countries, 141; Strategic Nuclear Forces, 141–42, 143, 144; technical capacities of, 145; territorial claims over Sevastopol, 16, 198, 199, 201, 208, 250, 259, 331n66; transfer of the nuclear warheads to, 139, 330n48
Russian diplomacy, 37–40, 147, 314n44
Russian Ministry of Foreign Affairs, 131, 202
Russia-Ukraine border, 16, 258, 304n55, 323n70, 330n55, 330n63, 331n65
Russia-Ukraine relations: 1997 agreement, 304n56; diplomatic scuffle in, 136; economic ties, 28, 311n80, 331n66; Minsk summit of 1993, 133–34; Moscow summit of 1993, 133, 134, 135, 136; Russia's vision of, 133–134, 263–265; talks on nuclear disarmament, 128, 154, 157–58, 202–3; territorial disputes, 16; Ukraine's vision of, 134; US role in, 8
Russo-Georgian War of 2008, 16, 299n1
Russo-Ukrainian war of 2014, 12, 16, 205, 263, 325n11, 330n55
Ryzhov, Iurii, 44, *44*, 303n46

Sakharov, Andrei, 43, *43*
SALT Treaties, 2
security guarantees: French government approach to, 254; Russian position on, 254, 329n33; Ukraine's position on, 252, 253, 255, 313n21; US approach to, 253, 312n15; world practice of, 252–53
Security Service of Ukraine (SBU): development of National Security Policy of Ukraine and, 47; directors of, 16, 45, 304n54, 317n113; name change, 46; nuclear

disarmament and, 40, 76, 78, 90, 103, 134, 162, 167, 310n52, 310n53; reform of the KGB into, 116; reports to the Working Group provided by, 104
Selivanov, Volodymyr, 47
Semenets', Serhii, 33, 46
Semipalatinsk Test Site, 21
Serebriannykov, Iurii, 46, 76
Sevastopol: Russia's territorial claims over, 16, 199, 201, 250, 259, 331n66
Shalikashvili, John, 121
Shaposhnikov, Yevgeny, 118, 119, *119*, 120
Shcherbak, Iurii, 44
Shekhovtsov, Oleksii, 76
Shevchenko, Oles, 132
Shevchenko, Vitalii, 250
Shishkin, Albert, 41
Shmarov, Valerii, 202–3, *203*, 211, 212
Shovkoshytnyi, V. F., 242
Shteinberg, Mark, 92
Shushkevych, Stanislav, 36, *38*, 127
Shvatskyi, Ievhen, 47
Sidak, Volodymyr, 47
Sikorski, Radek, 136
Sinovets, Polina, 299n1
Skoryk, Larysa, 43, 46
Slobodeniuk, Volodymyr, 46
Slocombe, Walter B., 195
Smirnov, Igor, 41
Sobchak, Anatolii, 44
solid-fuel missiles. *See also* SS-24 Scalpel missile systems; SS-25 Topol mobile ballistic missile systems
South Africa, Republic of (RSA): destruction of nuclear weapons, 21, 77–78
Soviet air defense rockets, *96*
Soviet Union: arms reduction treaties, 301n12; dissolution of, 18–19; division of military assets of, 118; national security policy, 303n46; nuclear tests in, 21, 234–35; nuclear weapons, 26, 89, 140; Strategic Rocket Forces, 302n19
SS-18 Satan missiles, 73, 127, 142, 240, 304n58, 307n14
SS-19 Stiletto missiles: cost of elimination of, 185; decommissioning of, 59, 152, 207, 258; engine case, *93*; maintenance of, 140; power of, 307n14
SS-24 Scalpel missile systems: cost of elimination of, 185; decommissioning of, 258; missile silo for, *103*; photographs of, *91, 106, 114*; power of, 90–91; production of, 304n58, 307n14; service of, 63, 142, 152; Ukraine's proposal on, 59, 164, 174
SS-25 Topol mobile ballistic missile systems, 36, 72, 118, 307n14
Stankevich, Sergei, 44
Starovoitova, Galina, 44
START I agreement: American media on, 128–29; cost of, 86–87, 185; implementation mechanisms, 68, 214–17; international significance of, 2; negotiations of, 202; parties of, 3; Russia's preconditions related to, 128; signing of, 29; Soviet collapse and, 69; supplemental protocol to, 31; terms of, 25, 67, 153, 166, 206, 243, 301n4; US Senate Committee hearings on, 193

START I ratification: by Belarus, 127; exchange of certificates on, 256; government public statements on, 210–11; Kravchuk's proposition on, 209–10; legal aspects of, 148–49; obstacles to, 148; parliamentary hearings on, 147–48, 188–89, 206–10; preparations for, 87, 120, 121, 122–23, 303n37, 308n27; press coverage of, 211–12, 213; resolution on, 208, 210–14; Russia's pressure on Ukraine for, 123, 128; Russia's reaction to, 212, 213; by Ukraine, 4, 69, 129; Ukraine's national interests and, 85, 153, 201; vote on, 210; Working Group and, 123, 207, 208–9

START II agreement: ratification of, 129, 212–13; Russia media on, 129; signing of, 3, 127, 129, 135; terms of, 127, 312n1; Ukraine's support of, 127; US and Russia's compliance with, 128

State Committee for Safeguarding the Population from the Effects of the Accident at Chernobyl NPP (Derzhkomchornobyl), 100

State Committee on the State of Emergency (GKChP), 301n7

State Nuclear and Radiation Safety Committee (Derzhatomnahliad), 87, 88, 120

Stevens, Christopher, 6

Stockholm International Peace Research Institute (SIPRI), 261, 330n56

Stoltenberg, Jens, 330n58

Strategic Nuclear Forces in Ukraine, 140, 141, 143

Strategic Nuclear Forces of the former USSR, 145, 146

Strelnykov, Volodymyr, 76

Streltsov, Mikhail, 127, 131

Stus, Volodymyr, 46

Sukhorukin, Vladimir, 107

Sukhorukov, Andrii, 46, 76

Sulzhynsky, Walter, 160

Sych, Iurii, 327n61

Systems Planning and Analysis, Inc. (SPA), 176–77

tactical nuclear weapons: compensation for, 98, 101, 138, 144, 153, 212, 227, 230, 231; Lisbon Protocol on, 208–9; transfer to Russia, 60, 61, 63, 64–66; in Ukraine's nuclear arsenal, 90, **92**, 305n69; Ukraine's policy regarding, 218–19, 242, 244, 246; Ukrainian press on, 309n48

Talbott, Strobe, 132, 133, 194, 198, 204, 205, 208, 223

Taman Division, 301n7

Tarasiuk, Borys: background and career of, 125, 160, 186, 215; diplomacy of, 130–31, 253; photograph of, 130; on security guarantee, 254; on Trilateral Statement, 326n14

Tashkent Pact, 74, 306n87

Techsnabexport company, 41

Telegraph Agency of the Soviet Union (TASS), 169

Tolstov, Serhii, 91, 197

Tolubko, Volodymyr, 18, 66, 76, 85, 119, 150, 162–63, 163

Treaty on Conventional Armed Forces in Europe (CFE), 27, 67, 302n12, 331n64

Treaty on Open Skies, 301n12

Trilateral Statement: information campaign on, 228–29; negotiations of, 326n14; parliamentary deliberations on, 229–32, 241; publication of, 225; ratification of, 4; Russian media on, 228; security guarantees issue,

231; signing of, 1, 4, 227; supporters of, 231; terms of, 225–27; Ukrainian law and, 225–27, 228

"Trilateral Statement of the Presidents of Ukraine, the US and Russia," 214

Tupolev Tu-95MS strategic bombers, 80, 91, 307n19

Tupolev Tu-160 ultramodern strategic bombers, 80, 91, 307n19

Tuzla Island: Russia's claim of, 16, 260–61, 304n55

TVEL (fuel rod assembly), 81, 93, 99; reprocessing of HEU into, 109–12

Tyma, Iurii, 247

Tymoshenko, Viktor, 163

Tymoshenko, Yulia, 304n56, 309n48

Udovenko, Hennadii, 136, 248, 254, 328n16

Ukraine: armed forces of, 12, 27, 163, 259, 265, 319n155; August Putsch and, 43; budget of, 86; Canada and, 319n154; constitution of, 10; currency depreciation, 20, 304n53, 325n3; declaration of sovereignty, 2; dependence on Russia, 4, 318n128; diplomacy of, 321n24; economic development of, 4, 12, 222, 300n6, 304n53, 325n3; energy sector of, 95; European Union and, 132, 173–74, 313n29; executive branch of, 10, 19, 20–21; foreign policy of, 161, 199–1, 221, 319n147; government of, 116, 117, 118, 191; information war and, 169; institutional conflict in, 10; legal recognition of property of, 150, 218; military doctrine, 161–64, 165, 318n123; military-industrial complex of, 45; national security policy, 15, 46, 48–51, 52, 161, 303n48; NATO membership debates, 8, 16, 132–33, 300n2, 313n21; non-nuclear status of, 2, 23–24, 25, 33, 70, 71; nuclear power industry, 308n36, 308n38, 310n51; oath of allegiance to, 146, 315n61; parallel with Bosnia, 194–5; path to Eurointegration of, 264; political development of, 9–10, 19–20, 301n1; ratification of arms reduction treaties, 74, 120–21, 302n12; reform program, 43–45, 221; right of ownership over nuclear weapons, 219; Russia's invasion of, 13, 14, 16; security dilemma of, 3, 13; Soviet debt and, 150; technical capabilities of, 58; territory of, 19. *See also* Russia–Ukraine relations

Ukraine-Hungary relations, 174–75

Ukraine's nuclear arsenal: conversion to nuclear fuel assemblies, 42; cost of maintenance of, 152, 163; debates on ownership of, 137–38, 150, 151–52, 157, 158, 159–60, 184–85; discussions about future of, 84, 155–57, 313n29; estimated quantity of, 22, 90–92, **92**; heavy bombers, 90–91; information exchange about, 84, 86, 90; limits of reduction of, 154; myths about, 155; proposal of international control of, 176; removal from battle duty, 152; status of, 62–63; storage program for, 155–56; strategic nuclear missiles, 80, 93; type of weapons, **92**

Ukraine's nuclear disarmament: Arbatov plan, 26–28, 37, 139; beginning of, 25; compensation for, 8, 94–5, 309n48, 328n66; consequences of, 13–14; cooperation with the West, 34–35, 82; cost of, 86, 175–76, 318–321n7; counterfactual scenarios of, 11–13; Defense Council meeting on, 56–61; expert opinion on, 89, 150–51; future implications of, 13–14; global security and, 261–62; idea of transfer weapons to Russia, 35, 36–37, 38, 78; Kostenko's arguments about, 7–11;

Kyiv's plan for, 33–37, 79–82, 305n73; lack of guiding documents about, 305n68; media coverage of, 170, 174, 306n6; national interests and, 52, 75, 149–50; parliamentary hearings on, 7–8, 33–34, 147–48; results of, 265–66; Russia's approach to, 28, 29, 37–40, 139, 203, 295, 325n111; security guarantees, 12, 130–33, 153; studies of, 5–7, 9, 16–17, 28; technical aspects related to, 61–62, 63; Ukraine's bargaining position on, 8, 10–11, 217–19; United States and, 8, 9, 29–31, 32, 66, 121, 153, 174, 196, 262–63; Western perception of, 14, 30–31; White House plan for, 29–32

Ukraine's nuclear fuel cycle, 100–1, 102, 103–4

Ukrainian-American relations, 29, 31–32, 174, 175, 192–3, 208, 221, 265–66, 315n55, 322n30, 326n23

Ukrainian Conservative Republican Party, 199

Ukrainian Insurgent Army (UPA), 16, 300n3

Ukrainian KGB: dissolution of, 44–45; leadership and personnel of, 43; reform of, 43–45, 116; telecommunication channels with the Kremlin, 116–17

Ukrainian Republican Party, 199, 200, 301n1

Ukrainian SSR (Ukrainian Soviet Socialist Republic), 17, 18

Ukratomenerhoprom, 308n38. See also Derzhkomatom

Unified Command Post (UCP), 108, 110, 112

United Armed Forces of the CIS, 118

United States: foreign policy of, 32; nuclear advantage of, 22, 261–62, 265; nuclear power centers of, 94, 99; purchase of highly enriched uranium (HEU), 40–41; Senate hearings on Ukraine, 195–6, 201; support of nuclear non-proliferation, 3, 31, 32

Ural Electrochemical Combine (Sverdlovsk-44), 41, 89

uranium enrichment, 57–58

uranium-plutonium oxide (MOX), 57

US-Russian Uranium Deal, 40–42, 93, 94, 98, 109

Valenia, I. I., 242

Varennikov, Valentin, 301n7

Vasylenko, Volodymyr, 24–25, 25

Vienna conventions, 79

Vitovych, Oleh, 247

VVER-1000 nuclear power reactors, 101, 103, 110, 111, 114

Vydrin, Dmytro, 148

Vyshnevs'kyi, Ivan, 56, 57–58, 108

Warsaw Pact, 307n10

Western media: bias of, 169; coverage of Ukraine's disarmament, 170–71, 172, 175, 190, 191, 193, 302n19

Westinghouse company, 95, 100, 101, 309n46, 309n47

Wisner, Frank, 8, 132, 253

Wolsack, Henk, 156

Woolsey, Robert James, Jr., 128, 194, 194, 302n19

Working Group for Addressing Issues Related to Ratification of START and to Ukraine's Attaining a Non-Nuclear Status: assessment of the Trilateral Statement, 230–31; creation of, 76, 303n37, 306n6; dismissal of, 121; effectiveness of, 123; expert committee of, 105; "For Service Use" policy brief, 188; meetings of, 160; members of, 76, 123; recommendations of, 78, 185–86, 189, 208–9; tasks of, 76–77

Yalta summit, 4, 11, 203

Yastrzhembsky, Sergey, 160

Yeltsin, Boris: Alma-Ata Protocol, 38; approach to democracy, 225; background and career of, 39; at Budapest summit, 256, 258; foreign policy of, 311n80, 321n16, 331n66; with George H. W. Bush, 30; in Masandra, 204; at Moscow summit of 1993, 105, 135; negotiations with Kravchuk, 124; photograph of, 39; at Russian-Ukrainian meeting, 133; signing of the Minsk Agreement, 36; at the Yalta summit, 134

Yushchenko, Victor, 35, 45, 247, 299n1, 313n21

Yuzhmash. See Pivdenmash

Zaiets', Ivan, 33, 46, 76, 182

Zakaria, Fareed, 196

Zelens'kyi, Viktor, 56, 105, 105, 107, 110, 308n27

Zenit space systems, 304n58

Zhirinovsky, Vladimir, 222, 222

Zlenko, Anatolii: arguments for nuclear disarmament, 56–58, 177–78, 181, 191, 240, 312n109, 320nn3–4; background and career of, 59; development of security policy, 47; diplomacy of, 40, 62, 223; on foreign policy of Ukraine, 200–1, 320n6; joint press conference with Kozyrev, 75; photograph of, 59; ratification of START I and, 211, 212; on security guarantees for Ukraine, 131, 313n21, 320n5; statement on Lisbon Protocol, 70–71; support of START II, 127; testimony before the Parliament on 3 February 1994, 237, 238, 239, 240; on Ukraine's ownership of nuclear weapons, 182; visit to Moscow, 75; Western media on, 191

Zviahils'kyi, Iefym, 223

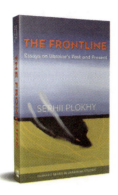

The Frontline: Essays on Ukraine's Past and Present

Serhii Plokhy

The Frontline presents a selection of essays drawn together for the first time to form a companion volume to Serhii Plokhy's *The Gates of Europe* and *Chernobyl*. Here he expands upon his analysis in earlier works of key events in Ukrainian history.

2021 (HC) / 2023 (PB) | 416 pp. / 420 pp.
10 color photos, 9 color maps

ISBN 9780674268821 (hardcover) $64.00

9780674268838 (paperback) $19.95

9780674268845 (epub)

9780674268852 (PDF)

Harvard Series in Ukrainian Studies, vol. 81

Read the
book online

The Moscow Factor: US Policy toward Sovereign Ukraine and the Kremlin

Eugene M. Fishel

This unique study that examines four key Ukraine-related policy decisions across two Republican and two Democratic U.S. administrations. Fishel asks whether, how, and under what circumstances Washington has considered Ukraine's status as a sovereign nation in its decision-making regarding relations with Moscow.

2022 | 324 pp., 2 figs.

ISBN 9780674279179 (hardcover) $59.95

9780674279186 (paperback) $29.95

9780674279421 (epub)

9780674279193 (PDF)

Harvard Series in Ukrainian Studies, vol. 82

Read the
book online

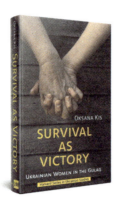

Survival as Victory

Oksana Kis

Hundreds of thousands of Ukrainian women were sentenced to the GULAG in the 1940s and 1950s. Only about half of them survived. With this book, Oksana Kis has produced the first anthropological study of daily life in the Soviet forced labor camps as experienced by Ukrainian women prisoners. Based on the written memoirs, autobiographies, and oral histories of over 150 survivors, this book fills a lacuna in the scholarship regarding Ukrainian experience.

2020 | 652 pp., 78 color and 10 b/w photos

ISBN 9780674258280 (hardcover) $94.00

9780674258327 (epub)

9780674258341 (PDF)

Harvard Series in Ukrainian Studies, vol. 79

Read the
book online

Ukrainian Nationalism in the Age of Extremes: An Intellectual Biography of Dmytro Dontsov

Trevor Erlacher

Ukrainian nationalism made worldwide news after the Euromaidan revolution and the outbreak of the Russo-Ukrainian war in 2014. Invoked by regional actors and international commentators, the "integral" Ukrainian nationalism of the 1930s has moved to the center of debates about Eastern Europe, but the history of this divisive ideology remains poorly understood.

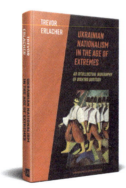

2021 | 658 pp., 34 photos, 5 illustr.

ISBN 9780674250932 (hardcover) $84.00

9780674250949 (epub)

9780674250956 (Kindle)

9780674250963 (PDF)

Harvard Series in Ukrainian Studies, vol. 80

Read the
book online

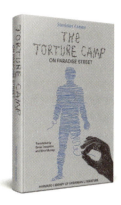

The Torture Camp on Paradise Street

Stanislav Aseyev

In *The Torture Camp on Paradise Street*, Ukrainian journalist and writer Stanislav Aseyev details his experience as a prisoner from 2015 to 2017 in a modern-day concentration camp overseen by the Federal Security Bureau of the Russian Federation (FSB) in the Russian-controlled city of Donetsk. This memoir recounts an endless ordeal of psychological and physical abuse, including torture and rape, inflicted upon the author and his fellow inmates over the course of nearly three years of illegal incarceration spent largely in the prison called Izoliatsiia (Isolation).

2023 | 299 pages, 1 map, 18 illustrations

ISBN 9780674291072 (hardcover) $59.95

9780674291089 (paperback) $29.95

9780674291102 (epub)

9780674291096 (PDF)

Harvard Series in Ukrainian Studies, vol. 5

Read the
book online

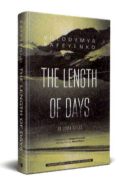

The Length of Days: An Urban Ballad

Volodymyr Rafeyenko

Translated by Sibelan Forrester

Afterword and interview with the author by Marci Shore

With elements of magical realism, Volodymyr Rafeyenko's novel combines a wicked sense of humor with political analysis, philosophy, poetry, and moral interrogation.

2023 | 348 pp.

ISBN 9780674291201 (hardcover) $39.95

9780674291218 (paperback) $19.95

9780674291232 (PDF)

9780674291225 (epub)

Harvard Library of Ukrainian Literature, vol. 6

Read the
book online

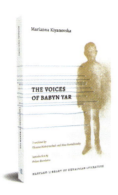

The Voices
of Babyn Yar

Marianna Kiyanovska

Translated by Oksana Maksymchuk
and Max Rosochinsky

Introduced by Polina Barskova

With this collection of stirring poems,
the award-winning Ukrainian poet
honors the victims of the Holocaust by
writing their stories of horror, death, and
survival in their own imagined voices.

2022 | 192 pp.

ISBN 9780674268760 (hardcover) $39.95

9780674268869 (paperback) $16.00

9780674268876 (epub)

9780674268883 (PDF)

Harvard Library of Ukrainian Literature, vol. 3

Read the
book online

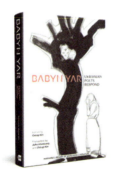

Babyn Yar: Ukrainian
Poets Respond

Edited by Ostap Kin

Translated by John Hennessy
and Ostap Kin

In 2021, the world commemorated the 80th
anniversary of the massacres of Jews at
Babyn Yar. The present collection brings
together for the first time the responses
to the tragic events of September 1941 by
Ukrainian Jewish and non-Jewish poets
of the Soviet and post-Soviet periods,
presented here in the original and in English
translation by Ostap Kin and John Hennessy.

2023 | 282 pp.

ISBN 9780674275591 (hardcover) $39.95

9780674271692 (paperback) $16.00

9780674271722 (epub)

9780674271739 (PDF)

Harvard Library of Ukrainian Literature, vol. 4

Read the
book online